Waste Valorisation

Wiley Series in Renewable Resources

Series Editor:

Christian V. Stevens, Faculty of Bioscience Engineering, Ghent University, Belgium

Titles in the Series:

Wood Modification: Chemical, Thermal and Other Processes
Callum A. S. Hill

Renewables-Based Technology: Sustainability Assessment
Jo Dewulf, Herman Van Langenhove

Biofuels
Wim Soetaert, Erik Vandamme

Handbook of Natural Colorants
Thomas Bechtold, Rita Mussak

Surfactants from Renewable Resources
Mikael Kjellin, Ingegärd Johansson

Industrial Applications of Natural Fibres: Structure, Properties and Technical Applications
Jörg Müssig

Thermochemical Processing of Biomass: Conversion into Fuels, Chemicals and Power
Robert C. Brown

Biorefinery Co-Products: Phytochemicals, Primary Metabolites and Value-Added Biomass Processing
Chantal Bergeron, Danielle Julie Carrier, Shri Ramaswamy

Aqueous Pretreatment of Plant Biomass for Biological and Chemical Conversion to Fuels and Chemicals
Charles E. Wyman

Bio-Based Plastics: Materials and Applications
Stephan Kabasci

Introduction to Wood and Natural Fiber Composites
Douglas D. Stokke, Qinglin Wu, Guangping Han

Cellulosic Energy Cropping Systems
Douglas L. Karlen

Introduction to Chemicals from Biomass, 2nd Edition
James H. Clark, Fabien Deswarte

Lignin and Lignans as Renewable Raw Materials: Chemistry, Technology and Applications
Francisco G. Calvo-Flores, Jose A. Dobado, Joaquín Isac-García, Francisco J. Martin-Martínez

Sustainability Assessment of Renewables-Based Products: Methods and Case Studies
Jo Dewulf, Steven De Meester, Rodrigo A. F. Alvarenga

Cellulose Nanocrystals: Properties, Production and Applications
Wadood Hamad

Fuels, Chemicals and Materials from the Oceans and Aquatic Sources
Francesca M. Kerton, Ning Yan

Bio-Based Solvents
François Jérôme, Rafael Luque

Nanoporous Catalysts for Biomass Conversion
Feng-Shou Xiao, Liang Wang

Thermochemical Processing of Biomass: Conversion into Fuels, Chemicals and Power, 2nd Edition
Robert C. Brown

The Chemical Biology of Plant Biostimulants
Danny Geelen, Lin Xu

Chitin and Chitosan: Properties and Applications
Lambertus A.M. van den Broek, Carmen G. Boeriu

Biorefinery of Inorganics: Recovering Mineral Nutrients from Biomass and Organic Waste
Erik Meers, Evi Michels, René Rietra, Gerard Velthof

Process Systems Engineering for Biofuels Development
Adrián Bonilla-Petriciolet, Gade P. Rangaiah

Waste Valorisation: Waste Streams in a Circular Economy
Carol Sze Ki Lin, Guneet Kaur, Chong Li, Xiaofeng Yang

Forthcoming Titles:

Biobased Packaging: Material, Environmental and Economic Aspects
Mohd Sapuan Salit, Rushdan Ahmad Ilyas

High-Performance Materials from Bio-based Feedstocks
Andrew J. Hunt, Nontipa Supanchaiyamat, Kaewta Jetsrisuparb, Jesper T.N. Knijnenburg

Waste Valorisation

Waste Streams in a Circular Economy

Edited by

CAROL SZE KI LIN
School of Energy and Environment, City University of Hong Kong,
Tat Chee Avenue, Kowloon Tong, Hong Kong

GUNEET KAUR
Department of Biology, Hong Kong Baptist University,
Kowloon Tong, Hong Kong
and
Department of Civil Engineering, York University, Toronto, Canada

CHONG LI
Shenzhen Branch, Guangdong Laboratory for Lingnan Modern Agriculture,
Genome Analysis Laboratory of the Ministry of Agriculture,
Agricultural Genomics Institute at Shenzhen,
Chinese Academy of Agricultural Sciences, Shenzhen, China

XIAOFENG YANG
School of Biology and Biological Engineering,
South China University of Technology, Guangzhou, China

WILEY

This edition first published 2021

Registered Offices
John Wiley & Sons, Inc., 111 River Street, Hoboken, NJ 07030, USA
John Wiley & Sons Ltd, The Atrium, Southern Gate, Chichester, West Sussex, PO19 8SQ, UK

Editorial Office
The Atrium, Southern Gate, Chichester, West Sussex, PO19 8SQ, UK

For details of our global editorial offices, customer services, and more information about Wiley products visit us at www.wiley .com.

Wiley also publishes its books in a variety of electronic formats and by print-on-demand. Some content that appears in standard print versions of this book may not be available in other formats.

Library of Congress Cataloging-in-Publication Data
Names Sze Ki Lin, Carol, 1977- editor. Kaur, Guneet, 1985- editor. Li, Chong, 1986- editor. Yang, Xiaofeng, 1987- editor.
Title Waste valorisation waste streams in a circular economy edited by Carol Sze Ki Lin, City University of Hong Kong, Hong Kong, Guneet Kaur Department of Biology, Hong Kong Baptist University, Kowloon Tong, Hong Kong and Department of Civil Engineering, York University, Toronto, Canada, Chong Li, Shenzhen Branch, Guangdong Laboratory for Lingnan Modern Agriculture, Genome Analysis Laboratory of the Ministry of Agriculture, Agricultural Genomics Institute at Shenzhen, Chinese Academy of Agricultural Sciences, Shenzhen, China, Xiaofeng Yang, South China University of Technology, Guangzhou.
Description First edition. Hoboken, NJ John Wiley & Sons, Inc., 2021. Series Wiley series in renewable resources Includes bibliographical references and index.
Identifiers LCCN 2020016810 (print) LCCN 2020016811 (ebook) ISBN 9781119502708 (cloth) ISBN 9781119502692 (adobe pdf) ISBN 9781119502746 (epub)
Subjects LCSH Biomass energy industries. Biomass energy–Economic aspects.
Classification LCC HD9502.5.B542 W37 2021 (print) LCC HD9502.5.B542 (ebook) DDC 338.4766288–dc23
LC record available at httpslccn.loc.gov2020016810
LC ebook record available at httpslccn.loc.gov2020016811

Cover Design: Wiley
Cover Images: Front cover design and images: X'ian Shangyuan Network Co. Ltd; Globe image © Ingram Publishing/Alamy Stock Photo

Set in 10/12pt TimesLTStd by SPi Global, Chennai, India

Printed and bound by CPI Group (UK) Ltd, Croydon, CR0 4YY

10 9 8 7 6 5 4 3 2 1

Contents

List of Contributors

Karthika Arumugam Department of Biotechnology, Kalasalingam University, India

Sandya Athukoralalage Chemical and Environmental Engineering, RMIT University, Australia

Ravindran Balasubramani Department of Environmental Energy and Engineering, Kyonggi University Youngtong-Gu, South Korea

Satinder Kaur Brar INRS-ETE, Université du Québec, Canada
Department of Civil Engineering, Lassonde School of Engineering, York University, Canada

Thamaraiselvi Chandran Department of Biotechnology, Mother Teresa Women's University, India

Soon Woong Chang Department of Environmental Energy and Engineering, Kyonggi University Youngtong-Gu, South Korea

Shauhrat S. Chopra School of Energy and Environment, City University of Hong Kong, Hong Kong

Namita Roy Choudhury Chemical and Environmental Engineering, RMIT University, Melbourne, Australia

Chengyu Dong Department of Civil and Environmental Engineering, Hong Kong Polytechnic University, Hong Kong

Chenyu Du School of Applied Sciences, University of Huddersfield, United Kingdom

Naba Kumar Dutta Chemical and Environmental Engineering, RMIT University, Australia

Cakin Ezgi School of Energy and Environment, City University of Hong Kong, Hong Kong

Yunchao Feng College of Energy; Fujian Engineering and Research Center of Clean and High-valued Technologies for Biomass; and Xiamen Key Laboratory of Clean and High-valued Applications of Biomass, Xiamen University, PR China

Pobitra Halder Chemical and Environmental Engineering, RMIT University, Australia

Zubeen Hathi School of Energy and Environment (SEE), City University of Hong Kong, Hong Kong

Hsien-Yi Hsu School of Energy and Environment, City University of Hong Kong, Hong Kong

Md Khairul Islam Department of Civil and Environmental Engineering, Hong Kong Polytechnic University, Hong Kong

Guneet Kaur Department of Biology, Hong Kong Baptist University, Kowloon Tong, Hong Kong and Department of Civil Engineering, York University, Canada

Meena Krishania Center of Innovative and Applied Bioprocessing, India

Deepak Kumar Department of Chemical Engineering, State University of New York College of Environmental Science and Forestry, USA

Sazal Kundu Chemical and Environmental Engineering, RMIT University, Australia

Shao-Yuan Leu Department of Civil and Environmental Engineering, Hong Kong Polytechnic University, Hong Kong

Chong Li Chinese Academy of Agricultural Sciences, China

Xiaotong Li School of Energy and Environment, City University of Hong Kong, Hong Kong

Carol Sze Ki Lin School of Energy and Environment, City University of Hong Kong, Hong Kong

Lu Lin College of Energy; Fujian Engineering and Research Center of Clean and High-valued Technologies for Biomass; and Xiamen Key Laboratory of Clean and High-valued Applications of Biomass, Xiamen University, PR China

Sishi Long College of Energy; Fujian Engineering and Research Center of Clean and High-valued Technologies for Biomass; and Xiamen Key Laboratory of Clean and High-valued Applications of Biomass, Xiamen University, PR China

Srinivas Mettu Sonochemistry Group, School of Chemistry, University of Melbourne, Parkville, Australia
Chemical and Environmental Engineering, RMIT University, Australia
School of Energy and Environment (SEE), City University of Hong Kong, Hong Kong

Jinhua Mou School of Energy and Environment (SEE), City University of Hong Kong, Hong Kong

Selvakumar Muniraj Water and Solid Waste Processing Lab, Department of Environmental Biotechnology, Bharathidasan University, India

Vasanthy Muthunarayanan Water and Solid Waste Processing Lab, Department of Environmental Biotechnology, India.

Khai Lun Ong School of Energy and Environment (SEE), City University of Hong Kong, Hong Kong

Parmjit S. Panesar Food Biotechnology Research Laboratory, Department of Food Engineering and Technology, Sant Longowal Institute of Engineering and Technology, India

Divyani Panwar Food Biotechnology Research Laboratory, Department of Food Engineering and Technology, Sant Longowal Institute of Engineering and Technology, India

Savankumar Patel Chemical and Environmental Engineering, RMIT University, Australia

Raffel Dharma Patria School of Energy and Environment, City University of Hong Kong, Hong Kong

Rajiv Periakaruppan Department of Biotechnology, Karpagam Academy of Higher Education, India

Joseph Sebastian INRS-ETE, Université du Québec, Canada

Kalpit Shah Chemical and Environmental Engineering, RMIT University, Australia

Archana Singh Institute of Environment and Sustainable Development, Banaras Hindu University, India

Gopal Shankar Singh Institute of Environment and Sustainable Development, Banaras Hindu University, India

Vijay Singh Agricultural and Biological Engineering Department, University of Illinois at Urbana-Champaign, USA

Gisha Singla Center of Innovative and Applied Bioprocessing, India

Karpagam Subramanian School of Energy and Environment, City University of Hong Kong, Hong Kong

Yong Sun College of Energy; Fujian Engineering and Research Center of Clean and High-valued Technologies for Biomass; and Xiamen Key Laboratory of Clean and High-valued Applications of Biomass, Xiamen University, PR China

Gayatri Suresh INRS-ETE, Université du Québec, Canada

Xing Tang College of Energy; Fujian Engineering and Research Center of Clean and High-valued Technologies for Biomass; and Xiamen Key Laboratory of Clean and High-valued Applications of Biomass, Xiamen University, PR China

Avinash Thakur Department of Chemical Engineering Sant Longowal Institute of Engineering and Technology, India

Huaimin Wang School of Energy and Environment, City University of Hong Kong, Hong Kong

Xiaofeng Yang South China University of Technology, China

Guihua Yan College of Energy; Fujian Engineering and Research Center of Clean and High-valued Technologies for Biomass; and Xiamen Key Laboratory of Clean and High-valued Applications of Biomass, Xiamen University, PR China

Shunyu Yao Sonochemistry Group, School of Chemistry, University of Melbourne, Parkville, Australia

Xianhai Zeng College of Energy; Fujian Engineering and Research Center of Clean and High-valued Technologies for Biomass; and Xiamen Key Laboratory of Clean and High-valued Applications of Biomass, Xiamen University, PR China

Series Preface

Renewable resources, their use and modification are involved in a multitude of important processes with a major influence on our everyday lives. Applications can be found in the energy sector; paints and coatings; and the chemical, pharmaceutical, and textile industries, to name but a few.

The area interconnects several scientific disciplines (agriculture, biochemistry, chemistry, technology, environmental sciences, forestry), which makes it very difficult to have an expert view on the complicated interaction. Therefore, the idea to create a series of scientific books, focusing on specific topics concerning renewable resources, has been very opportune and can help to clarify some of the underlying connections in this area.

In a very fast-changing world, trends are not only characteristic of fashion and political standpoints; science too is not free from hypes and buzzwords. The use of renewable resources is again more important nowadays; however, it is not part of a hype or a fashion. As the lively discussions among scientists continue about how many years we will still be able to use fossil fuels – opinions ranging from 50 to 500 years – they do agree that the reserve is limited, and that it is essential not only to search for new energy carriers but also for new material sources.

In this respect, the field of renewable resources is a crucial area in the search for alternatives for fossil-based raw materials and energy. In the field of energy supply, biomass- and renewables-based resources will be part of the solution alongside other alternatives such as solar energy, wind energy, hydraulic power, hydrogen technology, and nuclear energy. In the field of material sciences, the impact of renewable resources will probably be even bigger. Integral utilization of crops and the use of waste streams in certain industries will grow in importance, leading to a more sustainable way of producing materials. Although our society was much more (almost exclusively) based on renewable resources centuries ago, this disappeared in the Western world in the nineteenth century. Now it is time to focus again on this field of research. However, it should not mean a "retour à la nature," but should be a multidisciplinary effort on a highly technological level to perform research toward new opportunities, to develop new crops and products from renewable resources. This will be essential to guarantee an acceptable level of comfort for the growing number of people living on our planet. It is "the" challenge for the coming generations of scientists to develop more sustainable ways to create prosperity and to fight poverty and hunger in the world. A global approach is certainly favored.

This challenge can only be dealt with if scientists are attracted to this area and are recognized for their efforts in this interdisciplinary field. It is, therefore, also essential that consumers recognize the fate of renewable resources in a number of products. Furthermore, scientists do need to communicate and discuss the relevance of their work. The use and modification of renewable resources may not follow the path of the genetic engineering concept in view of consumer acceptance in Europe. Related to this aspect, the series will certainly help to increase the visibility of the importance of renewable resources. Being convinced of the value of the renewables approach for the industrial world, as well as for developing countries, I was myself delighted to collaborate on this series of books focusing on the different aspects of renewable resources. I hope that readers become aware of the complexity, the interaction and interconnections, and the challenges of this field, and that they will help to communicate on the importance of renewable resources.

I certainly want to thank the people of Wiley's Chichester office, especially David Hughes, Jenny Cossham, and Lyn Roberts, in seeing the need for such a series of books on renewable resources, for initiating and supporting it, and for helping to carry the project to the end.

Last, but not least, I want to thank my family, especially my wife Hilde and children Paulien and Pieter-Jan, for their patience, and for giving me the time to work on the series when other activities seemed to be more inviting.

Christian V. Stevens
Faculty of Bioscience Engineering, Ghent University, Belgium
Series Editor, "Renewable Resources"
June 2005

Preface

Global economic development is being confronted with the issues of limitation of natural resources, endless demands from the environment, and increased generation of waste streams. The exploration of circular economy principles to capture value, in the form of the energy, nutrients and materials embodied in these streams, would contribute to restoring the natural capital.

This book aims to give an overview of waste valorisation strategies under the concept of circular economy. The increasing amount of waste generated from human activities attracts global attention, and innovative biotechnologies for green and sustainable resource management are being closely and extensively studied in both academic and industrial-scale research. Technology-integrated biorefineries focus on the material recovery approach to resolve the twin problems of the global waste burden and resource/energy concerns in order to transform renewable resources from waste streams into value-added products. New research and explorations are rapidly emerging to lead the way toward a bio-based economy. Both theory and practice of waste valorisation is a topic of increasing importance in the 2030 Sustainable Development Agenda.

This book gives a detailed presentation of valorisation schemes for various types of waste streams. Sustainability analysis, circular economy development and future trends are also included. In this book, we will review the development of biochemical processes for utilisation of wastes as bioresources (Chapters 2 and 3), process integration for a waste-based biorefinery (Chapters 4–7), and closed loop recirculation of a waste-based biorefinery in a bio-based economy (Chapters 8–10).

As editors of this book, we sincerely hope it can successfully demonstrate the application of the integrated waste-based biorefinery, which has a strong potential to provide solutions to combat climate change. Based on these solutions on circular economy principles, cascaded benefits for food security, human health, and environmental protection are anticipated. We hope this book can promote and accelerate an innovative waste-based bioeconomy, and inspire more research and practice on waste management strategies in order to make a significant contribution to the UN Sustainable Development Goals.

We would like to take this opportunity to acknowledge all contributors for their great efforts in excellent and timely contributions and efficient collaboration. The co-editors would like to express their sincere appreciation and gratitude to the Series Editor Prof.

Christian Stevens from Ghent University, Belgium, for his kind recommendation for submission of a book proposal to the Wiley Series in Renewable Resources. Also, we would like to extend our gratitude to the Wiley Cover Team for their great effort in preparation of the book cover. Last but not least, we sincerely thank the publishing editor Sarah Higginbotham, project editor Aruna Pragasam, and assistant editor Emma Strickland, who patiently and kindly took us through the development of this book over the past three years to achieve this impressive final result, which would not have been possible without their support.

Sincere wishes for an enjoyable read!

Carol Sze Ki Lin, Guneet Kaur, Chong Li and Xiaofeng Yang
Hong Kong Special Administrative Region of the People's Republic of China
March 2020

1

Overview of Waste Valorisation Concepts from a Circular Economy Perspective

Jinhua Mou[1], Chong Li[2], Xiaofeng Yang[3], Guneet Kaur[4] and Carol Sze Ki Lin[1]

[1]*School of Energy and Environment, City University of Hong Kong, Tat Chee Avenue, Kowloon Tong, Hong Kong*
[2]*Shenzhen Branch, Guangdong Laboratory for Lingnan Modern Agriculture, Genome Analysis Laboratory of the Ministry of Agriculture, Agricultural Genomics Institute at Shenzhen, Chinese Academy of Agricultural Sciences, Shenzhen, China*
[3]*School of Biology and Biological Engineering, South China University of Technology, Guangzhou, China*
[4]*Department of Biology, Hong Kong Baptist University, Kowloon Tong, Hong Kong*

1.1 Introduction

Petroleum is one of the most important reserves used as a fundamental raw material for various industries. It has been in a predominant position in the world energy consumption structure since the 1970s. Petroleum-derived products, such as plastics, synthetic fiber, and synthetic rubber, are widely utilized in the agricultural, chemical, and pharmaceutical sectors, and have already become necessities in our daily life. Most industries, like the chemical industries and transportation, are heavily dependent upon petroleum and other fossil resources.

However, this fossil-based economy is facing two serious problems. On one hand, fossil resources are not renewable – they have limited availability on our planet and are

Waste Valorisation: Waste Streams in a Circular Economy, First Edition.
Edited by Carol Sze Ki Lin, Guneet Kaur, Chong Li and Xiaofeng Yang.

running out at a rapid rate. On the other hand, the industrial utilization of fossil resources has already caused many environmental problems, such as greenhouse effects, and air, water, and soil pollution. The serious energy and environmental crisis has aroused global concerns and reflections. The world needs to find substitutes for fossil resources to change the current energy-intensive and environmentally unfriendly economic model. Therefore, a low-consumption and high-value-added sustainable circular economy system needs to be established. Such a circular economy can replace the linear economy model of "make-use-dispose" with a "circular" model, in which the value of resources and products is maintained in the system for a long period. The efficient use of waste or side streams from production processes is another important aspect of a circular economy.

While traditional energy resources, such as petroleum and natural gas, are non-renewable and will be depleted in the near future, the substitution of emerging resources (energy and materials) and awareness of environmental protection outweigh seeking only economic profits and has become a significant worldwide issue. Besides the utilization of solar energy and wind energy, biomass energy, as an alternative form of energy derived from solar energy, has attracted an increasing amount of attention. Data from the US Energy Information Administration show that the percentage of biomass energy in total energy consumption has increased rapidly in recent years. For instance, in 2016, biomass energy contributed 5.8% to the source of US energy consumption. Biomass energy is a renewable (the only renewable carbon resource), clean (little pollution, low carbon emissions), and abundant resource (Field et al. 2008). It will greatly ease the energy and environment burden if biomass energy could be widely accepted and utilized in industries or in our daily life, and replace fossil resources as the preference in energy consumption. In fact, a global industrial revolution has already begun as the foundations of economic development change from hydrocarbon to carbohydrate, i.e., the transition from a petrol-based economy to a bio-based economy, which is a significant trend for sustainable development (Bozell and Petersen 2010) (Figure 1.1).

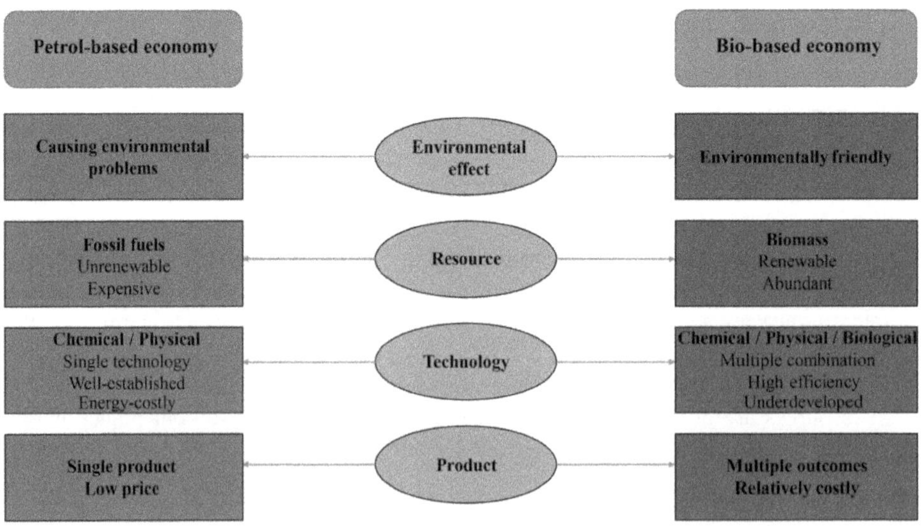

Figure 1.1 A comparison of petrol-based and bio-based economies.

With the rapid development of industries and improvement in living standards, the generation of waste is also increasing, which has already caused many environmental and social problems (Sharholy et al. 2008). Landfill and incineration are the most commonly used methods of waste management at present. However, they are not ideal solutions owing to the damage to the environment and human health. In particular, organic wastes, characterized by putrescibility, low heat value and high organic matter, could become a great threat to public health, because of the emission of toxic gases (e.g., oxole, dioxine) and the transmission of pathogenic micro-organisms when improperly treated by conventional methods (Polprasert and Koottatep 2017). But "wastes" can be regarded as valuable resources due to the functional components are present within them (Ong et al. 2018). It allows for their transformation into high-value products rather than being discarded as useless and unwanted. Overall, this provides solutions to both efficient waste management and provision of feedstock for industrially important products, which are fundamental solutions for sustainable development (Figure 1.2).

Together with increasing demand in both substance and spirit, the world today is facing many problems related to food security, energy consumption, and environmental protection. The development of biomass based industries could be one of the great efforts made in order to change this situation. Biorefinery, aimed at sustainable development through utilisation of renewable (and/or waste) resources and integration of high-efficiency technologies, can play an increasingly important role in achieving a green, circular, and sustainable economy.

In this book, we will take an overview of the development of biochemical processes for the utilization of wastes as a bioresource (Chapters 2 and 3), process integration for waste-based biorefinery (Chapters 4–7), and closed loop recirculation of waste-based biorefinery in a bio-based economy (Chapters 8–10) (Figure 1.3).

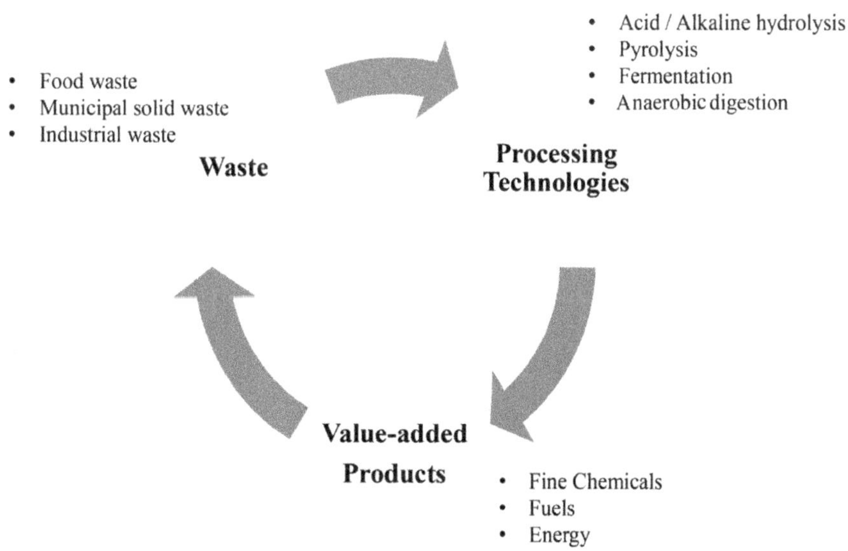

Figure 1.2 *Concept of waste valorisation to high-value products.*

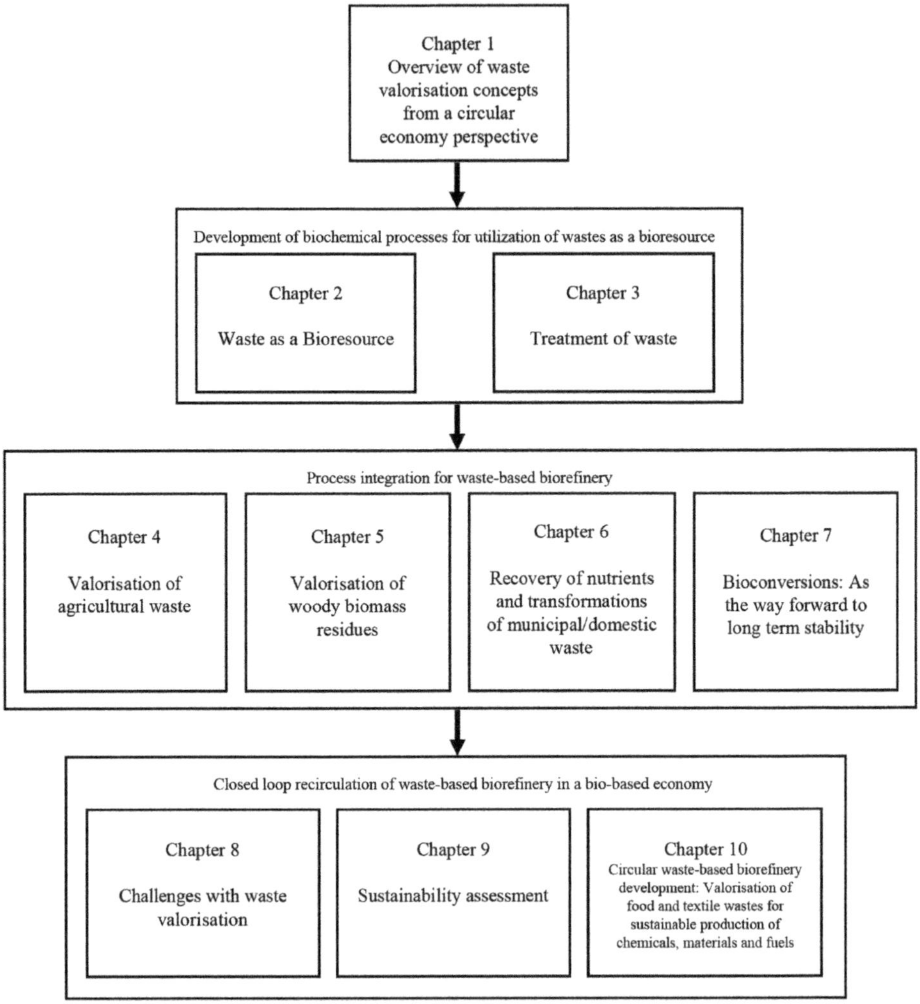

Figure 1.3 *Book layout.*

1.2 Development of (Bio)Chemical Process for Utilization of Waste as a Bioresource

Wastes are often defined as substances that are no longer useful to the holder. The rapid industrial and economic development of recent years has seen a huge amount of waste generated from human activities, which has caused many environmental and social problems. In fact, most of the wastes we are talking about have the potential for further processing and utilization. The main reasons hindering the effective recycling of the wastes are improper handling strategies and inadequate technologies in related industries. Sustainable waste management should be carried out for both environmental and economic benefits.

A waste stream is the flow of a specific waste, referring to the lifecycle from its source to recovery, recycling, or disposal. In general, waste streams are mainly divided into two

Table 1.1 Suitable waste streams for biorefineries.

Waste stream	Examples
Municipal/domestic waste	Food waste, Waste cooking oil, Sewage, Leather, Textiles
Animal waste	Fats, Blood, Meat, Manure
Forestry waste	Leaves, Straw, Wood
Agricultural waste	Crop waste, Rice straw, Citrus waste
Industrial waste	Pulp and paper industry waste, Sugar industry waste, Coffee industry waste

groups: material-related streams (e.g., metals, plastics, bio-waste) and product-related streams (e.g., e-waste, construction waste) (Bourguignon 2015). When talking about the biorefinery concept in waste valorisation, it aims to utilize waste as a bioresource, so we are more interested in the organic (or biodegradable) parts among the whole waste stream. Basically, the suitable sources for waste biorefineries are municipal/domestic waste, agricultural waste, industrial waste, forestry waste, and animal waste (Table 1.1).

In most cases, it is not efficient to convert the waste directly into products. Hence, before they are ready as substrates for biorefineries, certain pretreatments or modifications are required to make them accessible to the following downstream reactions, which may improve the conversion efficiency. Pretreatment methods are mainly classified into mechanical, physical, chemical and biological types, and each method has its advantages and disadvantages depending on the characteristics of the waste or specific need of the subsequent conversion (or production) process. Thus, it is an important consideration for the biorefinery management to decide the best pretreatment option. A brief summary of pretreatment methods is provided below.

1.2.1 Mechanical Pretreatment

The main purpose of mechanical pretreatment is the reduction of the particle size or crystallinity of the materials. It increases the surface area and reduces the degree of polymerization, which can benefit the downstream conversion process, for example, by improving enzymatic hydrolysis (Taherzadeh and Karimi 2008). Grinding, milling, and chipping are commonly used, although their power requirement is relatively high.

1.2.2 Physical Pretreatment

Physical pretreatment is usually proposed under specific temperature and pressure conditions. The extraction method could achieve effective isolation and fractionation of valuable products due to the unique properties of a solvent (Azmir et al. 2013). Ammonia fiber explosion makes hydrolysis of materials easier, increases the surface area for enzymatic conversion, and acts as a nitrogen source for microbes in downstream processing (Alizadeh et al. 2005).

1.2.3 Chemical Pretreatment

The main objectives of chemical pretreatment are breaking up any resistant structures, removal of unwanted fractions, and selective dissolution. For example, acid hydrolysis can effectively remove lignin and hemicellulose during wood biomass treatment, and hydrolyze cellulose into glucose. The major drawback, however, is the potential inhibition effect of the byproducts on the downstream process (Galbe and Zacchi 2012). Alkaline hydrolysis is a relatively mild process that can be used in biomass treatments such as delignification, while some obvious issues such as the corrosion problem during the process need to be tackled (Chen et al. 2013).

1.2.4 Biological Pretreatment

The best example of this pretreatment is fungal biodegradation, which is a widely used method for delignification of lignocellulosic biomass, wood chips and wheat straw for instance. White rot fungi are well studied and have shown a high degradation efficiency because of the extracellular enzymes they secrete, such as lignin peroxidases, manganese peroxidases, and laccases (Alvira et al. 2010). This biological approach is a more environmentally friendly pretreatment process that is safer and cheaper, but more time consuming in comparison to the conventional petrochemical production process.

1.3 Process Integration for Waste-Based Biorefinery

The concept of "biorefinery" was first put forward in *Science* in 1982 (Bungay 1982). Derived from petroleum refinery, biorefinery aims to produce a variety of valuable chemicals, materials, and energy by maximizing the conversion efficiency and minimizing cost and waste generation. What makes it different from petroleum-based biorefinery is the combination of facilities using biomass as feedstocks instead of fossil resources. Biorefinery overturns the stereotype of producing one product from biomass, as found in traditional chemical industries. Instead, it makes the most of every component of substrates and converts them into various products in order to achieve full resource utilization and value maximization. Biomass conversion is the core procedure of the biorefinery. Physical, chemical, and biological methods play important roles in this conversion process, such as mixing, separation, pyrolysis, modification, fermentation, and enzyme catalysis.

Based on the complexity of feedstock, combination of technologies and variety of outcomes, biorefineries are divided into three types: (I) single feedstock, single process and single major product; (II) single feedstock, multiple processes, and multiple major products; (III) multiple feedstocks, multiple processes, and multiple major products (Fernando et al. 2006). Biorefineries are capital-intensive projects, and it would be very difficult, and also costly, to increase the productivity or conversion efficiency when the designer put too much weight on just one conversion technology. Obviously, type III biorefineries are the most advanced and optimal, because they can efficiently produce multiple outcomes to meet market demands and achieve profitability. Also, they are capable of using different substrates when required after structural adjustment, which makes them flexible in industrial practice. However, there is still a long way to go for the commercial existence of such integrated biorefineries. It is not easy to achieve the combination of multiple inputs and considerable outputs on a large scale, and in one platform. Nevertheless, there are extensive works under way in many places on the design and feasibility of such facilities (Clark and Deswarte 2008).

Along with increasing population and urbanization, waste generation as well as its accompanying problems has become a global concern. Waste biorefineries are able to offer satisfying waste management solutions given the current circumstances. Although there is no standard classification for existing waste biorefineries, in most cases we divide waste biorefineries into different types according to the origin of feedstocks. For example, there are biorefineries based on food waste, agricultural waste, industrial waste, and wastewater.

1.3.1 Food Waste Biorefinery

Food waste is the food mass left before or after human consumption throughout the supply chain, including those arising from production, handling, storage and disposal. It is among the most widely generated biowastes in the world, amounting to about 1.3 billion tonnes globally every year (Paritosh et al. 2017). Food waste can be roughly divided into two groups: avoidable food waste and unavoidable food waste. Avoidable food waste is edible before being discarded, and we can usually decrease this waste by changing consumption strategies; while unavoidable food waste mainly originates from food (processing) industries and needs to be properly disposed of or recycled.

The organic fraction is the dominant component of food waste – carbohydrates, proteins and lipids that can be converted into simple compounds like glucose, amino acids and fatty acids through further conversion processes. Anaerobic fermentation is the most promising process for food waste valorisation to produce biogas and volatile fatty acids (Dahiya et al. 2018). Metabolite-rich effluents are generated during these specific processes. They have great potential to improve the conversion efficiency through integrating with other bio-processes to obtain additional products such as biopolymers, bioelectricity, and animal feed (Figure 1.4).

1.3.2 Agricultural Waste Biorefinery

Common agricultural residues or biomass include crop waste, wheat straw, seed, and wood chips. They are promising substrates for biogas and biofuel production. For instance, the

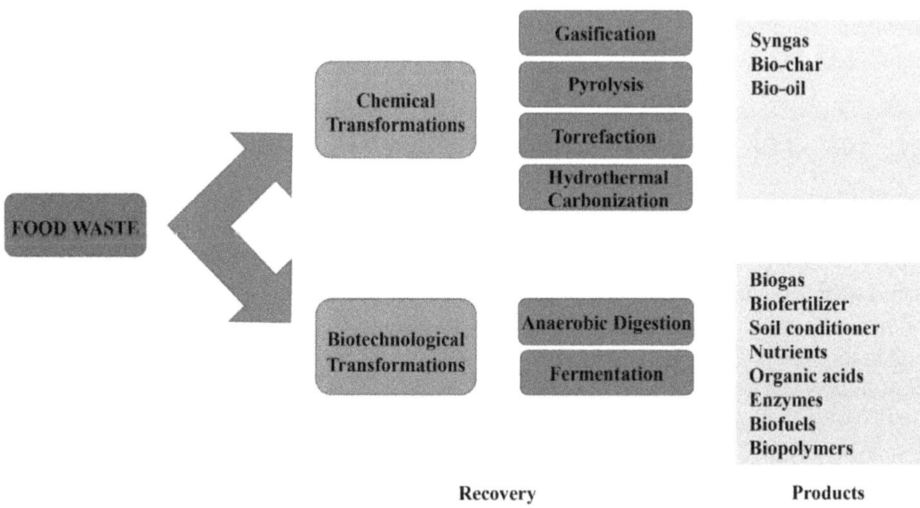

Figure 1.4 Valorisation of food waste to value-added bioproducts.

hydrolysate of wheat straw after hydrothermal pretreatment is rich in hemicellulose, which can be utilized in biohydrogen production, and bioethanol production is also achieved through the cellulose-rich solid fraction (Kaparaju et al. 2009). Leaves and wood waste are rich sources of lignin and cellulose. Thermochemical treatment, such as liquefaction, gasification, pyrolysis, and torrefaction, can help convert biomass into products like biochar and syngas (Lam et al. 2015).

1.3.3 Industrial Waste Biorefinery

The sludge produced by the paper and pulp industries has a high moisture content, and it is also rich in carbohydrates, which makes possible the production of cellulose-derived fuels and high-value chemicals like bioethanol, biohydrogen, and fatty acids by hydrolysis or fermentation techniques. Integration of two or more processes, a thermal process and bioprocess for instance, has proved to create higher energy outputs and chemical conversion yields. One study demonstrated that extra treatment of fermentation, together with pyrolysis, improved energy conversion compared with solo pyrolysis in paper sludge treatment (Gottumukkala et al. 2016). Also, the char produced from thermochemical conversion of sludge has the potential for use as activated carbon or biochar.

1.3.4 Wastewater Biorefinery

Wastewater management is inefficient in the world at present, as most of the wastewater is directly discharged into the natural water environment. Untreated wastewater can damage both the environment and human health. Hence, there is an urgent need for green wastewater treatment. Renewable energy production, hydrogen being a typical example, would be a promising approach for wastewater biorefineries. Microbial electrolysis cell technology has higher hydrogen productivity compared with conventional electrolysis or fermentation. Research shows that a cell utilizing wastewater as the substrate has a high removal rate for organic matter (Khan et al. 2017). If challenges of this technology could be resolved, such as low energy efficiency, and competition and growth inhibition between microbial communities, it would have a significant impact on wastewater treatment plants.

1.4 Closed Loop Recirculation in a Bio-based Economy

Landfill and incineration have been widely accepted as waste management strategies for a long time. However, drawbacks of these traditional waste management methods have already emerged. Landfilling is a serious waste of land resources, and without treatment, the buried wastes could result in serious problems in the future, such as heavy metal pollution, groundwater pollution, and fatal bacterial/viral diseases. Similarly, the disadvantages of incineration include public health problems and secondary environmental pollution (e.g., air pollution and acid rain).

In contrast, waste biorefineries have more opportunities in competition with traditional strategies. With the support of rich feedstocks and superior technologies, it realizes sustainable concepts by effective utilization of biomass resources and achieves numerous value-added outputs. Developing waste biorefineries could contribute to improving environmental conditions through a shift in the consumption model, from fossil fuels to renewable resources, and efficient waste management. Environmental benefits that waste biorefineries bring, such as the abatement of greenhouse gas emissions and increasing energy savings, would have a positive impact on economic development, especially in developing countries.

Undoubtedly, waste biorefinery could be the optimal solution for a cost-effective and environmentally friendly economy in such a rapid developing era. As a new concept, however, it still faces many challenges compared with traditional industries in a mature, petroleum-dominant economy.

Abundant availability of feedstock is one of the outstanding advantages for waste biorefinery. But in most cases, these resources are season-dependent and widely distributed. Collection, transportation and storage would be major issues for the early stages of waste management. In addition, technologies based on biology still need practical examination. They are usually more complicated than chemical technologies, which makes it difficult for bio-based substrates to achieve high conversion efficiencies and considerable productivity. Furthermore, despite the fact that many bioproducts are cost-competitive, low yields and low productivities increase their price and limit their popularity. This means that petroleum-derived products are still in the dominant position in the market, despite a high oil price, thanks to well developed and highly efficient processing technologies. The rapid realization of economic benefits further makes the fossil-based industries more favorable, while increasing competitive pressures for bio-based processes and products.

Nevertheless, the increasing demand for environmental protection and sustainable resource utilization, and improvements in efficiency and productivity of bio-based processes, will soon outweigh the quick yet non-sustainable benefits of the petroleum-based processes. This will promote the development of a bio-based economy that is both environmentally and economically sustainable.

1.5 Conclusions and Future Trends

Waste biorefinery is full of potential to achieve green and sustainable waste management and bring considerable environmental and economic benefits. Environmental benefits include abatement of greenhouse gas emissions, natural resources conservation, and elimination of other detrimental impacts on the environment caused by traditional waste disposal. As for economic benefits, waste biorefinery helps to break away from the fossil fuel-dependent situation and decrease the utilization of raw materials owing to widely available renewable waste resources. The development of innovative and optimal technologies could contribute to realizing cost-efficient outputs of energy recovery and value-added products from bio-based refineries in a circular economy. This closed-loop approach addresses sustainable development goals and will lead many economies to move towards an environmentally friendly development model, which is the reflection of a transitional trend to a circular economy.

The circular economy organizes economic activities into a feedback process of "resources – products – renewable resources," compared with the "make-use-dispose" process of the linear-increase economic model of traditional industries. It is characterized by low exploitation, high utilization, and low emissions, and all the substrates and energy under this concept can be rationally and circularly utilized for the long term. The circular economy minimizes the impact of economic activities on the environment, and combines green production with waste utilization, which in essence is a form of ecological economy. Core principles of this economic model facilitate the recycling and reuse of material though integrated sustainable approaches (Maina et al. 2017).

However, as upcoming green industries, waste biorefinery should be prepared for the opportunities and challenges when competing with traditional industries. The transition from the current petrol-dominant economy to a bio-based economy will require a systematic process involving many aspects. From the perspective of industrial chains, waste biorefinery shows its advantages during the adjustment of feedstocks, energy structure, and technology integration. Changing to renewable waste resources relieves the economic, environmental, and social burdens caused by the traditional linear economy. Integration of various processes enhances the efficiency of resource utilization and achieves considerable outcomes. However, there are still many problems, such as underdeveloped technologies, imperfect policies, and commercialization difficulties. Thus, in this significant period of social transformation toward a more sustainable (economic) system, those working in the area of waste biorefinery must be fully aware of its development needs, grasp opportunities, and strive in the competition. Finally, the objective is green and sustainable management, and the key aspect is to find a balance between innovation and regulation.

To allow optimal resource utilization and maximum economic benefit, a priority-hierarchy model could be established. This model adopts a waste biomass utilization sequence based on the ranking of the value chain. For example, high value-added industries, such as the chemical industrial chain, have the highest priority in the waste biomass distribution in order to increase the overall substrate utilization efficiency and economic value. Material and biofuel production would be the next. Subsequently, the residual biomass after distribution can be utilized in heating and electricity production. Additionally, "green policies" are needed to support and guide these industries, especially at the beginning of the research stage and/or industrialization. These policies would involve innovative research, green purchase, taxes, and green consumption, and they will contribute to promoting the development of bio-based industries, raise public awareness of bioproducts, and stimulate green behavior and consumption.

The above aspects of transitioning to a circular economy based on valorisation of resources in a biorefinery scheme are discussed in detail in subsequent chapters of this book.

References

Alizadeh, H., Teymouri, F., Gilbert, T.I., and Dale, B.E. (2005). Pretreatment of switchgrass by ammonia fiber explosion (AFEX). *Appl. Biochem. Biotechnol.* **124** (1–3): 1133–1141.

Alvira, P., Tomás-Pejó, E., Ballesteros, M., and Negro, M. (2010). Pretreatment technologies for an efficient bioethanol production process based on enzymatic hydrolysis: a review. *Bioresour. Technol.* **101** (13): 4851–4861.

Azmir, J., Zaidul, I., Rahman, M. et al. (2013). Techniques for extraction of bioactive compounds from plant materials: a review. *J. Food Eng.* **117** (4): 426–436.

Bourguignon, D. (2015). *Understanding Waste Streams. Treatment of Specific Waste*. European Parliamentary Research Service.

Bozell, J.J. and Petersen, G.R. (2010). Technology development for the production of biobased products from biorefinery carbohydrates – the US Department of Energy's "Top 10" revisited. *Green Chem.* **12** (4): 539–554.

Bungay, H.R. (1982). Biomass refining. *Science* **218** (4573): 643–646.

Chen, Y., Stevens, M.A., Zhu, Y. et al. (2013). Understanding of alkaline pretreatment parameters for corn Stover enzymatic saccharification. *Biotechnol. Biofuels* **6** (1): 8.

Clark, J.H. and Deswarte, F.E. (eds.) (2008). The biorefinery concept – an integrated approach. In: *Introduction to Chemicals from Biomass*, 2e, 1–29. Chichester: Wiley.

Dahiya, S., Kumar, A.N., Sravan, J.S. et al. (2018). Food waste biorefinery: sustainable strategy for circular bioeconomy. *Bioresour. Technol.* **248**: 2–12.

Fernando, S., Adhikari, S., Chandrapal, C., and Murali, N. (2006). Biorefineries: current status, challenges, and future direction. *Energy Fuel* **20** (4): 1727–1737.

Field, C.B., Campbell, J.E., and Lobell, D.B. (2008). Biomass energy: the scale of the potential resource. *Trends Ecol. Evol.* **23** (2): 65–72.

Galbe, M. and Zacchi, G. (2012). Pretreatment: the key to efficient utilization of lignocellulosic materials. *Biomass Bioenergy* **46**: 70–78.

Gottumukkala, L.D., Haigh, K., Collard, F.-X. et al. (2016). Opportunities and prospects of biorefinery-based valorisation of pulp and paper sludge. *Bioresour. Technol.* **215**: 37–49.

Kaparaju, P., Serrano, M., Thomsen, A.B. et al. (2009). Bioethanol, biohydrogen and biogas production from wheat straw in a biorefinery concept. *Bioresour. Technol.* **100** (9): 2562–2568.

Khan, M., Nizami, A., Rehan, M. et al. (2017). Microbial electrolysis cells for hydrogen production and urban wastewater treatment: a case study of Saudi Arabia. *Appl. Energy* **185**: 410–420.

Lam, W.C., Kwan, T.H., Budarin, V.L. et al. (2015). Pretreatment and thermochemical and biological processing of biomass. In: *Introduction to Chemicals from Biomass*, 2e (eds. J. Clark and F. Deswarte), 53–88. Chichester: Wiley.

Maina, S., Kachrimanidou, V., and Koutinas, A. (2017). A roadmap towards a circular and sustainable bioeconomy through waste valorisation. *Curr. Opin. Green Sustainable Che.* **8**: 18–23.

Ong, K.L., Kaur, G., Pensupa, N. et al. (2018). Trends in food waste valorisation for the production of chemicals, materials and fuels: case study South and Southeast Asia. *Bioresour. Technol.* **248**: 100–112.

Paritosh, K., Kushwaha, S.K., Yadav, M. et al. (2017). Food waste to energy: an overview of sustainable approaches for food waste management and nutrient recycling. *Biomed. Res. Int.*: 1–19. https://doi.org/10.1155/2017/2370927.

Polprasert, C. and Koottatep, T. (2017). *Organic Waste Recycling: Technology, Management and Sustainability*. IWA Publishing.

Sharholy, M., Ahmad, K., Mahmood, G., and Trivedi, R. (2008). Municipal solid waste management in Indian cities – a review. *Waste Manag.* **28** (2): 459–467.

Taherzadeh, M. and Karimi, K. (2008). Pretreatment of lignocellulosic wastes to improve ethanol and biogas production: a review. *Int. J. Mol. Sci.* **9** (9): 1621–1651.

2

Waste as a Bioresource

Gayatri Suresh[1], Joseph Sebastian[1] and Satinder Kaur Brar[1,2]

[1] *INRS-ETE, Université du Québec, 490 Rue de la Couronne, Québec G1K 9A9, Canada*
[2] *Department of Civil Engineering, Lassonde School of Engineering, York University, North York, Toronto, Ontario, Canada*

2.1 Introduction

Bioresources are biological, non-fossil resources that can be renewed naturally. These can be used by humans for the production of food, energy carriers, biomolecules, and in industries like construction, pulp and paper, and chemicals (Jadhav 2017; Narayana and Johnson 2017). Bioresources can be classified into four groups based on their origin – primary, secondary, tertiary, and quaternary bioresources (Körner 2015). Primary bioresources are generated for specific applications such as the production of food and high-value biobased products. These mainly include plants and animals grown for direct human use. Secondary bioresources are produced as residues of industrial processing, primary processing or area maintenance, and can be used for the production of food, high- and medium-value biobased products and energy, such as straw, husk, animal entrails, twigs, and so on. Tertiary bioresources are residuals that are mixed with fractions or impurities and have a lower value compared with secondary bioresources – examples are food processing residues, food wastes, and mixed slaughterhouse residues. Quaternary bioresources are generated as residues after the consumption of a biobased product – livestock excrement, used lumber, sewage sludge, industrial effluents, and so on. Both tertiary and quaternary bioresources can be used for the production of medium- and low-value biobased products and energy (Körner 2015).

Waste Valorisation: Waste Streams in a Circular Economy, First Edition.
Edited by Carol Sze Ki Lin, Guneet Kaur, Chong Li and Xiaofeng Yang.

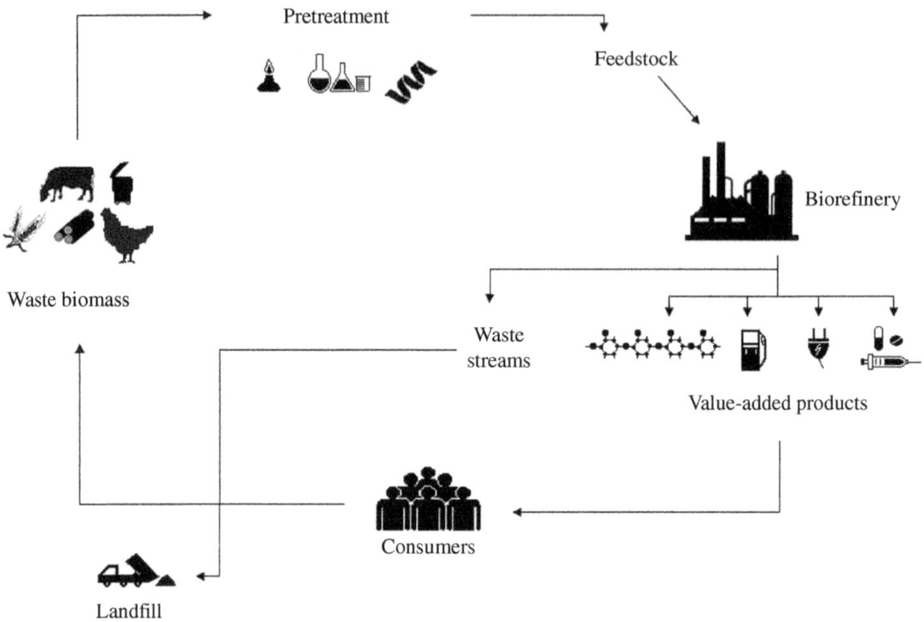

Figure 2.1 Waste biomass in a circular bioeconomy.

The use of bioresources for the manufacturing of products, materials, and energy plays an important role in the development of a sustainable bioeconomy (Adamowicz 2017). The continuous depletion of conventional, non-renewable resources such as fossil fuels, as well as the detrimental impact of their use on the environment, have necessitated the transition from a fossil-based economy to a bioeconomy (Saxena et al. 2009; ESEIA 2014; Sheldon 2014). For a bioeconomy, the concept of biorefineries comes into action. Unlike petroleum refineries, which use fossil fuels as input, biorefineries use biological materials as feedstock and subject them to specific conversion processes for the generation of power, energy, fuels, and chemicals (Figure 2.1) (Fernando et al. 2006). The sustainable use of bioresources would help in creating a resource-effective circular economy based on renewable products and energy, and additionally help in the reduction of waste, combating climate change and creating jobs (Rönnlund et al. 2014).

2.2 Waste Streams and Their Suitability as Feedstock for Valorisation: Is All Waste a Resource?

Globally, considerable quantities of municipal, agricultural, and industrial wastes are generated as a consequence of fulfilling the growing needs of an increasing global population. Organic matter accounts for almost 50% of the global waste generated, and is one of the main causes of greenhouse gas (GHG) emissions (Hoornweg and Bhada-Tata 2012). Organic waste streams, such as agro-industrial wastes, municipal solid wastes (MSWs), livestock and industrial wastes, do not directly compete with food crops for land, and hence these are a sustainable alternative to non-renewable resources (Ekman et al. 2013).

Table 2.1 Organic waste streams and their biological oxygen demand (BOD).

Organic waste streams	BOD (mg O_2/l of pollutant)
Raw domestic sewage	300–400
Vegetable washings	500–3000
Cattle slurry	10 000–20 000
Pig slurry	20 000–30 000
Milk	100 000–140 000
Dairy effluent	2500
Whey	34 000
Brewery wastes	775–1622
Distillery spent mash	12 000–34 000
Edible oil refinery	500–6700

Source: Adapted from Tarbox (2003), and http://www.ecifm.rdg.ac.uk/bod.htm.

Organic waste streams are rich in carbon and other nutrients (Table 2.1) and therefore these can be utilized as feedstock for material and energy recovery by implementing sustainable pathways to exploit their untapped potential. The use of carbohydrates as raw material eliminates the requirement for expensive oxidative processes, as well as it provides a viable microbial route for the production of valuable products (Ragauskas et al. 2006). Since microbes form an integral part of bioproduction strategies, only biodegradable wastes rich in organic matter (waste biomass such as agricultural wastes, forestry wastes, livestock wastes, spent biomass from biotechnology industries, and municipal wastes) can be valorized in a circular economy. Additionally, waste streams are available throughout the year and are cost-effective candidates for feedstock, providing a sustainable alternative to the conventional methods of waste management. For example, approximately 2.01 billion tonnes of MSW is generated annually, which amounts to a per-capita generation of 0.74 kg per day. This is expected to increase to approximately 3.40 billion tonnes per year by 2050 due to rapid urbanization and population growth (Statista 2018; The World Bank Group 2019). Additionally, it is estimated that if food wastage was a country, it would be the third largest emitter of GHGs in the world and would amount to 4.4 Gt CO_2 eq. per year (Bio-En Power Inc., 2018).

Conventional methods of management of organic waste streams include landfilling and incineration, which are not environmentally friendly as they can cause the release of toxic combustion products into the local environment which can lead to adverse health impacts (Hoornweg and Bhada-Tata 2012). Therefore, many studies have been carried out to identify sustainable technologies for waste management, which can be classified into four groups (Kaur et al. 2014; Lohri et al. 2017):

- Direct use as animal feed
- Biological treatments such as composting, anaerobic digestion, fermentation, etc.
- Physicochemical treatments such as transesterification and densification
- Thermochemical treatments such as gasification, pyrolysis, and liquefaction

The main challenge in using waste as a bioresource is that a viable commercial operation requires efficient use of feedstocks as well as product recovery. Hence, cost-effective methods need to be developed for purification of both feedstock as well as product. Currently, product recovery accounts for 60–80% of process cost of a chemical process, but this would be higher in the case of bioproduction strategy due to the presence of a plethora of microbial products (Statista 2018). Another challenge is the presence of recalcitrant and stable polymers, such as lignin, especially in lignocellulosic materials. Organic waste streams have high contents of lignocellulose, which is a highly crystalline complex of cellulose embedded in a matrix of lignin in hemicellulose, that makes the biomass resistant to enzymatic degradation during solid-state fermentation (SSF). Therefore, an important step to ensure the cost-effective use of this biomass is the conversion of complex lignocellulose into simple sugars that can easily be fermented by microorganisms.

2.3 (Bio)diversity and Variability of Waste Feedstock

Waste feedstocks can be broadly divided into four categories based on their source – agro-industrial wastes, MSWs, livestock wastes, and industrial wastes (Nizami et al. 2017), and they have been extensively studied for the generation of bioproducts (Table 2.2).

2.3.1 Agro-industrial Wastes

There has been a steady increase in the growth of agro-based industries over the past few years, and consequently in the amount of agro-industrial waste generated. Residues from agricultural crops are the primary source of agro-industrial wastes. Globally, cultivation of food crops such as maize, wheat, corn, and so on, produces 3.3 Gt of waste annually (Gopinath et al. 2016).

Agro-industrial wastes can be divided into two categories (Sadh et al. 2018):

- Agricultural residues from crop farming. Field residues such as stems, stalks, and leaves are generated after harvesting of crops. Process residues such as bagasse, molasses, husks, seeds, etc., are generated after the processing of crops into alternative resources.
- Industrial residues generated by the food processing industries during production and storage of raw material, and the industrial processing of food.

Lignocellulose forms a major component of agro-industrial wastes. While the exact lignocellulosic composition of the waste residues depends on their source (softwood, hardwood, grass, etc.), most agro-industrial residues comprise 40–50% cellulose; another 20–30% could possibly be used as a substrate for SSF by microorganisms for the production of bioenergy, as well as recovery or production of value-added products such as enzymes, vitamins, antioxidants, animal feed, antibiotics, and other chemicals (Sadh et al. 2018). Extensive research has been carried out on the use of agro-industrial wastes as feedstock for the production of second-generation biofuels such as bioethanol, biobutanol, methane, and biohydrogen (Schievano et al. 2009; Qureshi et al. 2010; Duhan et al. 2013; Iqbal et al. 2013; Shukor et al. 2014; Domínguez-Bocanegra et al. 2015; Saini et al. 2015; Kumar et al. 2016). The use of agro-industrial wastes for biofuel production would also cause a decrease in deforestation. Additionally, pretreated lignocellulosic biomass can also be converted to platform chemicals by appropriate processes. Depolymerization of lignin, cellulose, and hemicellulose leads to the formation of C-5 and C-6 sugars, which can be further used to produce a variety of platform chemicals such as glycerol, sorbitol, xylitol, aspartic acid, furfural, levulinic acid, malic acid, fumaric acid, succinic acid, lactic acid,

Table 2.2 Valorisation of different waste feedstocks into bioproducts.

	Feedstock	Product	Concentration	Reference
Agro-industrial waste	Jerusalem artichoke stalks	Biohydrogen	68.7 ml H_2/g_{TS}	Guo et al. (2014)
	Wheat straw hydrolysate	Butanol or (ABE)	12.0 (25.0) g l^{-1}	Jang et al. (2012)
	Lime-treated barley straw		18.0 (26.6) g l^{-1}	
	Lime-treated switchgrass hydrolysates		14.5 (26.3) g l^{-1}	
	Corn stover hydrolysate		10.4 (16.0) g l^{-1}	
Municipal sewage	Sewage-sludge	Biohydrogen	1.7 mol H_2/mol substrate	Nath & Das (2004)
	Wastewater sludge and MSW	Biogas	0.9 m^3 m^{-3} d^{-1}	
	Wastewater sludge	*B. thuringiensis* (biopesticide)	–	Vidyarthi et al. (2002)
	Wastewater sludge/crude glycerol	Oleaginous yeast *Yarrowia lipolytica SKY7*	–	Ram et al. (2018)
	Wastewater sludge	Extracellular polymeric substances (EPS)	1.36–2.12 g l^{-1}	More et al. (2014)
Livestock waste	Manure	Biogas	6.3 m^3 m^{-3} d^{-1}	Karakashev et al. (2005)
	Diary manure	Fumaric acid Chitin	31.0 g l^{-1} 0.21 g chitin/g biomass	Liao et al. (2008)
Industrial waste	Molasses	Biohydrogen	1.58 mol H_2/mol substrate	Nath & Das (2004)
	Wheat starch		1.9 mol H_2/mol substrate	Nath & Das (2004)
	Vegetable residues from restaurants		45.9 ml H_2/g_{TS}	Guo et al. (2014)
	Apple pomace ultrafiltration sludge	Fumaric acid	25.2 g l^{-1}	Das et al. (2015a)
	Brewery sludge		43.67 g l^{-1}	Das et al. (2015b)
	Apple pomace	Citric acid	766 g kg^{-1}	Vandenberghe (1999)
	Beet molasses		109 g l^{-1}	
	Coffee husk		150 g kg^{-1}	
	Starch industry wastewater	*B. thuringiensis* (biopesticide)	–	Ndao et al. (2017)

and glutamic acid (Isikgor and Becer 2015). Agricultural waste has also been reviewed as a low-cost substrate for the production of antibiotics by SSF (Asagbra et al. 2005; Vastrad and Neelagund 2011). Moreover, it can be used as feedstock for obtaining bioactive compounds such as polyphenols, catechins, essential oils, flavonoids, carotenoids growth factors, peptides, vitamins, and gallic acid, with antioxidant, antimicrobial, anti-inflammatory activities (Santana-Méridas et al. 2012).

Another field of research that has been gaining interest is the fabrication of green biocomposites, which are formed by biopolymers reinforced with natural fibers to give improved mechanical properties. With respect to this, agricultural waste has been evaluated as an alternative to glass and carbon fibers for reinforcement. Lignocellulosic residues like soya hull, corn stalk, wheat straw, and switchgrass have been studied as reinforcement for biodegradable polymers (Nagarajan et al. 2013; Nanda et al. 2013). Biocomposites reinforced with agricultural residues can have potential application in the development of an active packaging system (Valdés et al. 2014).

Wastes streams from food processing industries have a high content of sugars, therefore these can be used as low-cost substrates. Anaerobic digestion of these effluents gives the advantage of the generation of bioenergy. Carbon dioxide (CO_2), generated as a byproduct of wineries or breweries, can be harnessed for the production of fumaric and succinic acid (Koutinas et al. 2014). Studies have reported the extraction of flavonoids (e.g., anthocyanins, flavonones, flavanols, etc.) and phytochemicals (e.g., limonene and hesperidin) from wastes produced during fruit processing (Đilas et al. 2009; Balu et al. 2012). Wastes with a high lipid content, such as oilmeal cakes, can be fermented by microorganisms for the production of biosurfactants (Saravanan and Subramaniyan 2014).

2.3.2 Municipal Solid Wastes

MSWs encompass all solid or semi-solid wastes that are generated from residential, commercial, and industrial sources. These include organic wastes (food and garden wastes), plastics, glass, ceramics, electronic wastes, paper and cardboard, cans, and bottles. The conventional methods for the management of MSW are incineration and landfilling; however, both methods have their respective drawbacks (Matsakas et al. 2017). Incineration is a costly and energy-intensive process and it can also cause air pollution, while landfilling causes excessive release of methane, a GHG, into the atmosphere, with the potential spread of pathogens, and an unpleasant odor. Additionally, leachate from landfill sites can cause contamination of soil and water resources. Biogenic technologies, such as composting and anaerobic digestion, provide a sustainable, cost-effective and environmentally friendly alternative for MSW management.

Organic wastes generally account for a major proportion of MSW; however, the composition varies with location. Developing countries have a higher content of organic wastes in MSW (approximately 40–80%), compared with developed countries (approximately 30–40%) (Karak et al. 2012; Santibañez-Aguilar et al. 2013). During composting, this organic matter is decomposed and stabilized by microbial communities to produce organic-grade MSW compost that can be used as a soil conditioner or fertilizer. Composting also gives the advantage of reducing the volume, odorous compounds and pathogen load of the wastes. Several benefits have been reported for the addition of MSW compost to soils, such as improvement in organic matter content, water-holding capacity, and soil stability, as well as increased adsorption of pesticides in soil (Hargreaves et al. 2008).

MSW can also be used as feedstock for anaerobic digestion, during which bacterial degradation of the organic matter in MSW occurs in the absence of oxygen via reactions such as hydrolysis, acetogenesis, and methanogenesis, for the production of biogas (methane) (Wan et al. 2013). During anaerobic digestion, value-added products such as alcohols, carboxylic acids, aromatic compounds, terpenes, and ammonia are also produced (Papurello et al. 2012). Degradation of MSW in an anaerobic digester depends on its carbon to nitrogen ratio, moisture content, particle size, and digestion temperature.

Post-consumption food wastes account for a major part of the organic component of generated MSW. MSW also comprises of potential resources such as plastics, textile, polystyrene, and so on; however these are non-sustainable feedstocks as they are ultimately derived from crude oil (Tuck et al. 2012). Since the use of organic waste is the key for sustainable waste management and valorisation, the discussion in this section will be on post-consumption food wastes. These wastes contain carbohydrates, proteins, lipids, and minerals, and due to their composition, they have a very high potential for the generation of biofuel and bioelectricity, which can be produced during the food supply chain (Haddadi et al. 2018).

Hence, they can support the growth of desired microorganisms for bioproduction and biorefining strategies. The potential application of food wastes as feedstock for bioethanol production by microbial fermentation has been investigated. Proteins and minerals present in post consumption food wastes serve as a nutrient source for the microorganisms, which in turn ferment the sugars present in these wastes to ethanol (Matsakas et al. 2017). Previous studies have reported that the addition of hydrolytic enzymes such as glucoamylase, carbohydrase, and protease improve ethanol yield due to saccharification of the substrate (Moon et al. 2009; Walker et al. 2013; Uncu and Cekmecelioglu 2011; Kim et al. 2011; Huang et al. 2015; Uçkun Kıran and Liu 2015). Studies have also reported the extraction of high-value products and bioactive components from using food wastes as a feedstock. These include polyphenols, pectin, antioxidants, flavonoids, vitamins, carotenoids, pigments, aromas, colorants, and so on (Maina et al. 2017; Baiano 2014; Petkowicz et al. 2017; Mirabella et al. 2014). Fruit and vegetable peels, which account for the highest amount of kitchen wastes, have also been studied as a medium for the production of single cell proteins and poly(3-hydroxybutyric acid) and as biosorbents for water purification (Uncu and Cekmecelioglu 2011; Kim et al. 2011; Huang et al. 2015).

2.3.3 Livestock Wastes

The major wastes generated by the livestock industry are animal manures. Traditionally, raw manures have been used as organic fertilizers, due to their high content of Carbon, Nitrogen, and water, which provide nutrients for nitrifying and denitrifying soil bacteria, thus improving soil fertility. However, their use is under scrutiny due to several factors increasing their costs, such as waste transportation, malodor, and high GHG emissions (Teira-Esmatges and Flotats 2003). While the stabilization of the organic matter in waste can be carried out by composting, animal manure can also have concentrations of human pathogens (such as bacteria, protozoa, viruses, and prions), antibiotics, and micro- and macronutrients, which will have several detrimental effects when they are released into the environment, such as posing a hazard to human and animal health, enabling the spread of antibiotic resistance, causing eutrophication, and so on (Manyi-Loh et al. 2013). Keeping these in mind, currently, anaerobic digestion of animal manure provides an alternative for energy recovery of these wastes and generation of biogas. Conversion of animal waste to bioenergy also has significant economic impacts, such as generation of annual revenue, diversification of farm income, and moderation of commodity price impacts (Cantrell et al. 2008). Thermal treatment of animal wastes such as by pyrolysis and gasification are also under review for the production of electrical and thermal energy. Gasification, in particular, has been reported to give not only good yield but also a possibility of mineral and catalyst recovery from process effluents (Yanagida et al. 2009; Font-Palma 2012; Girotto and Cossu 2017). Animal-derived wastes are a good source of protein and can be used for the recovery of phosphorus to help deal with the depleting sources of phosphorus. Szögi et al. (2015) reported a feasible procedure for up to 90% recovery of phosphorus from pig manure solids. The recovered phosphorus can then be recycled and used as fertilizer. Additionally, several studies have been carried out to study the feasibility of using fats derived from animals as a feedstock for the production of biodiesel. Biodiesel consists of fatty acid methyl esters formed by the transesterification of methanol and triacylglycerols, and serves as a green alternative to petroleum. Currently, the biggest challenge faced for sustainable production of biodiesel is the high cost of feedstock (primarily vegetable oils), which can be addressed by the use of waste animal fats (such as chicken fat, lard, tallow, etc.) produced during meat processing and rendering, as low-cost substrates. Biodiesels produced from waste animal fats have the added advantage of having a higher Cetane number (due to the high ratio of saturated fatty acids in the feedstock), better oxidative stability and lower NO_x emissions (Villarreal et al. 2006). Chicken, tallow, and lard fats have been studied at the industrial scale for the production of biodiesel (Lammers et al. 2008a,2008b; Yang et al. 2012). A high content of free fatty acids in animal waste fats makes their conversion to biodiesel an expensive process (González-Pajuelo et al. 2004). However, this can be tackled by the inclusion of catalyzed/non-catalyzed/assisted transesterification of the animal waste fats, which decreases the energy consumption and cost of biodiesel production (Papanikolaou et al. 2002; Papanikolaou and Aggelis 2003; Mu et al. 2006; Hiremath et al. 2011).

Currently, poultry is the most highly produced livestock, with more than 500 billion units per year (Kamzolova et al. 2011), thus a lot of research has been conducted to study the potential of byproducts from chicken processing plants (such as carcasses, blood, bones, feathers, etc.) as a bioresource. These wastes are sources of proteins, particularly poultry blood (arginine, cysteine, lysine, and methionine) and feathers (primarily keratin, which consists primarily of glycine, proline, and serine), and therefore they can be used as feed supplements (Ashby et al. 2004; Ito et al. 2005; Jitrwung and Yargeau 2011). Poultry processing wastes can be converted to protein hydrolysates, which have various applications. After animal proteins are extracted and precipitated, they are hydrolyzed to produce a hydrolysate containing a mix of amino acids and peptides (Ibrahim and Steinbüchel 2009; Liang et al. 2010). Studies have reported that this hydrolysate can be used as antimicrobials, antioxidants, antihypertensives, emulsifiers, food supplements, a nutrient source in microbial growth media, flavor precursors, and so on (Saenge et al. 2011). Poultry byproducts can also be used for the recovery of proteases and polyunsaturated fatty acids (O'Grady and Morgan 2011; Lokesha et al. 2017). Chicken feathers, being rich in keratin (a fibrous protein), have been evaluated for their potential as reinforcements for the production of biocomposites that can be used in the packaging and textile industries (Ramachandran et al. 2007).

2.3.4 Industrial Wastes

Based on their source of generation, industrial waste streams can be classified into the following groups (Koutinas et al. 2014; Đilas et al. 2009; Balu et al. 2012; Technologico 2004).

2.3.4.1 *Wood Processing Industry Waste*

The wood industry is one of the largest producers of waste byproducts, with major byproducts being wood chips, wood particles, bark, and wood residues comprising of sawdust, processing rejects such as from edgings, veneers and trimmings. Wood chips, wood particles, and bark are used as feedstock for the paper and pulp industry, while the wood residues are mostly burnt for the production of heat (Koutinas et al. 2014). In recent times, however, wood residues have been explored for suitable usage. Wood residues can be used in the production of bedding for livestock, surfaces for playgrounds, and in landscaping. Wood pellet industries can use wood residues such as wood trimmings and shavings as feedstock (Thek and Obernberger 2012). Sawdust is used for the production of fiberboard and particle board (Peksa et al. 2007). Wood residues also serve as a substrate for the extraction of fuels and components such as turpentine, waxes, tannins, flavonoids, and steroids. Wood residues consist of celluloses and hemicelluloses that can be subjected to physical, chemical, or enzymatic pretreatments, and converted to C5 and C6 sugars which are easily fermentable for the production of value-added products such as ethanol (Galbe and Zacchi 2002; Silva et al. 2011) and organic acids such as lactic acid, fumaric acid, and succinic acid (Moldes et al. 2001; Lee et al. 2003; Wee et al. 2004; Kim et al. 2004; Rodríguez-López et al. 2012).

2.3.4.2 *Pulp and Paper Industry*

The pulp and paper industry accounts for a major part of the lignocellulosic wastes generated globally. Waste liquors from pulp and paper industries are mainly composed of lignin or lignosulfonates (depending on whether the chemical wood pulp has been produced by alkaline or acidic processing respectively), C5 and C6 sugars formed by hemicellulose degradation, and small amounts of acetic phenolics, 5-hydroxymethylfurfural, levulinic acid and furfural (Koutinas et al. 2014). Currently, these waste liquors are incinerated in broilers to produce energy and heat on-site for plant operation. However, research is being carried out on the potential of waste liquors for recovery of lignin, lignosulfonates, and phenolics (Varanasi et al. 2013). The use of waste liquor as a substrate for microbial fermentation is garnering interest, but it is limited by the presence of microbial inhibitors such as furfural, acetic acid and phenolics. Therefore, a pretreatment step is usually carried out before their use as a fermentation feedstock for the production of various products such as acetone, butanol, ethanol, cellulose, and xylitol (Villarreal et al. 2006; Carreira et al. 2011; Fernandes et al. 2012; Lu et al. 2013).

2.3.4.3 *Biodiesel Industry*

With the increasing global demand for biodiesel as a green fuel, it is imperative to have a sustainable system for biodiesel production, which can be achieved by the valorisation of the byproducts, especially crude glycerol, of this industry (Yang et al. 2012). The relatively high cost of biodiesel production can be offset by the utilization of crude glycerol, which accounts for almost 10% of the total byproducts formed (Yang et al. 2012). Crude glycerol finds its application as a low-cost ingredient in animal feed in both ruminant and non-ruminant diets, due to its high energy content, and conversion to glucose after absorption (Lammers et al. 2008a,2008b). Crude glycerol has also been reviewed for its potential as a feedstock for production of chemicals via bioconversions, such as 1,3-propanediol (González-Pajuelo et al. 2004; Mu et al. 2006; Hiremath et al. 2011), citric acid (Papanikolaou et al. 2002; Papanikolaou and Aggelis 2003; Kamzolova et al. 2011), biohydrogen (Ito et al. 2005; Jitrwung and Yargeau 2011), poly(hydroxyalkanoates) (Ashby et al. 2004; Ibrahim and Steinbüchel 2009), and lipids (Liang et al. 2010; Saenge et al. 2011; O'Grady and Morgan 2011). It has been observed that methanol is less toxic to microbes than ethanol, due to its lower lipid to water partition coefficient. Hence, it is recommended to use methanol instead of ethanol for biodiesel production in cases where crude glycerol produced is to be used as animal feed (Lokesha et al. 2017).

Residual oil meals from oil extraction also form a major component of biodiesel production wastes, and its composition varies with the feed material used and processing conditions. Carbohydrates, proteins, minerals, and phenolics form the main constituents of oilseed meals, and therefore can be used as a source as animal feeds, feed supplements, fertilizers substrate for enzyme production (primarily lipase and protease), for production of antibiotics, biopesticides, vitamins, antioxidants, and generation of energy by combustion (Ramachandran et al. 2007).

2.3.4.4 Bioethanol Industry

Bioethanol plays a crucial role, not only as a green alternative to fossil fuels, but also as an important platform chemical. Bioethanol production using cereal grains (such as corn and wheat) as feedstock and produces distillers' dried grains with solubles (DDGS) as a waste that contains proteins, lipids, fibers vitamins, and minerals (Rosentrater et al. 2012). The potential applications of DDGS include as animal feeds, production of biocomposites, bioenergy, biodiesel, gluten, biodegradable polymers, and platform chemicals (Arifeen et al. 2007; Saunders and Rosentrater 2009; Xu et al. 2010; Muniyasamy et al. 2013). Bioethanol produced from lignocellulosic biomass (such as sugarcane bagasse, or cereal straws) generates a waste residue with a high content of lignin, which can be extracted for further use (Pan et al. 2005). Additionally, this residue can also be used as a feedstock for microbial fermentation for the production of ligninolytic enzymes or other products of lignin degradation (Koutinas et al. 2014).

While agro-industrial residues, MSW, and industrial residues can be used as sustainable feedstock, the efficiency of the bioconversion process varies with the organic composition of these wastes. The efficient utilization of feedstock can be facilitated by the pretreatment of these wastes for their saccharification. In addition to increasing the content of fermentable sugars, pretreatment techniques should not produce any microbial inhibitory compounds that can interfere with the subsequent fermentation process, and should require the minimum input of energy (Anwar et al. 2014). Most importantly, since biomass pretreatment can account for more than 33% of the total cost, the techniques utilized for pretreatment should be cost-effective (Kim et al. 2013). While physical methods (such as milling, irradiation,

Table 2.3 *Comparison of the different methods of biomass pretreatment.*

	Physical pretreatment	Chemical pretreatment	Biological pretreatment
Production of inhibitors	Very low	High	Low, can even reduce toxic inhibitors
Environmentally friendly	Yes	No	Yes
Efficiency	Low	High	Low
Energy requirement	High	Moderate	Low
Cost-efficiency	Low	High	High

etc.) and chemical methods (such as alkaline or acid treatment, ozonation, etc.) of pretreatment have been used previously, the shift has been directed toward biological pretreatment of biomass using white-rot, soft-rot or brown-rot fungi, which can produce ligninolytic enzymes such as laccase and peroxidases (Kumar et al. 2009). Most of the microbial pretreatments rely on white-rot fungi, such as, *Phanerochaete chrysosporium, Ceriporia lacerata, Cyathus stercolerus, Ceriporiopsis subvermispora, Pycnoporuscinnarbarinus,* and *Pleurotusostreaus*. These microbes produce lignin peroxidases and manganese-dependent peroxidases which are lignin-degrading enzymes (Maurya et al. 2015). Unlike physical or chemical methods, biological pretreatment does not require high energy for lignin degradation and is an economically and environmentally feasible procedure (Lee et al. 2007; Chen et al. 2010). Moreover, the production of inhibitory compounds, formed during physical and chemical pretreatment, that affect product yield can be reduced. Table 2.3 gives a brief outline of the comparison of the different methods of biomass pretreatment (Yang et al. 2015).

One of the drawbacks with microbial pretreatment of waste is that it takes a long time due to slow microbial growth (Sun and Cheng 2002; Potumarthi et al. 2013). For example, Potumarthi et al. (2013) reported that when following the pretreatment of rice husk with fungus, the maximum reducing sugar release was observed on the 18th day. Such long durations for pretreatment cannot lead to a feasible and competitive bioprocess. This can be addressed by the direct use of enzymes for pretreatment of biomass, which has been extensively studied as a green technology for bioconversion of waste streams. The use of ligninolytic enzymes has been shown to significantly reduce the time for lignin depolymerization, compared with fungal pretreatment (Wan and Li 2012). Wastes with high amounts of sugars and proteins can be hydrolyzed with carbohydrases such as cellulases, amylases, glucanases, glucoamylases, and proteinases, to generate a feedstock for further fermentation (Ahuja et al. 2004). Enzymatic degradation of biomass can also lead to the recovery of valuable compounds such as phenolics, making the process economically feasible (Liguori and Faraco 2016). This can be a cost-effective process compared with the high input of energy and/or chemicals needed for physical and chemical pretreatment methods.

2.4 Drivers, Policies and Markets for Value-added Waste-derived Products

A circular economy is a regenerative and restorative system that focuses on a cyclical flow of materials and energy in the economy by emphasizing the reuse of products, components, and materials for as long as possible, while extracting the maximum value from them (Korhonen et al. 2018). The concept of a circular economy is garnering interest due to the present global drivers of sustainability, increased material and energy efficiency, decreased generation of wastes and GHG emissions (Kaur et al. 2018). Therefore, the use of wastes from different sources as a resource for the production of energy and other value-added products is a key point in the shift toward a circular economy. Moreover, the emphasis on transition to a bio-based economy from a petroleum-based economy has necessitated the need for development of alternate bioproduction strategies for commodity chemicals as well as polymers. Such bio-based strategies of production from lignocellulosic biomass have been discussed extensively in the review by Isikgor and Becer (2015). These strategies can be adapted for the utilization of other forms of feedstocks rich in microorganism-consumable and organic components.

An important step for the maintenance of a circular economy is the development of policies and regulations that recognize the value-added waste-derived products in the market, which would provide economic incentives for business investment, and thus maintain these products in the market. Up to the mid-nineteenth century, one of the earliest driving forces behind various polices of waste management in developed countries was the financial value of the waste as a resource. This led to the initiation of complementary policies such as "sustainable production and consumption" and the circular economy. In the 1970s, with increasing awareness of the impacts of waste management practices, environmental protection became an additional driver for the reduction of GHG emissions and subsequent resource management.

In the current scenario, the main drivers for the transition toward a bioeconomy are food security, resource security, climate change, and ecosystem services. The development of a bioeconomy is hindered due to the absence of comprehensive policies, which in turn are dependent on thriving biotechnological and chemical industries. Hence, it is important to develop policies that ensure efficient resource utilization for various sectors such as research, industry, and agriculture, via subsidies and taxes. These include measures such as competitive pricing of bioproducts, focused research for novel technologies, and stronger regulatory measures on the use of non-renewable resources (Amulya et al. 2016). The key point towards development of a sustainable bioeconomy is to reduce the dependence on fossil fuel. To this effect, Mandate M/429 was issued by the European Union (EU) to develop a standardization reference for biobased products, as well as increasing public awareness of these products. The mandate focuses on waste valorisation leading to decreased use of first-generation feedstocks, as well as reduction of waste volume (European Commission 2011).

Another driver for waste valorisation in the EU is the Landfill Directive (Council Directive 99/31/EC) which aims to reduce the amount of waste that is dumped in landfill sites, by increasing the landfill tax. In this scenario, policymakers support the production of value-added products from wastes as a means to achieve a "zero waste economy." According to the Europe 2020 strategy document, waste valorisation is the means toward achieving closed-loop manufacturing (Pfaltzgraff and Clark 2014). Countries like Japan, Sweden, and Germany have modified policy and legislation to encourage reduction and recycling of wastes for a closed loop economy. Legislations such as the Closed Substance Cycle and Waste Management Act in Germany, and focus on a recycling-based community by the Japanese Ministry of Economy Trade and Industry, are examples of policies implemented toward a circular economy (Lieder and Rashid 2016). More recently, the Paris Agreement on climate change has necessitated participating countries to reduce their greenhouse gas

Table 2.4 A brief summary of value-added products derived from wastes as feedstock.

Value-added products	Examples	Usage
Biofuels	• Biobutanol • Biohydrogen	• As chemical feedstock in the plastic industry, a more efficient fuel • In fuel cells to generate electricity
Biopolymers	• Bioplastics	• In food industry, packaging industry
Bioelectricity	• Microbial fuel cells	• Generation of electricity
Enzymes	• Hydrolytic enzymes • Oxidative enzymes	• Food industry, animal feed industry, pulp, and paper industry • Textile industry, detergent industry, as biocatalysts for bioremediation
Organic acids	• Volatile fatty acids	• Food and feed industry, supplements
Pharmaceuticals	• Hormones, vitamins, peptides, antioxidants	• Food, cosmetics, and biomedical industries
Food and feeds	• Mushroom cultivation • Single-cell proteins	• Human consumption, animal feed • As a protein-rich food/feed source

emissions with the sole purpose of keeping a global temperature rise in this century below 2°C. Therefore, the adoption of waste valorisation strategies has gained further significance and this transition is paramount to achieve the goal set during the accord.

With respect to the development of a sustainable circular bioeconomy, the concept of green chemistry has great importance. The three main elements of green chemistry are: (i) the use of renewable resources with waste elimination; (ii) avoiding toxic or hazardous reagents for product manufacturing; and (iii) use of renewable biomass as a sustainable alternative to fossil-based feedstock (Sheldon 2016). Keeping these in mind, the use of waste streams as feedstock leads to a sustainable system with feasible productivity. A variety of value-added products can be produced using waste streams as feedstock, a brief summary of which is given in Table 2.4 (Liguori et al. 2013).

2.5 Conclusions and Future Trends

The indiscriminate use of non-renewable resources combined with the increasing global population has led to the emergence of critical issues such as the energy crisis, environmental pollution, climate change, and scarcity of resources. Day-to-day human activities lead to the generation of large amounts of wastes, and it is estimated that currently about 90% of the resources used eventually end up as waste. This poses a significant challenge in the sustainable management and disposal of these wastes. Valorisation of biowaste into useful products solves the twin problem of finding an alternative to fossil-based resources, and the eco-friendly management and disposal of the wastes.

Sustainable waste valorisation is aimed at the generation of energy and fuel (to reduce the dependence on fossil-based residues), the production of high-value platform chemicals and other useful materials, with a significant decrease in GHG emissions, making waste valorisation a clean technology. Most biowastes are complex, heterogeneous mixtures of molecules, and therefore need to be separated before bioconversion. The biggest challenge with respect to the use of wastes as feedstock lies in its composition, which varies based on its source. Additionally, complex organic matter present in the wastes should be degraded into simple sugars by pretreatment of wastes, to enable their bioconversion by microorganisms.

Keeping in mind the limitations of the current practices for conversion of bioresources, future work can be aimed at developing frameworks according to the sourcing industry of wastes, development of effective and efficient technologies for waste pretreatment, as well as bioconversion. Apart from this, legislature and regulations must be modified to include current knowledge about waste valorisation to make the process feasible, and further research can focus on the integration of knowledge about waste valorisation into a targeted public policy.

Acknowledgements

The authors are sincerely thankful to the Natural Sciences and Engineering Research Council of Canada (Engage Grant 122842) for providing financial support.

References

Adamowicz, M. (2017). Bioeconomy – Concept, application and perspectives. *Prob. Agric. Econ.* **1** (350): 29–49.

Ahuja, S.K., Ferreira, G.M., and Moreira, A.R. (2004). Utilization of enzymes for environmental applications. *Crit. Rev. Biotechnol.* **24** (2–3): 125–154.

Amulya, K., Dahiya, S., and Mohan, S.V. (2016). Building a bio-based economy through waste remediation: innovation towards sustainable future. In: *Bioremediation and Bioeconomy* (ed. M.N.V. Prasad), 497–521. Elsevier.

Anwar, Z., Gulfraz, M., and Irshad, M. (2014). Agro-industrial lignocellulosic biomass a key to unlock the future bio-energy: a brief review. *J. Radiat. Res. Appl. Sci.* **7** (2): 163–173.

Arifeen, N., Wang, R., Kookos, I.K. et al. (2007). Process design and optimization of novel wheat-based continuous bioethanol production system. *Biotechnol. Prog.* **23** (6): 1394–1403.

Asagbra, A.E., Sanni, A.I., and Oyewole, O.B. (2005). Solid-state fermentation production of tetracycline by Streptomyces strains using some agricultural wastes as substrate. *World J. Microbiol. Biotechnol.* **21** (2): 107–114.

Ashby, R.D., Solaiman, D.K., and Foglia, T.A. (2004). Bacterial poly (hydroxyalkanoate) polymer production from the biodiesel co-product stream. *J. Polym. Environ.* **12** (3): 105–112.

Baiano, A. (2014). Recovery of biomolecules from food wastes—a review. *Molecules* **19** (9): 14821–14842.

Balu, A.M., Budarin, V., Shuttleworth, P.S. et al. (2012). Valorisation of orange peel residues: waste to biochemicals and nanoporous materials. *ChemSusChem* **5** (9): 1694–1697.

Bio-En Power Inc. (n.d.) Take advantage of a wide range of waste types. http://www.bio-enpower.com/feedstocks (accessed 1 April 2020).

Cantrell, K.B., Ducey, T., Ro, K.S., and Hunt, P.G. (2008). Livestock waste-to-bioenergy generation opportunities. *Bioresour. Technol.* **99** (17): 7941–7953.

Carreira, P., Mendes, J.A., Trovatti, E. et al. (2011). Utilization of residues from agro-forest industries in the production of high value bacterial cellulose. *Bioresour. Technol.* **102** (15): 7354–7360.

Chen, S., Zhang, X., Singh, D. et al. (2010). Biological pretreatment of lignocellulosics: potential, progress and challenges. *Biofuels* **1** (1): 177–199.

Das, R.K., Brar, S.K., and Verma, M. (2015a). A fermentative approach towards optimizing directed biosynthesis of fumaric acid by *Rhizopus oryzae* 1526 utilizing apple industry waste biomass. *Fungal Biol.* **119** (12): 1279–1290.

Das, R.K., Brar, S.K., and Verma, M. (2015b). Enhanced fumaric acid production from brewery wastewater by immobilization technique. *J. Chem. Technol. Biotechnol.* **90** (8): 1473–1479.

Đilas, S., Čanadanović-Brunet, J., and Ćetković, G. (2009). By-products of fruits processing as a source of phytochemicals. *Chem. Ind. Chem. Eng. Q.* **15** (4): 191–202.

Domínguez-Bocanegra, A.R., Torres-Muñoz, J.A., and López, R.A. (2015). Production of bioethanol from agro-industrial wastes. *Fuel* **149**: 85–89.

Duhan, J.S., Kumar, A., and Tanwar, S.K. (2013). Bioethanol production from starchy part of tuberous plant (potato) using *Saccharomyces cerevisiae* MTCC-170. *Afr. J. Microbiol. Res.* **7** (46): 5253–5260.

Ekman, A., Campos, M., Lindahl, S. et al. (2013). Bioresource utilisation by sustainable technologies in new value-added biorefinery concepts – two case studies from food and forest industry. *J. Clean. Prod.* **57**: 46–58.

ESEIA Secretariat (2014). *Innovation Challenge towards Rational use of Bio-Resources in Europe*. http://www.eseia.eu/files/attachments/10457/453058_eseia_Discourse_Book_May_2014.pdf (accessed 1 April 2020).

European Commission (2011). *Technical Committee TC411. M/492 Mandate Addressed to CEN, CEN-ELEC and ETSI for the Development of Horizontal European Standards and Other Standardisation Deliverables for Bio-Based Products*. Brussels: European Commission.

Fernandes, D.L., Pereira, S.R., Serafim, L.S. et al. (2012). Second generation bioethanol from lignocellulosics: processing of hardwood sulphite spent liquor. In: *Bioethanol* (eds. M.A.P. Lima and A.P.P. Natalense), 123–152. InTech.

Fernando, S., Adhikari, S., Chandrapal, C., and Murali, N. (2006). Biorefineries: current status, challenges, and future direction. *Energy Fuel* **20** (4): 1727–1737.

Font-Palma, C. (2012). Characterisation, kinetics and modelling of gasification of poultry manure and litter: an overview. *Energy Convers. Manag.* **53** (1): 92–98.

Galbe, M. and Zacchi, G. (2002). A review of the production of ethanol from softwood. *Appl. Microbiol. Biotechnol.* **59** (6): 618–628.

Girotto, F. and Cossu, R. (2017). Animal waste: opportunities and challenges. In: *Sustainable Agriculture Reviews* (ed. E. Lichtfouse), 1–13. Springer.

González-Pajuelo, M., Andrade, J., and Vasconcelos, I. (2004). Production of 1, 3-propanediol by *Clostridium butyricum* VPI 3266 using a synthetic medium and raw glycerol. *J. Ind. Microbiol. Biotechnol.* **31** (9): 442–446.

Gopinath, K., Sankaranarayanan, A., and Nivedhitha, L. (2016). Platform chemical biorefinery and agroindustrial waste management. In: *Platform Chemical Biorefinery* (eds. S.K. Brar, S.J. Sarma and K. Pakshirajan), 379–391. Elsevier.

Guo, X.M., Trably, E., Latrille, E. et al. (2014). Predictive and explicative models of fermentative hydrogen production from solid organic waste: role of butyrate and lactate pathways. *Int. J. Hydrog. Energy* **39** (14): 7476–7485.

Haddadi, M.H., Aiyelabegan, H.T., and Negahdari, B. (2018). Advanced biotechnology in biorefinery: a new insight into municipal waste management to the production of high-value products. *Int. J. Environ. Sci. Technol.* **15** (3): 675–686.

Hargreaves, J.C., Adl, M.S., and Warman, P.R. (2008). A review of the use of composted municipal solid waste in agriculture. *Agric. Ecosyst. Environ.* **123** (1): 1–14.

Hiremath, A., Kannabiran, M., and Rangaswamy, V. (2011). 1, 3-Propanediol production from crude glycerol from jatropha biodiesel process. *New Biotechnol.* **28** (1): 19–23.

Hoornweg, D. and Bhada-Tata, P. (2012). *What a Waste: A Global Review of Solid Waste Management.* Washington, DC: World Bank.

Huang, H., Qureshi, N., Chen, M.-H. et al. (2015). Ethanol production from food waste at high solids content with vacuum recovery technology. *J. Agric. Food Chem.* **63** (10): 2760–2766.

Ibrahim, M.H. and Steinbüchel, A. (2009). Poly (3-hydroxybutyrate) production from glycerol by *Zobellella denitrificans* MW1 via high-cell-density fed-batch fermentation and simplified solvent extraction. *Appl. Environ. Microbiol.* **75** (19): 6222–6231.

Iqbal, H.M.N., Kyazze, G., and Keshavarz, T. (2013). Advances in the valorisation of lignocellulosic materials by biotechnology: an overview. *Bioresources* **8** (2): 3157–3176.

Isikgor, F.H. and Becer, C.R. (2015). Lignocellulosic biomass: a sustainable platform for the production of bio-based chemicals and polymers. *Polym. Chem.* **6** (25): 4497–4559.

Ito, T., Nakashimada, Y., Senba, K. et al. (2005). Hydrogen and ethanol production from glycerol-containing wastes discharged after biodiesel manufacturing process. *J. Biosci. Bioeng.* **100** (3): 260–265.

Jadhav, D. (2017). Bio-resources & its utilization in health sector of India. *Int. J. Environ. Sci. Nat. Res.* **1** (4): 555568.

Jang, Y.-S., Malaviya, A., Cho, C. et al. (2012). Butanol production from renewable biomass by clostridia. *Bioresour. Technol.* **123**: 653–663.

Jitrwung, R. and Yargeau, V. (2011). Optimization of media composition for the production of biohydrogen from waste glycerol. *Int. J. Hydrog. Energy* **36** (16): 9602–9611.

Kamzolova, S.V., Fatykhova, A.R., Dedyukhina, E.G. et al. (2011). Citric acid production by yeast grown on glycerol-containing waste from biodiesel industry. *Food Technol. Biotechnol.* **49** (1): 65–74.

Karak, T., Bhagat, R., and Bhattacharyya, P. (2012). Municipal solid waste generation, composition, and management: the world scenario. *Crit. Rev. Environ. Sci. Technol.* **42** (15): 1509–1630.

Karakashev, D., Batstone, D.J., and Angelidaki, I. (2005). Influence of environmental conditions on methanogenic compositions in anaerobic biogas reactors. *Appl. Environ. Microbiol.* **71** (1): 331–338.

Kaur, S., Dhillon, G.S., Sarma, S.J. et al. (2014). Waste biomass: a prospective renewable resource for development of bio-based economy/processes. In: *Biotransformation of Waste Biomass into High Value Biochemicals* (eds. S.K. Brar, G.S. Dhillon and M. Fernandes), 3–28. Springer.

Kaur, G., Uisan, K., Ong, K.L., and Ki Lin, C.S. (2018). Recent trends in green and sustainable chemistry & waste valorisation: rethinking plastics in a circular economy. *Current Opinion in Green and Sustainable Chemistry* **9**: 30–39.

Kim, D.Y., Yim, S.C., Lee, P.C. et al. (2004). Batch and continuous fermentation of succinic acid from wood hydrolysate by *Mannheimia succiniciproducens* MBEL55E. *Enzym. Microb. Technol.* **35** (6–7): 648–653.

Kim, J.H., Lee, J.C., and Pak, D. (2011). Feasibility of producing ethanol from food waste. *Waste Manag.* **31** (9): 2121–2125.

Kim, S.B., Cui, C., Lee, J.H. et al. (2013). Rapid analysis of barley straw before and after dilute sulfuric acid pretreatment by photoluminescence. *Bioresour. Technol.* **146**: 789–793.

Korhonen, J., Honkasalo, A., and Seppälä, J. (2018). Circular economy: the concept and its limitations. *Ecol. Econ.* **143**: 37–46.

Körner, I. (2015). Civilization biorefineries: efficient utilization of residue-based bioresources. In: *Industrial Biorefineries and White Biotechnology* (eds. A. Pandey et al.), 295–340. Elsevier.

Koutinas, A.A., Vlysidis, A., Pleissner, D. et al. (2014). Valorisation of industrial waste and by-product streams via fermentation for the production of chemicals and biopolymers. *Chem. Soc. Rev.* **43** (8): 2587–2627.

Kumar, P., Barrett, D.M., Delwiche, M.J., and Stroeve, P. (2009). Methods for pretreatment of lignocellulosic biomass for efficient hydrolysis and biofuel production. *Ind. Eng. Chem. Res.* **48** (8): 3713–3729.

Kumar, A., Sadh, P., Kha, S., and Duhan, J. (2016). Bio-ethanol production from sweet potato using co-culture of saccharolytic molds (*Aspergillus* spp.) and *Saccharomyces cerevisiae* MTCC170. *J. Adv. Biotechnol.* **6** (1): 822–828.

Lammers, P., Kerr, B., Weber, T. et al. (2008a). Digestible and metabolizable energy of crude glycerol for growing pigs. *J. Anim. Sci.* **86** (3): 602–608.

Lammers, P., Kerr, B., Honeyman, M. et al. (2008b). Nitrogen-corrected apparent metabolizable energy value of crude glycerol for laying hens. *Poult. Sci.* **87** (1): 104–107.

Lee, P., Lee, S., Hong, S. et al. (2003). Biological conversion of wood hydrolysate to succinic acid by *Anaerobiospirillum succiniciproducens*. *Biotechnol. Lett.* **25** (2): 111–114.

Lee, J.-W., Gwak, K.-S., Park, J.-Y. et al. (2007). Biological pretreatment of softwood *Pinus densiflora* by three white rot fungi. *J. Microbiol.* **45** (6): 485–491.

Liang, Y., Sarkany, N., Cui, Y., and Blackburn, J.W. (2010). Batch stage study of lipid production from crude glycerol derived from yellow grease or animal fats through microalgal fermentation. *Bioresour. Technol.* **101** (17): 6745–6750.

Liao, W., Liu, Y., Frear, C., and Chen, S. (2008). Co-production of fumaric acid and chitin from a nitrogen-rich lignocellulosic material – dairy manure – using a pelletized filamentous fungus *Rhizopus oryzae* ATCC 20344. *Bioresour. Technol.* **99** (13): 5859–5866.

Lieder, M. and Rashid, A. (2016). Towards circular economy implementation: a comprehensive review in context of manufacturing industry. *J. Clean. Prod.* **115**: 36–51.

Liguori, R. and Faraco, V. (2016). Biological processes for advancing lignocellulosic waste biorefinery by advocating circular economy. *Bioresour. Technol.* **215**: 13–20.

Liguori, R., Amore, A., and Faraco, V. (2013). Waste valorisation by biotechnological conversion into added value products. *Appl. Microbiol. Biotechnol.* **97** (14): 6129–6147.

Lohri, C.R., Diener, S., Zabaleta, I. et al. (2017). Treatment technologies for urban solid biowaste to create value products: a review with focus on low- and middle-income settings. *Rev. Environ. Sci. Biotechnol.* **16** (1): 81–130.

Lokesha, E., Marappan, G., Kumar, R.D. et al. (2017). Crude glycerol: by-product of biodiesel industries as an alternative energy source for livestock feeding. *J. Exp. Bio. Agric. Sci.* **5** (6): 755–766.

Lu, C., Dong, J., and Yang, S.-T. (2013). Butanol production from wood pulping hydrolysate in an integrated fermentation–gas stripping process. *Bioresour. Technol.* **143**: 467–475.

Maina, S., Kachrimanidou, V., and Koutinas, A. (2017). A roadmap towards a circular and sustainable bioeconomy through waste valorisation. *Curr. Opin. Green Sustainable Chem.* **8**: 18–23.

Manyi-Loh, C.E., Mamphweli, S.N., Meyer, E.L. et al. (2013). Microbial anaerobic digestion (bio-digesters) as an approach to the decontamination of animal wastes in pollution control and the generation of renewable energy. *Int. J. Environ. Res. Public Health* **10** (9): 4390–4417.

Matsakas, L., Gao, Q., Jansson, S. et al. (2017). Green conversion of municipal solid wastes into fuels and chemicals. *Electron. J. Biotechnol.* **26**: 69–83.

Maurya, D.P., Singla, A., and Negi, S. (2015). An overview of key pretreatment processes for biological conversion of lignocellulosic biomass to bioethanol. *3 Biotech* **5** (5): 597–609.

Mirabella, N., Castellani, V., and Sala, S. (2014). Current options for the valorisation of food manufacturing waste: a review. *J. Clean. Prod.* **65**: 28–41.

Moldes, A.B., Alonso, J.L., and Parajo, J.C. (2001). Strategies to improve the bioconversion of processed wood into lactic acid by simultaneous saccharification and fermentation. *J. Chem. Technol. Biotechnol.* **76** (3): 279–284.

Moon, H.C., Song, I.S., Kim, J.C. et al. (2009). Enzymatic hydrolysis of food waste and ethanol fermentation. *Int. J. Energy Res.* **33** (2): 164–172.

More, T., Mahmoudi, A., Yan, S., and Tyagi, R.D. (2014). Extracellular polymeric substances production kinetics of 13 sludge isolates using wastewater sludge as raw material and its flocculation potential. *Environ. Technol.* **36** (23): 3022–3035.

Mu, Y., Teng, H., Zhang, D.-J. et al. (2006). Microbial production of 1, 3-propanediol by *Klebsiella pneumoniae* using crude glycerol from biodiesel preparations. *Biotechnol. Lett.* **28** (21): 1755–1759.

Muniyasamy, S., Reddy, M.M., Misra, M., and Mohanty, A. (2013). Biodegradable green composites from bioethanol co-product and poly (butylene adipate-co-terephthalate). *Ind. Crop. Prod.* **43**: 812–819.

Nagarajan, V., Mohanty, A.K., and Misra, M. (2013). Sustainable green composites: value addition to agricultural residues and perennial grasses. *ACS Sustain. Chem. Eng.* **1** (3): 325–333.

Nanda, M.R., Misra, M., and Mohanty, A.K. (2013). Performance evaluation of biofibers and their hybrids as reinforcements in bioplastic composites. *Macromol. Mater. Eng.* **298** (7): 779–788.

Narayana, D. and Johnson, T. (2017). Emerging scenario on industrial use of bio resources and development of models for computing access and benefit sharing (ABS) mechanisms. *J. Biodivers. Bio-prospect. Dev.* **4**: 165.

Nath, K. and Das, D. (2004). Improvement of fermentative hydrogen production: various approaches. *Appl. Microbiol. Biotechnol.* **65** (5): 520–529.

Ndao, A., Sellamuthu, B., Gnepe, J.R. et al. (2017). Pilot-scale biopesticide production by *Bacillus thuringiensis* subsp. *kurstaki* using starch industry wastewater as raw material. *J. Environ. Sci. Health B* **52** (9): 623–630.

Nizami, A.S., Rehan, M., Waqas, M. et al. (2017). Waste biorefineries: enabling circular economies in developing countries. *Bioresour. Technol.* **241**: 1101–1117.

O'Grady, J. and Morgan, J.A. (2011). Heterotrophic growth and lipid production of *Chlorella protothecoides* on glycerol. *Bioprocess Biosyst. Eng.* **34** (1): 121–125.

Pan, X., Arato, C., Gilkes, N. et al. (2005). Biorefining of softwoods using ethanol organosolv pulping: preliminary evaluation of process streams for manufacture of fuel-grade ethanol and co-products. *Biotechnol. Bioeng.* **90** (4): 473–481.

Papanikolaou, S. and Aggelis, G. (2003). Modelling aspects of the biotechnological valorisation of raw glycerol: production of citric acid by *Yarrowia lipolytica* and 1, 3-propanediol by *Clostridium butyricum. J. Chem. Technol. Biotechnol.* **78** (5): 542–547.

Papanikolaou, S., Muniglia, L., Chevalot, I. et al. (2002). *Yarrowia lipolytica* as a potential producer of citric acid from raw glycerol. *J. Appl. Microbiol.* **92** (4): 737–744.

Papurello, D., Soukoulis, C., Schuhfried, E. et al. (2012). Monitoring of volatile compound emissions during dry anaerobic digestion of the organic fraction of municipal solid waste by proton transfer reaction time-of-flight mass spectrometry. *Bioresour. Technol.* **126**: 254–265.

Peksa, M., Dolzan, P., Grassi, A. et al. (2007). *Global Wood Pellets Markets and Industry: Policy Drivers, Market Status and Raw Material Potential*. Utrecht University.

Petkowicz, C., Vriesmann, L., and Williams, P. (2017). Pectins from food waste: extraction, characterization and properties of watermelon rind pectin. *Food Hydrocoll.* **65**: 57–67.

Pfaltzgraff, L.A. and Clark, J.H. (2014). Green chemistry, biorefineries and second generation strategies for re-use of waste: an overview. In: *Advances in Biorefineries* (ed. K.W. Waldron), 3–33. Elsevier.

Potumarthi, R., Baadhe, R.R., Nayak, P., and Jetty, A. (2013). Simultaneous pretreatment and sacchariffication of rice husk by *Phanerochete chrysosporium* for improved production of reducing sugars. *Bioresour. Technol.* **128**: 113–117.

Qureshi, N., Saha, B.C., Dien, B. et al. (2010). Production of butanol (a biofuel) from agricultural residues: part I–Use of barley straw hydrolysate. *Biomass Bioenergy* **34** (4): 559–565.

Ragauskas, A.J., Williams, C.K., Davison, B.H. et al. (2006). The path forward for biofuels and biomaterials. *Science* **311** (5760): 484–489.

Ram, S., Tyagi, R., and Drogui, P. (2018). Effect of sludge concentration and crude glycerol matrix as a substrate on the production of single-cell oil by oleaginous yeast *Yarrowia lipolytica* SKY7. *Fermentation* **4** (2): 24.

Ramachandran, S., Singh, S.K., Larroche, C. et al. (2007). Oil cakes and their biotechnological applications – a review. *Bioresour. Technol.* **98** (10): 2000–2009.

Rodríguez-López, J., Sánchez, A.J., Gómez, D.M. et al. (2012). Fermentative production of fumaric acid from *Eucalyptus globulus* wood hydrolyzates. *J. Chem. Technol. Biotechnol.* **87** (7): 1036–1040.

Rönnlund, I., Pursula, T., Bröckl, M. et al. (2014). *Creating value from bioresources: Innovation in Nordic Bioeconomy*. Nordic Innovation.

Rosentrater, K.A., Ileleji, K., and Johnston, D.B. (2012). Manufacturing of fuel ethanol and distillers grains-current and evolving processes. In: *Distillers Grains: Production, Properties, and Utilization*, 73–102. Boca Raton, FL: CRC Press.

Sadh, P.K., Duhan, S., and Duhan, J.S. (2018). Agro-industrial wastes and their utilization using solid state fermentation: a review. *Bioresou. Bioprocess.* **5** (1): 1.

Saenge, C., Cheirsilp, B., Suksaroge, T.T., and Bourtoom, T. (2011). Potential use of oleaginous red yeast *Rhodotorula glutinis* for the bioconversion of crude glycerol from biodiesel plant to lipids and carotenoids. *Process Biochem.* **46** (1): 210–218.

Saini, J.K., Saini, R., and Tewari, L. (2015). Lignocellulosic agriculture wastes as biomass feedstocks for second-generation bioethanol production: concepts and recent developments. *3 Biotech* **5** (4): 337–353.

Santana-Méridas, O., González-Coloma, A., and Sánchez-Vioque, R. (2012). Agricultural residues as a source of bioactive natural products. *Phytochem. Rev.* **11** (4): 447–466.

Santibañez-Aguilar, J.E., Ponce-Ortega, J.M., Betzabe González-Campos, J. et al. (2013). Optimal planning for the sustainable utilization of municipal solid waste. *Waste Manag.* **33** (12): 2607–2622.

Saravanan, V. and Subramaniyan, V. (2014). Production of biosurfactant by *Pseudomonas aeruginosa* PB3A using agro-industrial wastes as a carbon source. *Malaysian J. Microbiol.* **10** (1): 57–62.

Saunders, J.A. and Rosentrater, K.A. (2009). Survey of US fuel ethanol plants. *Bioresour. Technol.* **100** (13): 3277–3284.

Saxena, R., Adhikari, D., and Goyal, H. (2009). Biomass-based energy fuel through biochemical routes: a review. *Renew. Sust. Energ. Rev.* **13** (1): 167–178.

Schievano, A., D'Imporzano, G., and Adani, F. (2009). Substituting energy crops with organic wastes and agro-industrial residues for biogas production. *J. Environ. Manag.* **90** (8): 2537–2541.

Sheldon, R.A. (2014). Green and sustainable manufacture of chemicals from biomass: state of the art. *Green Chem.* **16** (3): 950–963.

Sheldon, R.A. (2016). Green chemistry, catalysis and valorisation of waste biomass. *J. Mol. Catal. A Chem.* **422**: 3–12.

Shukor, H., Al-Shorgani, N.K.N., Abdeshahian, P. et al. (2014). Production of butanol by *Clostridium saccharoperbutylacetonicum* N1-4 from palm kernel cake in acetone–butanol–ethanol fermentation using an empirical model. *Bioresour. Technol.* **170**: 565–573.

Silva, N.L.C., Betancur, G.J.V., Vasquez, M.P. et al. (2011). Ethanol production from residual wood chips of cellulose industry: acid pretreatment investigation, hemicellulosic hydrolysate fermentation, and remaining solid fraction fermentation by SSF process. *Appl. Biochem. Biotechnol.* **163** (7): 928–936.

Statista (2018). Distribution of municipal solid waste generated worldwide in 2016, by material type. https://www.statista.com/statistics/916666/global-generation-of-municipal-solid-waste-share-by-material (accessed 1 April 2020).

Sun, Y. and Cheng, J. (2002). Hydrolysis of lignocellulosic materials for ethanol production: a review. *Bioresour. Technol.* **83** (1): 1–11.

Szögi, A.A., Vanotti, M.B., and Hunt, P.G. (2015). Phosphorus recovery from pig manure solids prior to land application. *J. Environ. Manag.* **157**: 1–7.

Tarbox, M. (2003). Effluents from food processing: composition and analysis. In: *Encyclopedia of Food Sciences and Nutrition*, 2e (ed. B. Caballero), 1991–1996. Oxford: Academic Press.

Technologico, G.C. (2004). *Handbook for the Prevention and Minimization of Waste and Valorisation of By-products in European Agro-Food Industries*. AWARENET.

Teira-Esmatges, M.R. and Flotats, X. (2003). A method for livestock waste management planning in NE Spain. *Waste Manag.* **23** (10): 917–932.

The World Bank Group. (2019). *Solid Waste Management*. http://www.worldbank.org/en/topic/urbandevelopment/brief/solid-waste-management (accessed 1 April 2020).

Thek, G. and Obernberger, I. (2012). *The Pellet Handbook: The Production and Thermal Utilization of Biomass Pellets*. Routledge.

Tuck, C.O., Pérez, E., Horváth, I.T. et al. (2012). Valorisation of biomass: deriving more value from waste. *Science* **337** (6095): 695.

Uçkun Kiran, E. and Liu, Y. (2015). Bioethanol production from mixed food waste by an effective enzymatic pretreatment. *Fuel* **159**: 463–469.

Uncu, O.N. and Cekmecelioglu, D. (2011). Cost-effective approach to ethanol production and optimization by response surface methodology. *Waste Manag.* **31** (4): 636–643.

Valdés, A., Mellinas, A.C., Ramos, M. et al. (2014). Natural additives and agricultural wastes in biopolymer formulations for food packaging. *Front. Chem.* **2**: 6.

Vandenberghe, L.P.S., Soccol, C.R., Pandey, A., and Lebeault, J.-M. (1999). Microbial production of citric acid. *Braz. Arch. Biol. Technol.* **42** (3): 263–276.

Varanasi, P., Singh, P., Auer, M. et al. (2013). Survey of renewable chemicals produced from lignocellulosic biomass during ionic liquid pretreatment. *Biotechnol. Biofuels* **6** (1): 14.

Vastrad, B. and Neelagund, S. (2011). Optimization and production of neomycin from different agro industrial wastes in solid state fermentation. *Int. J. Pharm. Sci. Drug Res.* **3** (2): 104–111.

Vidyarthi, A.S., Tyagi, R.D., Valero, J.R., and Surampalli, R.Y. (2002). Studies on the production of *B. thuringiensis* based biopesticides using wastewater sludge as a raw material. *Water Res.* **36** (19): 4850–4860.

Villarreal, M., Prata, A., Felipe, M., and Silva, J.A.E. (2006). Detoxification procedures of eucalyptus hemicellulose hydrolysate for xylitol production by *Candida guilliermondii*. *Enzym. Microb. Technol.* **40** (1): 17–24.

Walker, K., Vadlani, P., Madl, R. et al. (2013). Ethanol fermentation from food processing waste. *Environ. Prog. Sustain. Energy* **32** (4): 1280–1283.

Wan, C. and Li, Y. (2012). Fungal pretreatment of lignocellulosic biomass. *Biotechnol. Adv.* **30** (6): 1447–1457.

Wan, S., Sun, L., Douieb, Y. et al. (2013). Anaerobic digestion of municipal solid waste composed of food waste, wastepaper, and plastic in a single-stage system: performance and microbial community structure characterization. *Bioresour. Technol.* **146**: 619–627.

Wee, Y.-J., Yun, J.-S., Park, D.-H., and Ryu, H.-W. (2004). Biotechnological production of L (+)-lactic acid from wood hydrolyzate by batch fermentation of *Enterococcus faecalis*. *Biotechnol. Lett.* **26** (1): 71–74.

Xu, Y., Wang, R.-H., Koutinas, A., and Webb, C. (2010). Microbial biodegradable plastic production from a wheat-based biorefining strategy. *Process Biochem.* **45** (2): 153–163.

Yanagida, T., Minowa, T., Shimizu, Y. et al. (2009). Recovery of activated carbon catalyst, calcium, nitrogen and phosphate from effluent following supercritical water gasification of poultry manure. *Bioresour. Technol.* **100** (20): 4884–4886.

Yang, F., Hanna, M.A., and Sun, R. (2012). Value-added uses for crude glycerol – a byproduct of biodiesel production. *Biotechnol. Biofuels* **5** (1): 13.

Yang, X., Choi, H.S., Park, C., and Kim, S.W. (2015). Current states and prospects of organic waste utilization for biorefineries. *Renew. Sust. Energ. Rev.* **49**: 335–349.

3

Treatment of Waste

Ravindran Balasubramani[1], Vasanthy Muthunarayanan[2], Karthika Arumugam[3], Rajiv Periakaruppan[4], Archana Singh[5], Soon Woong Chang[1], Thamaraiselvi Chandran[6], Gopal Shankar Singh[5] and Selvakumar Muniraj[2]

[1] *Department of Environmental Energy and Engineering, Kyonggi University Youngtong-Gu, Gyeonggi-Do, Suwon 16227, South Korea*
[2] *Water and Solid Waste Processing Lab, Department of Environmental Biotechnology, Bharathidasan University, Tamil Nadu, Tiruchirappalli 620 024, India*
[3] *Department of Biotechnology, Kalasalingam University, Tamil Nadu, Srivilliputtur 626 126, India*
[4] *Department of Biotechnology, Karpagam Academy of Higher Education, Eachanari Post, Tamil Nadu, Coimbatore 641021, India*
[5] *Institute of Environment and Sustainable Development, Banaras Hindu University, Varanasi, India*
[6] *Department of Biotechnology, Mother Teresa Women's University, Tamil Nadu, Kodaikanal 624 101, India*

3.1 Introduction

Waste is defined as a used-up, unwanted, and unusable material thrown into the trash bin (Leblanc 2019). Main sources of waste include domestic, municipal, hazardous, agricultural, and industrial wastes, which are represented in Figure 3.1. Based on their physical, chemical, and biological properties, the waste materials are classified into degradable and non-degradable types. Improper disposal of solid waste results in severe impacts to the environment, such as air, soil, and water pollution, greenhouse gas emissions, infectious diseases, and so on (Leblanc 2019). Therefore, necessary steps should be taken to manage solid waste generation. These steps are collection, separation, segregation and processing,

Waste Valorisation: Waste Streams in a Circular Economy, First Edition.
Edited by Carol Sze Ki Lin, Guneet Kaur, Chong Li and Xiaofeng Yang.

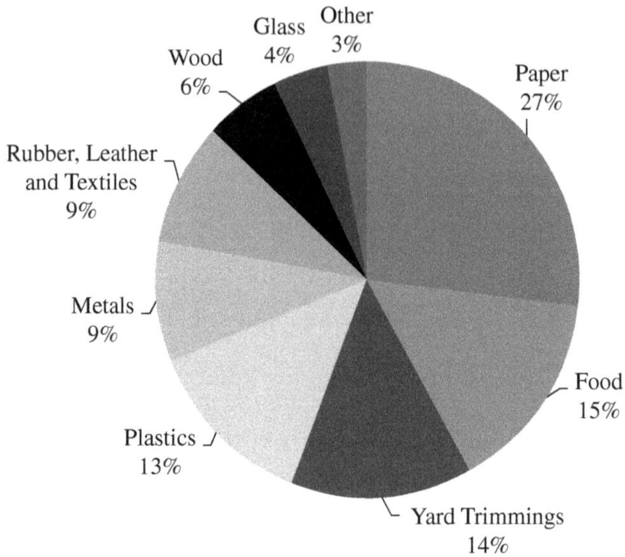

Figure 3.1 Contributions of all types of waste to total solid waste generation.

transfer and transport, reuse, landfill, and energy generation. Management and treatment of waste is discussed as follows.

3.2 Solid Waste Management

The solid waste management rules of each nation follow the collection, treatment, and disposal. The state and central board of a specific country is the monitoring and granting authorization to local bodies for processing and disposal of solid waste. It could be done by conventional methods like land filling, incineration, or utilization of the waste for energy consumption (Liyanage et al. 2015). The method of management depends on the type of waste. The management of different types of wastes varies and is discussed in the following sections as follows.

3.2.1 E-waste Management

Electronic waste, or e-waste, comprises old, end-of-life electrical and electronic appliances such as telephones, cellular handsets, computers, laptops, television sets, refrigerators, washing machines, air-conditioners, fluorescent and other mercury-containing lamps, and so on. The E-waste management rules apply to every manufacturer, producer, consumer, bulk consumer, collection center, dealer, e-retailer, refurbisher, dismantler, and recycler involved in the manufacture, sale, transfer, purchase, collection, storage and processing of e-waste or electrical and electronic equipment (Li et al. 2015). E-waste management could be done by returning the electronic item to the certified e-waste recycler to find appropriate recycling options.

3.2.2 Hazardous Waste Management

Hazardous waste refers to wastes that are toxic, flammable, explosive, or corrosive waste that could cause danger to health or environment (Safari et al. 2018). Hazardous waste management must follow the correct steps, namely prevention, minimization, reuse, recycling, recovery, utilization including co-processing, and safe disposal. Treatment of hazardous waste can be done either by incineration or pyrolysis.

3.2.3 Biomedical Waste Management

Biomedical waste means any waste generated during the diagnosis, treatment or immunization of human beings or animals, or research activities, or during the production or testing of biological specimens. A common biomedical waste treatment and disposal facility (CBMWTF) shall be responsible for the safe handling and disposal of biomedical waste. Biomedical waste could be treated using autoclaves, microwave digesters, and chemical disinfectants.

3.2.4 Plastic Waste Management

Plastic waste include everyday items like used plastic bags and water bottles, broken furniture, product packaging, broken home appliances, and so on. Plastic waste could be managed by reducing, reusing, and recycling the plastic materials.

3.2.5 Solid Waste Management Options

Waste treatment techniques could be followed in order to reduce the volume and toxicity of solid waste. These could help to transform the waste into a more convenient form before disposal (Aryan et al. 2019). Based on the form, composition and quantity of solid waste material, different treatment processes and disposal methods are chosen.

Hence, the main objective of solid waste management is to reduce the impact of waste materials on human health and environment. There are several types of waste treatment steps that are followed. They are as follows:

- Waste bins are placed at particular sites where the waste generation is sufficient so that handling, storage, and processing facilitate easiest way for the waste treatment.
- Waste transfer and transport involve moving waste material from local, regional to waste disposal site through large transport vehicle.
- High defined equipment and techniques are employed to recover, reusable, or recyclable materials from waste stream and to improve the effectiveness of waste management.
- Finally, waste is disposed by various methods like thermal (Landfill, Incineration, Gasification, Pyrolysis, open dumping), biological (composting, vermicomposting), and mechanical treatment (Separation, Compaction, etc.).

This chapter discusses the classification of waste, factors influencing specific treatment process and the techniques available for the conversion of particular waste type.

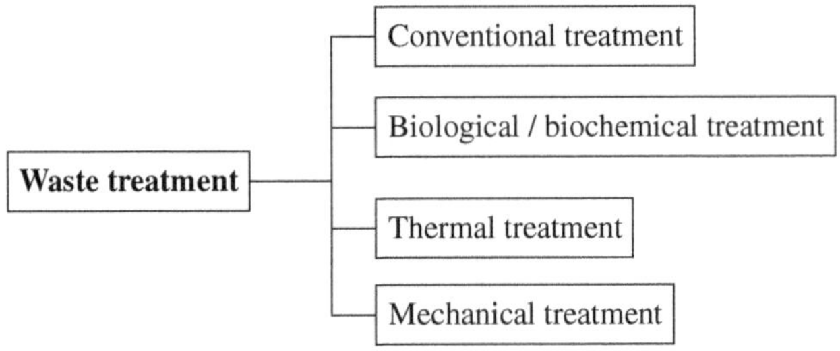

Figure 3.2 Classification of solid waste treatment processes.

3.3 General Approach for Waste Treatment and Conversion to Value-added Products: Biochemical, Mechanical, and Thermochemical

Solid waste management is necessary to protect both public health and the environment. A solid waste management program guides people to manage solid waste properly. Proper waste disposal brings a safer world with less potential harm to both humans and animals. Commonly, three major treatment methods, namely physical, biological, and thermal methods, are followed for managing waste (Figure 3.2).

3.3.1 Conventional Treatment

The conventional treatment of solid wastes includes open dumps and landfills. They are highly preferred disposal methods as they are of low cost and need no prior investment (Figure 3.3).

3.3.1.1 Open Dumps

Dumping is the common method used everywhere to treat household waste. The waste is dumped in open sites, often a vast area without residents nearby. Various animals like

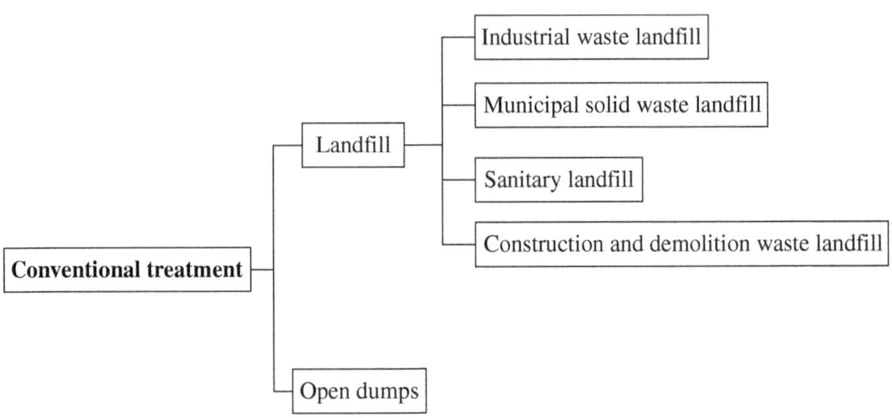

Figure 3.3 Types of conventional waste treatment methods.

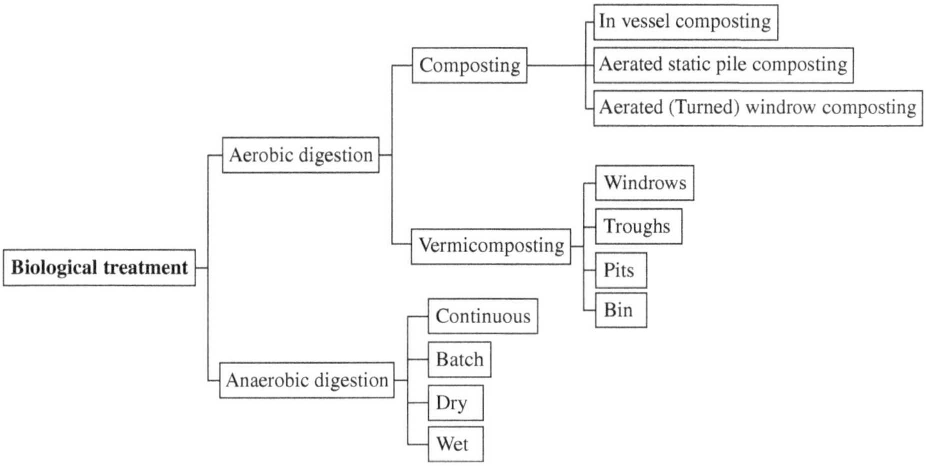

Figure 3.4 *Classification of biological/biochemical treatment.*

rats, mice, birds, and worms are involved the degradation of waste. Rodents, insects, and other vermin attracted to open dump sites may pose health risks. Such open dumps become a source of physical, chemical, and biological hazards. Contaminants can end up in our groundwater, rivers, and lakes that are also sources of drinking water.

3.3.1.2 Landfill

Landfills is designed carefully for the purpose of treating household and nonhazardous wastes. In this process, the waste is buried in a well-defined layer in one specific site to prevent groundwater contamination (Brunner 2019). It is one of the oldest methods of waste treatment. However, hazardous and industrial solid waste can also make its way into landfill. Gases escaping from landfills contain toxic pollutants that can cause cancer, asthma, and other serious health effects. These gases have been reported to carry toxic chemicals such as paint thinner, solvents, pesticides, and other hazardous volatile organic compounds. All dumps also leak toxic leachate; even "state-of-the-art" landfills will eventually leak and pollute nearby groundwater.

3.3.2 Biological/Biochemical Treatment

Biological treatment of waste is the process of recycling waste material in order to produce biofertilizer, sludge, compost, and biogas by means of aerobic/anaerobic processes (Figure 3.4).

3.3.2.1 Aerobic Digestion

Aerobic digestion refers to biological decomposition of organic matter that takes place in the presence of oxygen with the help of microorganisms. The resultant product, namely organic compost, could be used as organic fertilizer. This process is applicable for both small- and large-scale application.

3.3.2.2 Composting

Composting is the best example of a controlled aerobic process. The decomposition takes place by the action of small invertebrates and microorganisms. The most common composting techniques include static pile composting, vermicomposting, windrow composting, and in-vessel composting. Solid organic waste material, sewage sludge material, food waste, and garden waste could be treated using composting.

The composting process occurs naturally in the environment, although efficient composting is needed to control several factors to avoid environmental problems such as odors and dust, and for obtaining quality agricultural manure. Generally, the composting process is divided into two phases: the bio-oxidative phase and the maturing phase, also called the curing phase. The bio-oxidative phase again includes several stages: initial activation, the thermophilic phase, and the mesophilic or maturation phase. The organic waste degradation occurs during the thermophilic phase, as microorganisms degrade the available compounds in the organic waste (Keener et al. 2000). In this phase, high-temperature generation in the composting pile is observed due to the microbial degradation of organic waste. Generally, high temperatures are required (i.e., 55°C) to destroy pathogens in the organic solid waste, and composting piles often reach temperatures as high as 70°C during the degradation of animal manure and green waste (Singh and Kalamdhad 2014; Tang et al. 2011; Cáceres et al. 2015). In addition, in this stage the organic compounds degrade to CO_2 and NH_3 (Jeong et al. 2017; Kim et al. 2017), with uptake of O_2. Finally, when the decomposable organic fraction in the waste is exhausted, the temperature starts decreasing in the composting pile and the maturation phase commences, with stabilization and humification of the organic matter, producing mature compost with humic characteristics (Ravindran and Sekaran 2010). Hence, this compost can be characterized as the stable and sanitized product of the composting process, making composting suitable for sustainable agriculture and resource management (Ravindran et al. 2019; Gajalakshmi and Abbasi 2008; Haug 1993).

Over recent decades, researchers have been focusing on the study of the complex interactions between physical, chemical, and biological factors that takes place during natural composting, using controlled conditions for the composting process. There are several factors affecting the composting process, such as nutrient balance, pH, porosity, and moisture of the composting mix, together with O_2 and water content, and temperature. Thus, the control of these parameters are established to be key factors for optimization of the composting process, as they determine the optimal conditions for organic matter degradation and microbial growth (Agnew and Leonard 2003; Richard et al. 2002). The effectiveness of compost material, with regard to its beneficial effects on agricultural as a nutrient source, depends on the quality of the compost. The quality criteria for compost relate to the nutrient content, organic matter stabilization, and presence of heavy metals or toxic compounds.

On the other hand, the traditional composting process also results in partial decomposition, which has phytotoxic effects in the final product (Ravindran et al. 2014). In order to overcome the conventional composting limitations, several researchers have employed an in-vessel composting strategy for different types of waste such as flyash with organic waste (Mandpe et al. 2019); swine manure (Awasthi et al. 2019; Ravindran et al. 2019); municipal solid waste (MSW) (Malinowski et al. 2019); and sewage sludge (Jain et al. 2019). An in-vessel composting system provides a controlled environment that eliminates some of the obstacles inherent in traditional composting systems for producing nutrient-rich matured compost.

3.3.2.3 Vermicomposting

Vermicomposting is widely used for managing solid waste. They stabilize organic matter present in the solid waste with the help of earthworms and microorganisms (Williams 2013). The end product can be applied as organic fertilizer in the agriculture field in order to reduce the harm caused by hazardous chemical fertilizers. The entire organic fraction of solid waste and sewage waste can be treated using vermicomposting.

The vermicomposting process for organic waste disposal is a suitable approach for decreasing environment pollution by degrading organic waste. Vermicomposting, similar to composting, is biological decomposition with an addition of earthworms for degradation of organic waste for the production of stabilized bioorganic fertilizer, namely vermicompost, for agricultural purposes (Lim et al. 2015). Earthworms are the main drivers, and along with the influence of microorganisms, they give a faster rate of biodegradation of organic waste than the composting process alone (Sim and Wu 2010). Researchers have attempted vermicomposting with many types of organic solid waste, such as manure from livestock, food and vegetable waste, industrial waste, sewage sludge and agricultural waste. However, as several researchers explained in their research, for an efficient and successful vermicomposting process, an appropriate organic waste or feedstock for earthworms is needed (Ravindran et al. 2008, 2014; Yadav and Garg 2011; Rajiv et al. 2013, 2014). For example, cow dung is provided as an amendment in the vermicomposting process as it is the best nutrient-rich animal waste for the growth of earthworms (Edwards 2004). In addition, one of the major drawbacks of utilizing the vermicomposting process is that it does not involve a thermophilic stage, unlike composting. Hence, reduction of pathogens in the organic waste was found to be a difficult process. However, a study conducted by Sim and Wu (2010), explained that pathogens could be killed or reduced through the action of enzymes present in the intestines of earthworms, making the organic waste a less conducive environment for pathogens (Sim and Wu 2010). But the low temperatures may mean that the vermicomposting process cannot meet the required low levels of pathogens in the resulting organic fertilizer (Fornes et al. 2012). Additionally, it can also lead to secondary air pollution such as GHG emissions, thus reducing the positive benefits of the process. This is due to the biodegradation process that produces CO_2, methane, and nitrous oxide when undergoing vermicomposting (Friedrich and Trois 2013). However, vermicomposting is considered to be a simple, clean, and sustainable technology for production of biofertilizer to be applied to agricultural land.

3.3.2.4 Anaerobic Digestion

Anaerobic treatment of solid waste is a natural process which takes place in the absence of oxygen to degrade the organic matter in the solid waste. It results in the production of biogas, which can be converted into heat or other energy forms. The remaining sludge can be further composted in such a way to recover the nutrients.

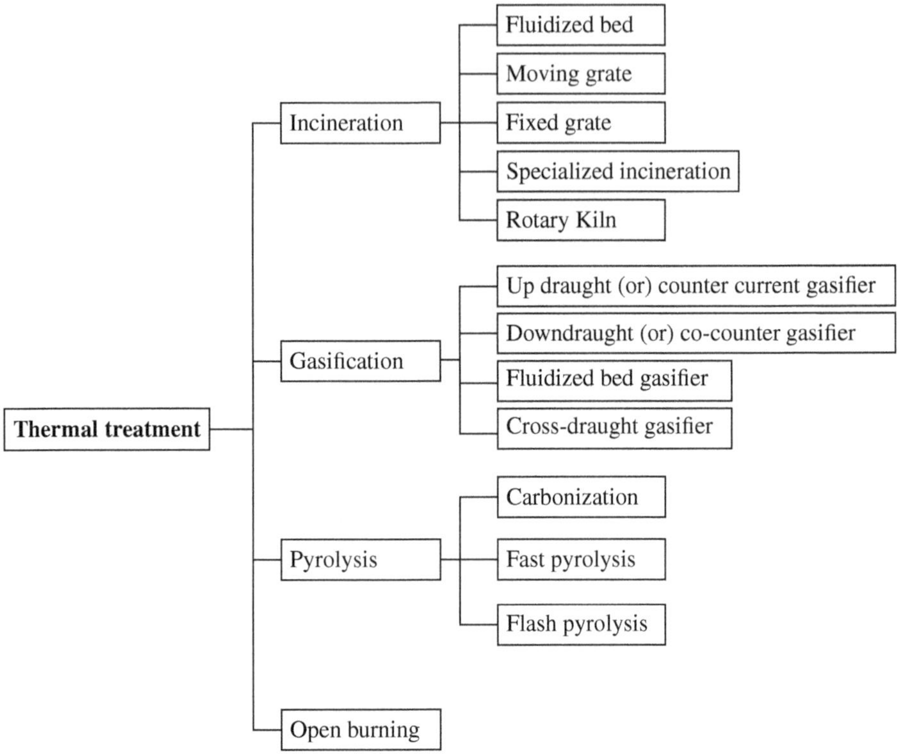

Figure 3.5 Thermal treatment classification.

3.3.3 Thermal Methods

Thermal treatment comprises of waste incineration, pyrolysis, hydration, and drying. The thermal treatment of waste allows for the possibility of volume reduction, energy recovery, and production of marketable secondary raw materials, such as bottom ash, which could be mixed with glass and could be made into inert glass. Further, bottom ash is also used for the production of tiles and bricks (Figure 3.5).

3.3.4 Open Burning

Open burning is the process of burning all types of degradable and non-degradable wastes in an open space, where smoke and other emissions are released directly into the air. This type of burning is commonly practiced everywhere to dispose of waste. Biomedical waste could also be treated using the open burning process.

3.3.5 Mechanical Treatment

Mechanical treatment is the process of sorting or separating recyclable materials from solid waste. Various steps are followed for the mechanical treatment process (Matellanes et al. 2019) (Figure 3.6).

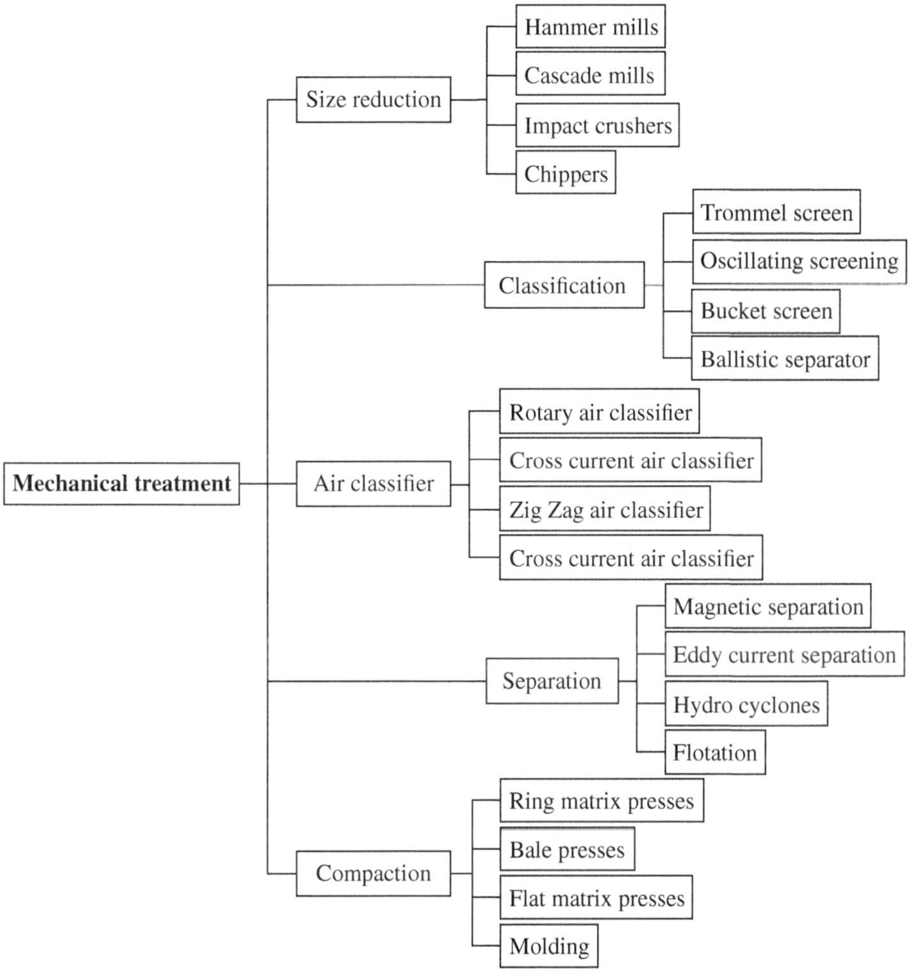

Figure 3.6 *Mechanical treatment classification.*

Size reduction is the breaking down of waste material into small pieces. Size reduction increases the specific surface area of waste material. The following mechanisms are followed for reducing the size of material.

For size reduction, air classifiers are employed to assist the process of screening and classifying the waste material based on their properties, such as size, shape, and density. This process helps to identify the various aggregates separately to follow a specific waste treatment.

Separation or sorting refers to the process where waste materials are collected individually based on their physical characteristics. This technique is very expensive, and hence not suitable for a heterogeneous waste stream. A compaction process is basically done by binding via adhesion and cohesion forces, solid bridging, capillary force attraction, and interlocking bridging, so that the dust particles from waste can be converted into solid granulates. This process plays a significant role in waste management, due to the low cost for transportation and storage of waste.

3.4 Factors Influencing Selection of an Appropriate Valorisation Technique for Specific Waste Types

The growth of urban green areas globally produces large amounts of green waste. Hence, it has become an environmental problem in developed and developing countries (Zhang and Sun 2017). In addition, green wastes comprise the largest fraction of MSW (Wei et al. 2017). Green waste comprises tree bark and wood, green and dead leaves, soil and grass clippings, young trees and shrubs pruning, and is produced from parks, gardens, reserves, and domestic dwellings (Reyes-Torres et al. 2018). Green waste is unattractive and expensive because of its low bulk density and low value, respectively.

Use of wastes for energy production is a cost-effective and sustainable process. Paper wastes can be valorized into valuable energy routes through advanced thermal modification techniques, hence providing not only a solution for disposal of waste, but also a means to producing important sustainable energy at paper mill sites (Ouadi et al. 2017). Important factors for waste valorisation as energy or materials are an awareness of the waste stream composition and its availability (e.g., seasonality).

Pretreatment of waste is another important factor in making the wastes suitable for further valorisation. In this respect, pretreatment of lignocellulosic wastes relies on various factors: (i) lignin polymer heterogeneity; (ii) lignocellulosic biomass nature; (iii) requirement of higher energy to produce a lower energy product; and (iv) production of toxic inhibitor compounds in recycling of chemicals used (Corrado et al. 2019).

3.4.1 Case Study of Paper Waste Recycling

Worldwide, paper waste recycling has taken place for many years. It was observed that around 419.7 million metric tonnes of paper and cardboard were produced globally in 2017, and more than 50% of that production was attributable to packaging paper. Nowadays, in a more environmentally concerned globe, the recycling of paper waste has growing importance (Bajpai 2013). In this respect, the European paper recycling value chain will increase the rate of paper recycling to about 74% by 2020, from around 71.5% in 2015 (www.paperforrecycling.eu). The American Forest & Paper Association announced a goal to increase the paper recycling rate to around 70% by 2020, from 63.5% in 2010 (Bajpai 2013). Mostly, paper producers are reutilizing fibers for paper production and associated products. Secondary papers are actually produced from used paper; hence the methods suit the rapid advancements in deinking methods. With huge amounts of paper consumed each year, paper waste is available in abundance, producing a huge amount of recycled paper and related products of different qualities and levels. Paper consumption mainly for printing is expected to increase 3–4 times in the next 10–15 years (Scott 2019).

3.4.2 Deinking Process

Once the paper waste is collected and sorted into various categories, it is modified into pulps by water-logging in large vats with a repulping process. After separation of inks from coarse fibers, chemicals were added to avoid their attachment to paper fibers. Then, other additives and inks are isolated from the coarse pulp utilizing a screening series, and fibers of paper further cleansed repeatedly using chemicals and heat. Then, the pulp and chemicals including calcium soap are blended in a flotation device using air bubbles. Any residual ink float to the surface for removal. This is known as the deinking method. So, the clean pulps might be processed into different papers and related products, depending on its quality (Bajpai 2015, 2016).

Thus the conventional methods of waste disposal, such as landfill or incineration, are diverted to reduce land use and harmful emissions. Recycled fiber use decreases energy and water requirements in comparison with virgin wood use. Moreover, paper recycling methods produce less CO_2, at around 20–50% that of consuming virgin fibers (Bajpai 2013). But not all aspects of the paper recycling methods are good. The deinking process generates significant amounts of sludge and waste, known as deinking residues or deinking paper sludge, which commonly ends in landfill, land spreading or incineration. Increasing costs of landfill or land spreading, and introducing strict laws associated with waste disposal, gives the motivation for efficient valorisation of these solid wastes, which are notable environmental and economic burdens for paper recycling industries. Furthermore, as the world worries about pollution, waste generation, and landfills, the need for maximum efficiency and competitiveness will become more apparent. key industrial sectors will be important in finding value in deinking residue, which presently remains as low-value treated waste or underutilized waste (Bajpai 2013).

3.4.3 Paper Deinking Residue

Four major categories of waste are usually produced from the pulp and paper industry:

(i) Discards, for instance, fiber, staples, glass and sizing agents and other metals.
(ii) Deinking sludge/residue comprising inks, short fibers, and calcium carbonate and other additives.
(iii) Primary sludge produced from fillers and flotation of fines.
(iv) Secondary sludge with more microbial matter produced in the wastewater treatment process.

It is notable that the waste stream characteristics are dissimilar from each other (Gottumukkala et al. 2016).

Around 5–40% of raw materials in the recycling of paper come from deinking sludge and the rest is good quality recovered paper (Zhang and Matharu 2018). The average formation rate of excluded fibers and deinking sludge is about 15%, measured by the raw component input on a dried basis (Bajpai 2013). Around 11 million tons of paper wastes are produced by Europe through the paper industries (Zhang and Matharu 2018). Deinked recycled fibers account for approximately 70% of wastes, and deinked residues amount to about 2.3 million tons of waste per year (Monte et al. 2009).

Paper can be recycled six to seven times theoretically. But, in actuality, paper can only be recycled two to three times, since the process of recycling the cellulosic fibers results in shorter fibers after each recycling process. Eventually these fibers must be discarded as deinked residues because of their short length (Gottumukkala et al. 2016). Therefore, the deinked residues are compound waste material and chiefly comprises short and distorted cellulosic fibers and inorganic mineral filler, such as clay, silica, calcium carbonate, coatings, and ink particles, depending on the type of paper being recycled (Monte et al. 2009).

Thermochemical technologies, including pyrolysis and microwave heating, should be considered for conversion of paper deinking residues into value-added products with a diverse range of potential purposes. The combination of pyrolysis and microwave heating make the recycling process more efficient and faster under mild situations. Microwave-assisted pyrolysis at low temperatures is useful for valorisation of deinking residues in the biorefinery, and has a low adverse effect on the environment.

3.5 Conventional and Novel Techniques: Overall Comparison in Terms of Energy Consumption, Waste Stream Generation and Cost

Since 1960, waste generation has drastically increased because of different intrinsic and extrinsic factors. Therefore, the management of such waste has become more challenging for both the public and environmentalists in recent years. The treatment of such waste must be effectively safe and, most importantly, it should be technically sound. A much greater revolution has been observed during the last few decades in treating and processing environmental waste. Different methods of waste management have been proposed by researchers so far, of which a few are discussed in detail in this chapter.

3.5.1 Pyrolysis

Pyrolysis is another conventional method of disposal of carbon-based materials via internal and external sources of heat, typically at temperatures ranging from 450–750°C, anaerobic conditions that would result in production of a carbonaceous char, oils, and combustible gases. This eliminates the volatile portions of the organic materials, resulting in a composition of gas called syngas consisting primarily of H_2, CO, CO_2, CH_4, and complex hydrocarbons.

3.5.2 Gasification

Gasification is a process of anaerobic waste treatment that involves the conversion of all hydrocarbon compounds to CO_2 and H_2O through the application of thermal heat by the injection of steam at 700–1600°C. This partial oxidation process produces a composite gas comprising primarily hydrogen (H_2) and carbon monoxide (CO).

3.5.3 Incineration

Incineration involves the process of combustion of solid waste chemical elements (carbon, hydrogen, sulfur) in an aerobic environment, at temperatures greater than 850°C. As a result, this process produces combustion gases consisting of CO, CO_2, NO_x, H_2O, SO_2, ash, and heat. Thus the inorganic content of the waste is converted to ash.

The main goal of this heat treatment of waste is to reduce the volume of the treated waste with simultaneous extraction of energy. The recovered energy can be used for various purposes such as heating, steam production, or generation of electricity. The net energy that can be produced per ton of solid waste approximately ranges from about 0.7 MW/h of electricity to about 2 MW/h for district heating.

3.6 Energy Consumption, Waste Stream Generation, and Costs of Conventional and Novel Waste Treatment Technologies

An economic comparison of these techniques should not only compare the investment against the collection costs, but also consider and evaluate the area required, as well as operational and maintenance costs.

The treatment and disposal of waste at the conventional level includes their cost in terms of the cost paid to the garbage collector, the social cost of waste (disposed of in landfills), and also includes the amount of toxic gases such as methane emitted to air, and contamination of groundwater. Other environmental (intangible) costs include unwanted odor and noise from heavy vehicles. In conventional methods, the land values are added cost. There are some municipalities that charge a rent, which is a major limitation to conventional waste treatment methods such as landfill. The construction of collection vehicles is included, as well as their operation maintenance of the same to prevent loss of liquid or solid waste material, in order to minimize health and safety hazards. Prevention of odors or fly breeding can be an added economic cost.

For modern waste treatment methods, the operational cost is lowered by storing waste in a single place prior to transporting it to its final treatment destination. Therefore, fewer large-volume trucks are required to transfer the waste to the final disposal facility, thereby reducing transportation cost. This method saves fuel as well as manpower. By using modern methods such as waste to energy, the waste could be converted into useful material and be used for resource recovery, thereby greatly reducing the final waste volume. The obtained resources could be utilized to generate revenue to fund waste management. Recent studies have proven that advanced methods of waste treatment are the most cost-effective method to achieve the objectives of waste management disposal in developing countries (Brunner and Fellner 2007).

The recovery of energy from non-recyclable wastes is the need of the hour. This is referred to as WtE – the waste to energy concept. This is normally made possible by subjecting the waste to combustion, followed by utilization of the produced chemical energy into heat and/or power. In comparison with commercial fuels, MSW is reported to have lower energy content (Changkook and Donghoon 2013). However, there are technologies available for WtE conversion from MSW, such as grate-type combustion, fluidized bed combustion, and pyrolysis/ash-melting (Caneghem et al. 2012). Other technologies available include production of solid recovered fuel, and gasification (Caputo and Pelagagge 2002; Arena 2012). The solid residue obtained after the recovery of heat could be used for construction (ISWA 2006). This type of sludge disposal is reported be less detrimental to the environment (Astrup 2008).

3.7 Conclusions and Future Trends

Urbanization, along with the population explosion and industrialization, has posed challenges to mankind, including sustainable solid waste management and tackling the energy crisis. It is necessary to select a solid waste treatment methodology based on the characteristics of the waste. For the management of non-biodegradable wastes, thermochemical methods, namely incineration, gasification and pyrolysis, could be utilized. For the organic fraction, aerobic and anaerobic digestion could be used with a recycling strategy. Currently, waste to energy schemes are gaining much attention, along with the concept of the circular economy. This focuses on a "win-win" philosophy which states that a prosperous economy and a healthy environment can co-exist. The main aim of the circular economy is to maintain the value of materials, resources, and products. A healthy WtE supply chain certainly helps to ensure a circular economy (Massarutto 2015; Trindade et al. 2018; Palacio et al. 2019). However, as mentioned above, the circular economy has now gained much importance as it insists on recovery of the waste, depending on the quality of waste and the adapted technology. But it is also reported that an optimized plant could generate good amounts of energy (Massarutto 2015), as there now exists a wide range of WtE technologies. Each of the thermochemical and biochemical conversion processes has their own pros and cons. Hence such energy recovery could be framed based on the economic aspects and based on the waste available for energy recovery. The investment costs depend on the complexity of the technology along with the auxiliary processes needed to be installed. It is of course clear that biological treatments are always more cost-effective than thermochemical treatment. While the capital and operational costs of incineration and pyrolysis are reported to be around 400–700 and 40–80 US$/ton of MSW/yr respectively, and gasification accounts for 250–850 US$/ton of MSW/yr capital cost, but landfilling with gas recovery accounts for a capital cost of 10–30 US$/ton of MSW/yr and an operational cost of 1–3 US$/ton of MSW/yr (Kumar and Samadder 2017). While the process of incineration facilitates the recovery of the energy value of waste, pyrolysis and gasification help to recover the chemical value of waste by using the derived chemical products as inputs for other processes. Conventional methods such as sanitary landfills could be practiced, provided proper plastering and care is taken. However, open dumping must no longer be practiced. Hence, the current focus must be on maximizing recycling, safe waste management, along with energy production from solid wastes, ensuring sustainability and environmental protection.

Acknowledgement

The authors from Bharathidasan University (VM and SM) acknowledge the UGC – SAP, UGC – Non SAP, DST-FIST, DST – PURSE and MHRD – RUSA for providing the necessary facilities in the department and University.

References

Agnew, J.M. and Leonard, J.J. (2003). The physical properties of compost. *Compost Sci. Util.* **1**: 238–264.

Arena, U. (2012). Process and technological aspects of municipal solid waste gasification. A review. *Waste Manag.* **32** (4): 625–639.

Aryan, Y., Yadav, P., and Samadder, S.R. (2019). Life Cycle Assessment of the existing and proposed plastic waste management options in India: A case study. *J. Clean. Prod.* **211**: 1268–1283. https://doi.org/10.1016/j.jclepro.2018.11.236.

Astrup, T. (2008). *Management of APC residues from W-t-E plants. An overview of management options and treatment methods.* ISWA Working Group on Thermal Treatment of Waste. https://www.iswa.org/

uploads/tx_iswaknowledgebase/Management_of_APC_residues_from_W-t-E_Plants_2008_01.pdf (accessed 26 April 2020).

Awasthi, M.K., Liu, T., Chen, H. et al. (2019). The behavior of antibiotic resistance genes and their associations with bacterial community during poultry manure composting. *Bioresour. Technol.* **280**: 70–78.

Bajpai, P. (2013). *Recycling and Deinking of Recovered Paper*. Elsevier.

Bajpai, P. (2015). *Management of Pulp and Paper Mill Waste*. Switzerland: Springer International Publishing.

Bajpai, P. (2016). *Pulp and Paper Industry: Energy Conservation*. Amsterdam: Elsevier.

Brunner, P.H. (2019). WTE: Thermal waste treatment for sustainable waste management. In: *Recovery of Materials and Energy from Urban Wastes*, Encyclopedia of Sustainability Science and Technology Series (eds. N.J. Themelis and A.C. Bourtsalas). New York: Springer.

Brunner, P.H. and Fellner, J. (2007). Setting priorities for waste management strategies in developing countries. *Waste Manag. Res.* **25** (3): 234–240.

Cáceres, R., Coromina, N., Malińska, K., and Marfà, O. (2015). Evolution 729 of process control parameters during extended co-composting of green waste and solid fraction of cattle slurry to obtain growing media. *Bioresour. Technol.* **179**: 396–406.

Caneghem, J.V., Brems, A., Lievens, P. et al. (2012). Fluidized bed waste incinerators: Design, operational and environmental issues. *Prog. Energy Combust. Sci.* **38**: 551–582.

Caputo, A.C. and Pelagagge, P.M. (2002). RDF production plants: I. Design and costs. *Appl. Therm. Eng.* **22** (4): 423–437.

Changkook, R. and Donghoon, S. (2013). Combined heat and power from municipal solid waste: current status and issues in South Korea. *Energies* **6** (1): 45–57.

Corrado, S., Caldeira, C., Eriksson, M. et al. (2019). Food waste accounting: methodologies, challenges and opportunities. *Glob. Food Sec.* **20**: 93–100.

Edwards, C.A. (ed.) (2004). *Earthworm Ecology*, 2e 448 pp. Boca Raton, FL: CRC Press.

Fornes, F., Mendoza-Hernández, D., García-de-la-Fuente, R. et al. (2012). Composting versus vermicomposting: a comparative study of organic matter evolution through straight and combined processes. *Bioresour. Technol.* **118**: 296–305.

Friedrich, E. and Trois, C. (2013). GHG emission factors developed for the recycling and composting of municipal waste in South African municipalities. *Waste Manag.* **33**: 2520–2531.

Gajalakshmi, S. and Abbasi, S.A. (2008). Solid waste management by composting: state of the art. *Crit. Rev. Env. Sci. Technol.* **38**: 311–400.

Gottumukkala, L.D., Haigh, K., Collard, F.-X. et al. (2016). Opportunities and prospects of biorefinery-based valorisation of pulp and paper sludge. *Bioresour. Technol.* **215**: 37–49.

Haug, R.T. (1993). *The Practical Handbook of Compost Engineering*. Boca Raton, FL: Lewis Publishers.

ISWA. (2006). *Management of Bottom Ash from WTE Plants*. ISWA Working Group on Thermal Treatment of Waste. https://www.iswa.org/uploads/tx_iswaknowledgebase/Bottom_ash_from_WTE_2006_01.pdf (accessed 26 April 2020).

Jain, M.S., Paul, S., and Kalamdhad, A.S. (2019). Interplay of physical and chemical properties during in-vessel degradation of sewage sludge. *Waste Manag.* **98**: 58–68.

Jeong, K.H., Kim, J.K., Ravindran, B. et al. (2017). Evaluation of pilot-scale in-vessel composting for Hanwoo manure. *Bioresour. Technol.* **245**: 201–206.

Keener, H.M., Dick, W.A., and Hoitink, H.A.J. (2000). Composting and beneficial utilization of composted by- product materials. In: *Land Application of Agricultural, Industrial, and Municipal By-Products* (ed. W.A. Dick), 315–341. Madison: Soil Science Society of America, Inc.

Kim, J.K., Lee, D.J., Ravindran, B. et al. (2017). Evaluation of integrated ammonia recovery technology and nutrient status with an in-vessel composting process for swine manure. *Bioresour. Technol.* **245**: 365–371.

Kumar, A. and Samadder, S.R. (2017). A review on technological options of waste to energy for effective management of municipal solid waste. *Waste Manag.* **69**: 407–422.

Leblanc, R. (2019). *An Introduction to Solid Waste Management*. The Balance Small Business. https://www.thebalancesmb.com/an-introduction-to-solid-waste-management-2878102 (accessed 25 April 2020).

Li, J., Xianlai, Zeng, and Stevels, Ab. (2015). Ecodesign in Consumer Electronics: Past, Present, and Future. *Crit. Rev. Environmental Science and Technology* **45** (8): 840–860, DOI: 10.1080/10643389.2014.900245.

Lim, S.L., Wu, T.Y., Lim, P.N., and Shak, K.P.Y. (2015). The use of vermicompost in organic farming: overview, effects on soil and economics. *J. Sci. Food Agric.* **95**: 1143–1156.

Liyanage, B., Gurusinghe, R., Herat, S., and Tateda, M. (2015). Case study: Finding better solutions for municipal solid waste management in a semi local authority in Sri Lanka. *Open J. Civ. Eng.* **5**: 63–73.

Malinowski, M., Wolny-Koładka, K., and Vaverková, M.D. (2019). Effect of biochar addition on the OFMSW composting process under real conditions. *Waste Manag.* **84**: 364–372.

Mandpe, A., Paliya, S., Kumar, S., and Kumar, R. (2019). Fly ash as an additive for enhancing microbial and enzymatic activities in in-vessel composting of organic wastes. *Bioresour. Technol.* **293**: 122047.

Massarutto, A. (2015). Economic aspects of thermal treatment of solid waste in a sustainable WM system. *Waste Manag.* **37**: 45–57.

Matellanes, G., Victoria, M., Vázquez, G. et al. (2019). Assessing the influence of biomass properties on the gasification process using multivariate data analysis. *Energy Convers. Manag.* **184**: 649–660.

Monte, M.C., Fuente, E., Blanco, A., and Negro, C. (2009). Waste management from pulp and paper production in the European Union. *Waste Manag.* **29** (1): 293–308.

Ouadi, M., Jaeger, N., Greenhalf, C. et al. (2017). Thermo-catalytic reforming of municipal solid waste. *Waste Manag.* **68**: 198–206.

Palacio, J., Santos, J., Reno, M. et al. (2019). Municipal solid waste management and energy recovery. In: *Energy Conversion – Current Technologies and Future Trends* (ed. I.H. Al Bahadly), 127–146. Intech Open https://doi.org/10.5772/intechopen/79235.

Rajiv, P., Rajeshwari, S., Yadav, R.H., and Rajendran, V. (2013). Vermiremediation: detoxification of parthenin toxin from parthenium weeds. *J. Hazard. Mater.* **262**: 489–495.

Rajiv, P., Rajeshwari, S., and Rajendran, V. (2014). Impact of parthenium weeds on earthworms (*Eudrilus eugeniae*) during vermicomposting. *Environ. Sci. Pollut. Res. Int.* **21** (21): 12364–12371.

Ravindran, B. and Sekaran, G. (2010). Bacterial composting of animal fleshing generated from tannery industries. *Waste Manag.* **30**: 2622–2630.

Ravindran, B., Dinesh, S.L., John Kennedy, L., and Sekaran, G. (2008). Vermicomposting of solid waste generated from leather industries using epigeic earthworm *Eisenia foetida*. *Appl. Biochem. Biotechnol.* **151** (2–3): 480–488.

Ravindran, B., Contreras-Ramos, S.M., Wong, J.W.C. et al. (2014). Nutrient and enzymatic changes of hydrolysed tannery solid waste treated with epigeic earthworm. *Environ. Sci. Pollut. Res. Int.* **21** (1): 641–651.

Ravindran, B., Nguyen, D.D., Chaudhary, D.K. et al. (2019). Influence of biochar on physico-chemical and microbial community during swine manure composting process. *J. Environ. Manag.* **232**: 592–599.

Reyes-Torres, M., Oviedo-Ocaña, E.R., Dominguez, I. et al. (2018). A systematic review on the composting of green waste: Feedstock quality and optimization strategies. *Waste Manag.* **77**: 486–499.

Richard, T.L., Hamelers, H.V.M., Veeken, A.H.M., and Silva, T. (2002). Moisture relationships in composting processes. *Compost Sci. Util.* **10**: 286–302.

Safari, Y., Karimyan, K., Gupta, V.K. et al. (2018). A study of staff's awareness and attitudes towards the importance of household hazardous wastes (HHW) management (A Case Study of Kermanshah University of Medical Sciences, Kermanshah, Iran). *Data Brief* **19**: 1490–1497. https://doi.org/10.1016/j.dib.2018.06.022.

Scott, G.M. (2019). Recovered paper. In: *Waste*, 2e (eds. T.M. Letcher and D.A. Vallero), 291–305. Academic Press.

Sim, E.Y.S. and Wu, T.Y. (2010). The potential reuse of biodegradable municipal solid 1077 wastes (MSW) as feedstocks in vermicomposting. *J. Sci. Food Agric.* **90**: 2153–2162.

Singh, J. and Kalamdhad, A.S. (2014). Potential for composting of green phumdi biomass of Loktak lake. *Ecol. Eng.* **67**: 119–126.

Tang, Z., Yu, G., Liu, D. et al. (2011). Different analysis techniques for fluorescence excitation-emission matrix spectroscopy to assess compost maturity. *Chemosphere* **82**: 1202–1208.

Trindade, A., Palacio, J., Gonzalez, A. et al. (2018). Advanced exergy analysis and environmental assessment of the steam cycle of an incineration system of municipal solid waste with energy recovery. *Energy Convers. Manag.* **157**: 195–214.

Wei, Y., Li, J., Shi, D. et al. (2017). Environmental challenges impeding the composting of biodegradable municipal solid waste: A critical review. *Resour. Conserv. Recycl.* **122**: 51–65.

Williams, P.T. (2013). *Waste Treatment and Disposal*. Wiley.

Yadav, A. and Garg, V.K. (2011). Industrial wastes and sludges management by 1201 vermicomposting. *Rev. Environ. Sci. Biotechnol.* **10**: 243–276.

Zhang, Z. and Matharu, A.S. (2018). Thermochemical valorisation of paper deinking residue through microwave-assisted pyrolysis. *Waste Biorefin.* **23**: 671–692.

Zhang, L. and Sun, X. (2017). Addition of fish pond sediment and rock phosphate enhances the composting of green waste. *Bioresour. Technol.* **233**: 116–126.

4

Valorisation of Agricultural Waste Residues

Srinivas Mettu[1, 2, 3], Pobitra Halder[2], Savankumar Patel[2], Sazal Kundu[2], Kalpit Shah[2], Shunyu Yao[1], Zubeen Hathi[3], Khai Lun Ong[3], Sandya Athukoralalage[2], Namita Roy Choudhury[2], Naba Kumar Dutta[2] and Carol Sze Ki Lin[3]

[1]*Sonochemistry Group, School of Chemistry, University of Melbourne, Parkville, Melbourne, Victoria 3010, Australia*
[2]*Chemical and Environmental Engineering, RMIT University, Melbourne, Victoria 3000, Australia*
[3]*School of Energy and Environment (SEE), City University of Hong Kong, Tat Chee Avenue, Kowloon, Hong Kong*

4.1 Introduction

Climate change effects are being felt throughout the world, as evidenced by extreme weather conditions and increases in average temperatures. The recent extensive and unmanageable bushfire situation in Australia (December 2019 and January 2020 in New South Wales and Victoria states) exemplifies the effects of climate change directly affecting the lives of many rural and urban communities. Millions of animals have died in these bushfires, while the surviving animals face total habitat loss and food scarcity. The total number of animals that are affected by these bushfires is still increasing and it might reach more than 1 billion animals according to some estimates (WWF-Australia 2020). The secondary consequences of these bushfires have affected major cities in Australia such as Sydney and Melbourne. The smoke from the bushfires blanketed the two cities, causing hazardous air quality issues for the general public and visibility issues for high-rise construction workers.

Waste Valorisation: Waste Streams in a Circular Economy, First Edition.
Edited by Carol Sze Ki Lin, Guneet Kaur, Chong Li and Xiaofeng Yang.

The utilization of fossil fuels to meet the energy demand of the world is a major contributor to global warming (Intergovernmental Panel on Climate Change 2014). There has been a huge concern recently about pollution by microparticles of plastics on land, air, and water bodies (Boucher and Friot 2017). Microplastic pollution in oceans and other water bodies is wreaking havoc on native wildlife (Moore 2008; Müller et al. 2012; Vegter et al. 2014). Every year around 100 million tons of plastic gets discarded and ends up in landfills, accumulating in the environment (Geyer et al. 2017). Apart from mitigating climate change, there is a need for eliminating fossil-fuel derived synthetic polymer use, reduce landfilling, and reducing food and agricultural waste. It is imperative that the world either reduces or eliminates its dependence on fossil fuels, as well as invest in novel technologies that can meet the growing energy demand.

The circular bioeconomy concept has attracted significant interest recently due to its promise to reduce our reliance on fossil-derived fuels in almost every possible aspect of life (Kaur et al. 2018). The increasing demand for food production to feed the rapidly increasing world population is creating huge amounts of agricultural waste (agro-waste). If the agro-waste is properly integrated into the energy and materials economy of the world, it has significant promise to achieve the goal of eliminating use of fossil fuels and synthetic polymers. Synthetic polymers have been a great help to the world with their excellent physical and chemical properties. However, their robustness has created problems as they take a long time to degrade, unlike natural biopolymers. Hence there is a need to replace synthetic polymers with renewable biopolymers. Most of the discarded plastics that ends up in landfills is single-use plastics that is used as a packaging material. Single-use plastics for packaging is a major contributor to land, air, and water pollution as it accounts for a major portion of total plastics use (about 40% of total plastic use in Europe in 2016, according to PlasticsEurope). The biodegradability of the biopolymers produced from agricultural waste has the potential to reduce the accumulation of non-degradable synthetic polymers in the environment.

The agricultural industry is very diverse across the world, producing waste biomass of a diverse composition. The major components of waste streams from the agricultural industry (farming) comprise cellulose, lignin, fiber, carbohydrates, polysaccharides, and plant-based proteins. Existing conventional techniques of disposal, such as landfill, incineration, and composting, even though they are well-developed waste treatment methods, either produce low-value commodity products or toxic gases and bad odors. There is a great emphasis now on reclaiming the valuable fine chemicals, organic solvents, fuels, medicinal compounds, biopolymers, and proteins from these agro-waste streams. The green chemistry principles have attracted great interest in reclaiming such valuable compounds from agricultural wastes without using toxic chemicals or generating toxic waste streams from the processing techniques. Processing of agro-waste using strong chemicals derived from fossil fuels is not considered as a green alternative process, since it relies on fossil fuels. However, if the chemicals to be used in the processing are also derived from renewable bioresources, then the whole production cycle of valuable products from agro-waste is recognized as "sustainable and green." Hence, there is a need for a detailed and fresh look at the agro-waste being utilized in the circular economy concept. This chapter attempts to produce a comprehensive analysis and account of the existing and novel uses of agro-waste.

4.2 Agricultural Waste Definition, Composition, Variability, and Associated Policies and Regulations

The United Nations (UN) defines agricultural waste as: "the waste produced as a result of various agricultural operations. It includes manure and other wastes from farms, poultry houses and slaughterhouses; harvest waste, fertilizer run-off from fields, pesticides that enter into water, air or soils, salt and silt drained from fields" (United Nations 1997). The above definition classifies as unusable or unwanted these waste products from a series of agricultural premises, such as some crop materials, and relevant products from farming activities; most agricultural waste is not classified as controlled wastes. They are discarded after use and most of them are worthless and defective. Table 4.1 shows a summary of agricultural land usage and the corresponding total waste generated in some chosen countries (Letcher 2011).

Agricultural waste consists of food processing waste, livestock waste, crop waste, hazardous, and toxic agricultural waste; some examples are shown in Figure 4.1. The wastes generated based on different types of agricultural activities are shown in Figure 4.2, along with current valorisation approaches formulated in a European Horizon 2020 research and development project.

Lately, agricultural waste management (AWM) is gaining momentum to achieve sustainable development while following ecological agricultural methods. This issue has been recognized by policymakers (Hai 2010). For example, the 3R (Reduce, Reuse, Recycle) policy is being explored in Vietnam (Hai 2010). In the UK, agricultural waste was previously excluded from household waste management regulations, but now agro-waste is included. In Canada's Prince Edward Island, specific and detailed agro-waste policy management strategies have been proposed (Government of Prince Edward Island 1999). The Australian government's policies are specified in the Water Pollution Control Act 2002 and the Agricultural and Veterinary Chemicals (Control of Use) Act 1992. Agricultural-caused pollution is one of the biggest global environment quality problems, therefore it should be addressed in the same way as other pollution sources (such as industrial, mining, marine, etc.) to enhance the global environmental quality. By doing this, we can avoid contamination of the environment as well as propagation of hazardous materials.

Recently, the spike in agriculture-related environmental pollution has forced federal and state authorities to protect the public interest. Natural agricultural waste, which comes from

Table 4.1 *Agricultural area and its total generated waste.*

Country	Area (million hectares)	Waste generated (million tonnes annually)
Australia	425.44	16.2
China	552.83	56.2
Egypt	3.53	27
Germany	16.9	13.85
India	179.9	500
Indonesia	48.5	8.65
UK	17.6	43
USA	411.16	1100

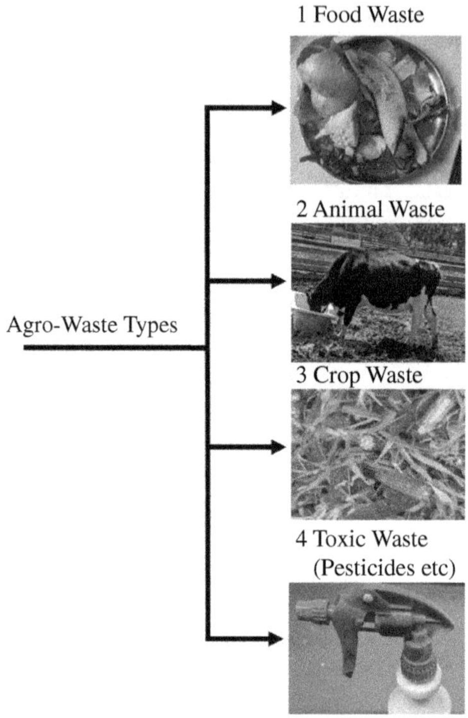

Figure 4.1 *Different types of agricultural waste, modified and adapted from Obi et al. (2016). Source for (3): http://Cropwatch.unl.edu.* Source: Photo by F. John Hay, University of Nebraska –Lincoln, used with permission.

sources in nature (such as deep forest, rainforest, barren land without human beings, etc.), is tough to control, due to its diversity, distance, and less controllable resources. The treatment of agricultural waste has switched from total agricultural wastes towards controllable pollution. Therefore, many new legal restraints that look beyond old policies are enforced on agricultural waste processing.

The basic policy of agricultural wastewater control in the United States (US Federal Water Pollution Control Act 2002) specifies "water pollution control activities in the Public Health Service of the Federal Security Agency and in the Federal Works Agency, and for other purposes." The declaration goals include: (i) "Congress has the right to recognize, preserve, and protect the primary responsibilities and rights of States to prevent, reduce, eliminate pollution, and exercise control of pollution in the waterways of the nation"; (ii) "The Administrator, in cooperation with the nation, shall develop and publish regulations specifying minimum guidelines for public participation in such processes to control water pollution"; (iii) "A national policy for prevention, control of water pollution should be established".

It is inevitable that chemicals will be used in the cultivation of crops and livestock. There are many countries where their major industry is based on animal husbandry. There is usually a legislative Act to protect people, the environment, and animals. For instance, in Australia the basic policy of agricultural chemical control can be found in the "Agricultural

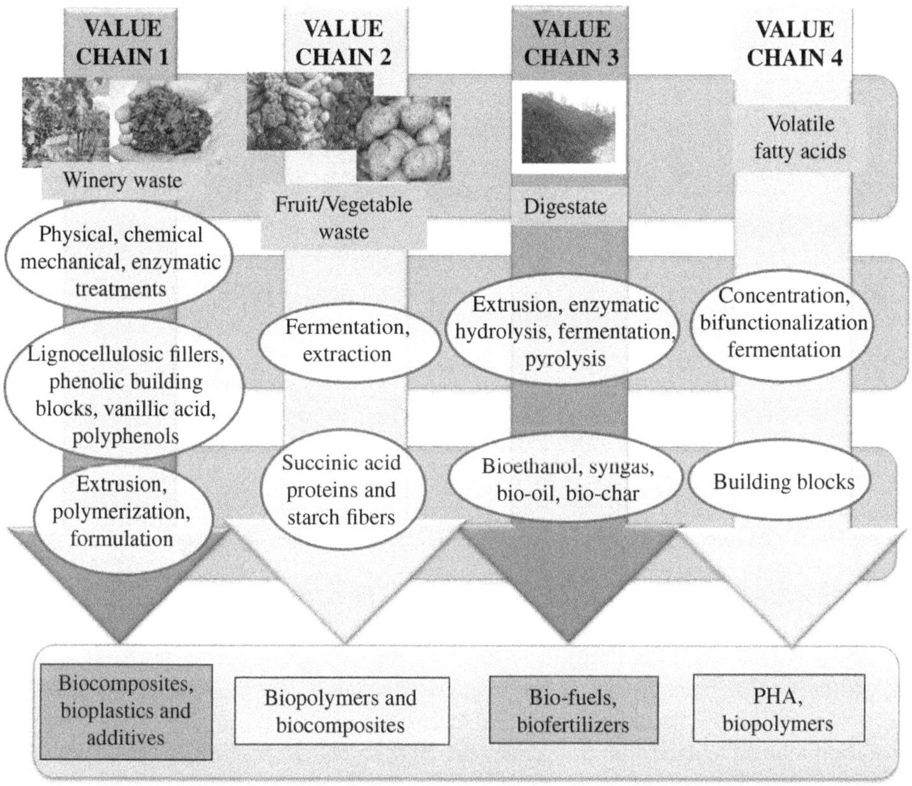

Figure 4.2 *Upstream and downstream processes for valorisation of agro-waste formulated in a European Horizon 2020 research and development project titled "NoAW – No Agro-Waste: Innovative approaches to turn agricultural waste into ecological and economic assets," as a part of the workpackage entitled "Eco-design of innovative cascading agro-waste conversion into molecules and materials."* Source: Modified from Gontard et al. (2018), reproduced with permission of Taylor & Francis.

and Veterinary Chemicals (Control of Use) Act 1992," and the purposes include: (i) "Protecting the health of the general public, welfare of animals and the environment"; (ii) "To impose controls in relation to the production of agricultural produce to avoid the contamination of food for human consumption"; (iii) "Promoting uniformity of regulation throughout Australia and reduce agricultural waste generation".

In the following sections, we summarize the conventional techniques used to valorize agro-waste, before delving into novel techniques and product streams.

4.2.1 Agricultural Waste from Farming

Agriculture is the mixture of science and art of cultivating plants and livestock, and it is without a doubt that agricultural activities are the widest utilizers of freshwater resources. There has been a great need to increase productivity against the decreasing availability of and competition for land and water from other uses. In addition, there is a need to meet the demands of rapid population increase. When cultivating crops, though fertilizer is needed,

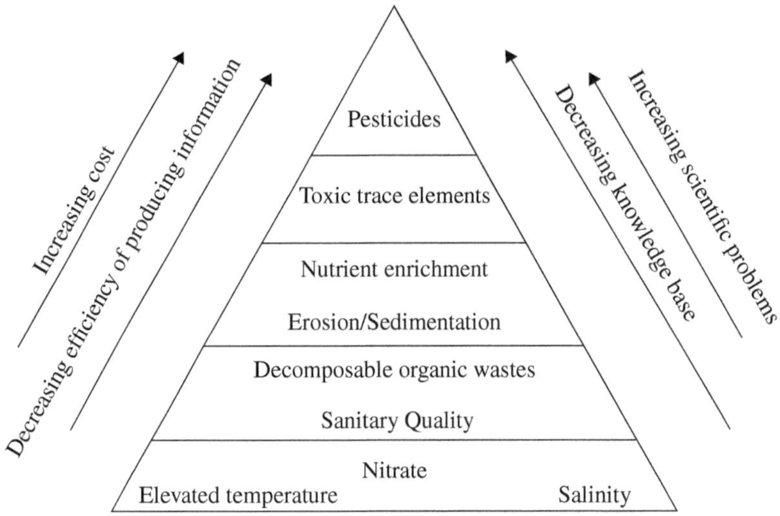

Figure 4.3 *Hierarchical, interdependent, and complex nature of water quality problems arising from agriculture (Rickert 1993).*

the excessive nutrition availability from unused fertilizer indirectly sustains the generation and growth of pests. This situation requires farmers to use pesticides, herbicides, or insecticides to kill the pests and protect the crops for healthy growth. Such a situation is quite common in the agricultural industry, resulting in the overuse of pesticides by farmers. Traces of residual pesticides in discarded pesticide bottles lead to the contamination of water sources and livestock indirectly, harming public health through the consumption of meat produced from such livestock. These hierarchical and interdependent, complex water quality problems arising from agriculture were analyzed by Rickert (1993), as shown in Figure 4.3.

In addition, pesticide packaging as well as canned materials used during food processing can also cause environmental problems if not well-kept or buried properly. Residual pesticides may slowly leak into the soil and waterways, contaminating the soil and drinking water sources. Empty cans from packaged foods, if not properly recycled or disposed of, may leak paints and other coating materials into the soil and waterways. The environmental pollution and hazards to aquatic wildlife is a serious issue that needs to be considered when disposing of packaging materials from agricultural and food processing industries.

The increase in production of cereals is also creating a lot of agro-waste. Cereals are grasses cultivated for the edible components of their grain, and provide more food energy than any other crops; thus they are also becoming a huge component of agricultural wastes. Table 4.2 summarizes the composition of general agricultural crop waste.

4.2.2 Agricultural Wastes from Livestock

Waste from livestock includes gases, liquids, and solids. Livestock digestive systems emit smelly odors, consisting of NH_4, CH_4, H_2S, and so on. The different waste odors depend

Table 4.2 Typical agricultural waste composition.

Composition no.	Source	Composition
1	Wheat	Bran, straw (Arvanitoyannis and Tserkezou 2008)
2	Barley	Husk, straw (Arvanitoyannis and Tserkezou 2008)
3	Oats	Straw (Arvanitoyannis and Tserkezou 2008)
4	Corn	Cobs, stalk (Arvanitoyannis and Tserkezou 2008)
5	Millet	Stover (Phonbumrung and Khemsawas 1998)
6	Rice	Husk, bran (Phonbumrung and Khemsawas 1998)
7	Sugarcane	Sugarcane tops, bagasse (Phonbumrung and Khemsawas 1998)

on different digestive processes and different animals, different foods, slaughterhouse temperatures, humidity, and activity of microorganisms.

Wastewater is also generated from a variety of sources, for example, urine, cage washing water, water for maintaining cages or slaughterhouses, and so on. Liquid livestock waste accounts for 75–95% of the total livestock waste, and also brings about water pollution (Hai 2010). Livestock manure in the cage or slaughterhouse, as well as the decay process of manure, can build up organic materials in the solid wastes formed. Those wastes and smelly ordures may propagate pathogens to humans and cause air quality deterioration.

4.2.3 Agricultural Waste Availability

Most agricultural wastes are from livestock, so this significant source makes it a promising candidate for the biorefinery field. For the rest of agricultural wastes, such as seeds, bran, straws, stalks, and bagasse, these residues are composed of lignin, cellulose, and hemicellulose (Taherzadeh and Karimi 2007). All these agricultural wastes are sustainable, available, and renewable resources amounting to about a billion tons a year (Forster-Carneiro et al. 2013). Typical agricultural residues consist of cellulose, hemi-celluloses, and lignin (Ghatak 2002). The easy and abundant availability of agro-waste in many countries at low cost makes them a very good option. The agricultural residues (lignin, cellulose, and hemicellulose) could be processed into some products, for instance, renewable energy, animal feed, film barriers, solvent chemical additives, biodegradable plastics, and textile fibers.

As CO_2 and many other greenhouse gases are progressively accumulating in the atmosphere, there is an urgent requirement to reduce these emissions to slow down the global warming process. Agro-waste can be used as a source for bioenergy. However, novel technologies need to be developed that either eliminate or lower the emissions during energy production. The dry crop residue produced while cultivating different crops varies a lot, as shown in Table 4.3 for the United States (Wolf and Snyder 2013; Wolf et al. 1980; Lal 2005).

Such a huge energy content in agricultural residue makes it a significant source for potential biorefinery feedstock. The availability of agro-residues in the world is around 10^{10} Mt, which corresponds to an enormous amount of energy of 47×10^{18} J (Ghatak 2011). In terms of this energy value, cereal residue has an estimated two-thirds of its value, at approximately 3.8×10^9 Mt (Lal 2008). Wheat straw and rice husk are commonly utilized as raw material for papermaking in China and India. There are also several agricultural wastes that have promising potential as biorefinery feedstock, for example, barley husk, oat straw, corn cobs,

Table 4.3 The dry crop residue produced in the US.

Crop	Residue dry weight (wt%)
Barley	4.3
Corn	10.1
Cotton	6.7
Oats	5.6
Peanuts	5.6
Rice	6.7
Sorghum	8.4
Tobacco	4.0
Tomatoes	5.0
Sugar beet	5.6

Source: Data from Wolf and Snyder (2013); Wolf et al. (1980); Lal (2005).

and sugarcane bagasse (Cherubini and Ulgiati 2010; Stephen et al. 2010; Matsumura et al. 2005; Boyden 2001).

In addition, another type of agricultural waste – livestock waste, for example, animal manure – can be directly used as combustible fuel and further converted into biogas. Solid waste residues can still be used as fertilizers. Existing literature shows that significant progress has already been made, where fermentation from animal waste has shown increased efficiencies in producing high-quality biogas and methane, hence the possibility of self-supporting biogas and methane production (Forster-Carneiro et al. 2004, 2010).

4.3 Conventional Techniques – Anaerobic Digestion, Pyrolysis, Gasification, and Solvent Treatment/Extraction

4.3.1 Anaerobic Digestion

The idea of anaerobic digestion (AD) of a diverse range of wastes has been pitched as a lucrative tactic to address waste management issues and simultaneously to produce biogas along with quality digestate (Khoshnevisan et al. 2019). The digestate produced is reported to have higher levels of recalcitrant rich in essential organic solids that could be used for bio-augmentation processes (Vaneeckhaute et al. 2013). On the other hand, extensive usage and prolonged storage of digestate leads to eutrophication, an increase in total salinity and acidity of soil (Li et al. 2017). Disposal of such toxic solid waste is an uphill battle for all nations, whether they are developing or developed countries. Additionally, collection and disposal of municipal solid waste (MSW) is a daunting issue on a global front. It is therefore no surprise that valorisation of agricultural waste has received tremendous attention over the last decade. Strategies to counter such global issues must be legally, financially, and environmentally sustainable. From a conventional perspective, composting and anaerobic digestion of agricultural waste have always been suitable, sustainable, and feasible measures to counter this challenge. There is a dramatic rise in the generation of agricultural waste (mainly organic in nature) every year (Abdel-Shafy and Mansour 2018). The generated waste is mainly handled by incineration, as animal feed, or disposed of without pretreatment, which has severe effects on the environment and all life forms on or off the land. Organic agricultural waste is rich in proteins, sugars, and several minerals that can be used to make useful products after proper segregation and pretreatment (Torales 2013). The generated waste and byproducts, if left underutilized, might lead to a loss of potential capital investment as well as contributing to rising disposal costs. AD is a highly favorable valorisation method to turn generated wastes into renewable energy, having higher percentages of methane gas, which can be used as fuel.

Reportedly, AD has quite a rich history going back as far as the tenth century BCE, when biogas was used for water heating purposes (Lin et al. 2011). Currently, over 50 million biogas units, including 42 million in China and 4.9 million in India, have been mounted and are fully operational throughout the world (www.worldbiogasassociation.org/). Back in 2012, De Meester et al. studied the overall sustainability of AD from an environmental perspective involving three major parameters: electricity generation from organic waste; the possibility of applying varying digester feed to agricultural land, and lastly, the significance of using digestate in the form of fertilizers. Outcomes of this study revealed that biomass generated an energy efficiency of 30%. From a rational perspective, this results in resource conservation of more than 90% through bio-based electricity in comparison with its conventional counterpart. Moreover, in comparison with aerobic treatment processes, AD does not require the energy needed for aeration, generates less than half the sludge and produces resourceful biogas having no less than 50% methane. AD consists of four major phases in which complete breakdown occurs along with final transformation of biomass to valuable biogas:

(i) Hydrolysis, a process where complex organic hydrocarbons are broken down to simpler peptides, di-, tri- or oligosaccharides and long-chain volatile fatty acids (LCVFA).
(ii) Acidogenesis, in which oligopeptides and di-, tri- saccharides combine to form carbon dioxide and hydrogen along with ammonium ions and hydrogen sulfide. Oligosaccharides and LCVFA combine to give short-chain volatile fatty acids (SCVFA), alcohol and lactic acid.
(iii) Acetogenesis, a process where acetate along with carbon dioxide and hydrogen gas are formed from lactate, SCVFA, and alcohols.
(iv) Methanogenesis, in which acetic acid is reduced to methane and carbon dioxide through hydrogenotrophs and methylotrophs.

Production of biogas depends on several technical aspects, such as the substrate loading rate, C/N ratio, mixing, pH and temperature, with temperature being the most crucial as methanogens are reported to be very sensitive to thermal changes (Torales 2013).

There are numerous types of digesters and systems that combine these models, since each of these combinations has a specific advantage of their own. For example, at a small scale, digesters having a simple design and operative features combined with high robustness and low maintenance costs could be more appealing for farmers, and can be installed in larger numbers. These digesters should essentially be of simple design and easy to operate, they could be dry or wet, single or multistage, meso or thermophilic, etc.; two main models of this type are a Chinese model and an Indian model used in rural parts of India (Rajendran et al. 2012). Altogether, AD is a highly promising valorisation technology that can potentially convert almost all the waste generated into value-added products.

Anaerobic digestion processes can be categorized as batch or continuous, based on the continuity of the process; psychrophilic, mesophilic, or thermophilic, based on the operating temperature; closed lagoon, completely-mixed, or plug-flow, based on the reactor design; and wet or dry based on the solid content in the feedstock (Nizami et al. 2013). Dry or solid-state AD is considered the most appropriate for describing the degradation process of feedstocks containing more than 15% dry matter, particularly agricultural residues (André et al. 2018). Solid-state AD offers a number of advantages over wet AD (Guendouz et al. 2008; Xu et al. 2014). For instance, solid-state AD is associated with low energy requirements, small reactor volume and no mixing requirement; the digestate from the solid-state process is easier to handle than effluent from a wet process. However, the solid-state process involves high retention time and is associated with reduced mass transfer, which requires further improvement.

In AD the waste organics are processed to produce useful biogas under a low-oxygen environment using various microorganisms. This process significantly reduces the biological oxygen demand and chemical oxygen demand (Chen et al. 2008). The organic matter undergoes a number of metabolic stages such as hydrolysis, acidogenesis, acetogenesis and methanogenesis (Figure 4.4) and produces methane-rich biogas (Halder et al. 2016). Typical biogas consists of methane (about 50–75%), CO_2 (about 25–50%), and hydrogen sulfide (about 0–3%) (Khurshid 2009).

Methane-rich biogas has several economic benefits, and it is widely used for electricity production and heat generation. Additionally, the use of upgraded biogas (often called bio-methane), particularly after removing carbon dioxide, as an alternative to natural gas in the grids and for vehicles has gained substantial attention in recent times. The residual digestate can be used for land application as a fertilizer as well as solid fuel. The cost-effectiveness of the AD process depends on the capital and maintenance costs of the plant. Vasco-Correa et al. (2018) presented an economic evaluation of small-scale biogas plants for different regions. The cost of a biogas plant in Asia was relatively consistent, whereas in Africa and in Latin America, the variability in cost was attributed to the variation in labor and construction materials costs and a lack of established technology (Vasco-Correa et al. 2018).

Agricultural residues are considered as a viable input source for the AD process because of their abundance, renewability, and pattern of current use. For instance, Austria is currently producing 106.8 MW of electricity from around 368 AD plants based on agricultural biomass feedstock (Kral et al. 2016). Germany is producing at least 50% of their total biogas from 700 plants using agricultural energy crops biomass (Sukhesh and Rao 2018). It has been estimated that Europe has about 37.2 billion m^3 of biogas potential from agricultural residues, including 27.2 billion m^3 and 10.0 billion m^3 from crops and straw, respectively (Scarlat et al. 2018). In Europe, more than 60% of the plants are producing biogas mainly

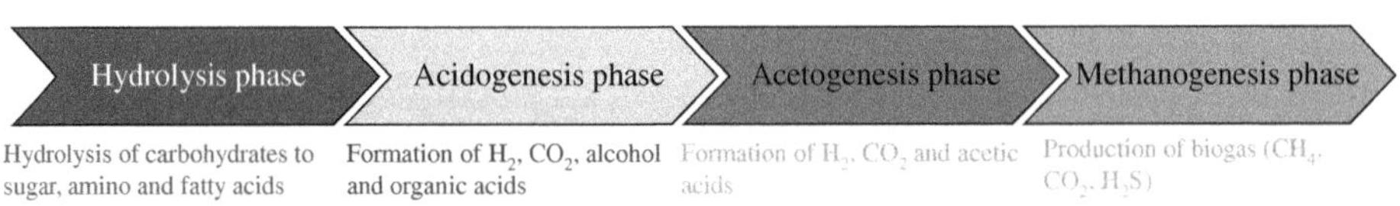

Figure 4.4 *Various metabolic phases during anaerobic digestion. Source: Halder et al. (2016), reproduced with permission of Elsevier.*

from crop silages via solid-state anaerobic digestion (Kalamaras and Kotsopoulos 2014; Karthikeyan and Visvanathan 2013). In Australia, about 22 plants are producing biogas from agricultural residues, and are mainly used for combined heat as well as for power generation. Currently, there are three commercial-scale agricultural biogas plants are in operation as follows:

- In 2011, Air Liquide and Swedish Biogas jointly established the first commercial-scale agricultural residues-based AD plant in Sweden. The plant uses 72 600 tons per year of local vegetable waste to produce biogas of 4000 tons per year of biogas.
- Biogest Biogas has started producing biogas from a mixture of agricultural residues such as maize silage, sugar beet, and cattle slurry and manure. The plant was established in the UK under the Bio-methane project in 2014. The output of the plant includes 4000 MW of electricity and 7.2 million m^3 of biogas which is further converted into 4.3 million m^3 of biomethane.
- Recently, in 2019, Verbio started producing about 10 MW of biogas using straw from an AD plant built in Germany under the "Verbiogas Pinnow" project.

Biogas production from agricultural residues via AD is significantly affected by several factors, including reactor design, operating conditions, and feedstock composition. The high carbon-to-nitrogen ratio of agricultural residues reduces its conversion rate (Demirbas and Ozturk 2005). Therefore, co-digestion with livestock manure or food waste and the addition of ammonia can release nitrogen during the process, leading to an increase in biogas yield. The effect of solid content in agricultural residues on biogas yield has been extensively studied. For example, Suksong et al. (2017) observed a higher methane yield at 16% solid content in palm oil residues, compared with 25–35% solid content. Xu et al. (2014) noticed an increase in methane production from corn stover digestion up to 20% solid content, while further increase in solid content reduced the methane yield. The effect of temperature on biogas yield varies because microbial growth depends on temperature. Yan et al. (2015) observed maximum biogas yield at 35°C from rice straw digestion, while Liu et al. (2018) obtained maximum yield at 44°C from corn stover digestion. The anaerobic digestion of corn stover yielded about 36% more biogas at thermophilic conditions (55°C) compared with that at mesophilic conditions (37°C) (Y. Li et al. 2011a). Additionally, the application of rumen microorganisms as inoculum enhanced the degradation of the volatile solids of corn stover during anaerobic fermentation, which increased volatile fatty acids, resulting in improved biogas yield (Hu and Yu 2005). Zhou et al. (2017) observed low methane yield from rice straw digestion at low inoculum percentage, due to the limitation of mass transfer. Vats et al. (2019) reported about 22% more biogas production when sugarcane bagasse and food waste were digested together, compared with that from a single feedstock. Apart from these, the recalcitrance of the complex lignocellulosic structure of agricultural residue biomass to hydrolysis is considered a bottleneck for anaerobic digestion which needs to be overcome in order to enhance the conversion rate (Veluchamy and Kalamdhad 2017). The presence of lignin in agricultural biomass forms three-dimensional rigid crosslinking structures, which slow down the hydrolysis process, thereby requiring a longer retention time. Several researchers have linked the biogas yield from lignocellulosic biomass to the lignin content, and observed that biogas production decreases with lignin content (Sukhesh and Rao 2018). Therefore, pretreatment of the feedstock prior to AD has attracted significant interest from the scientific community, with a view to removing

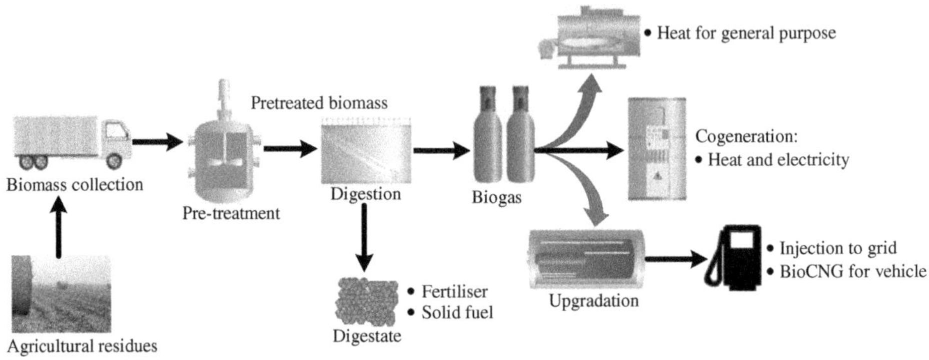

Figure 4.5 Value chain of biogas integrated with pretreatment.

lignin or deconstructing the lignocellulosic structure. Figure 4.5 depicts the integration of pretreatment and the value chain of biogas.

So far, various pretreatment techniques such as chemical, physical, thermal, and biological pretreatments have been investigated. The incorporation of pretreatment before anaerobic digestion significantly improved the biogas yield compared with untreated agricultural residue (Zheng et al. 2014). Figure 4.6 shows different pretreatment

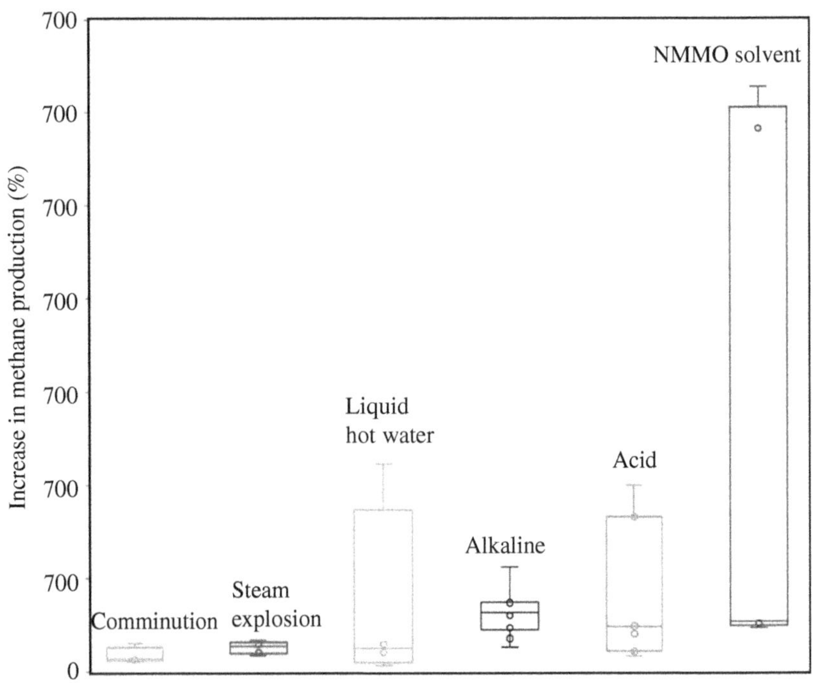

Figure 4.6 Effects of different pretreatments on the variability in methane production from anaerobic digestion of agricultural biomass. Source: Adapted from Zheng et al. (2014) and Teghammar et al. (2012).

methods to produce methane from agricultural residues via anaerobic digestion. The N-methylmorpholine-N-oxide monohydrate (NMMO) solvent pretreatment outperformed the other pretreatments for methane production from agricultural residues within a wide range from 47% to 628%. For example, rice straw and triticale straw pretreated with NMMO solvent showed around 628% and 583% increase in production of methane respectively, compared with untreated biomass. However, the methane production increased by 1200% after NMMO pretreatment of softwood such as chips and milled spruce (Teghammar et al. 2012). Methane production depends on pretreatment, composition of feedstock, and operational conditions.

4.3.2 Solvent Treatment

Processing of lignocellulosic biomass, particularly pulping, has been widely investigated from the late nineteenth century using various solvents such as alcohols, ethylene glycol, glycerol, acetone, acetic acid, and formic acid, either in the presence or absence of catalysts (Li et al. 2012). However, recent interest is to explore the biorefinery route for platform chemicals and value-added products, rather than being confined to the pulping process. Solvent treatment alters the lignocellulosic structure of agricultural biomass and disintegrates it into its constituent parts, which is a favorable step for further processing of the feedstock. Figure 4.7 shows the solvent treatment integrated process steps for bioethanol production from agricultural biomass. Solvent treatment removes lignin and increases the specific surface area of the pretreated materials, leading to improved enzymatic hydrolysis efficiency and pyrolysis selectivity (Z. Zhang et al. 2016; Halder et al. 2019a). In addition to these, catalytic solvent treatment of agricultural biomass can convert the hemicellulose into high-value chemicals, including furfurals, hydroxymethyl furfural, and xylitol (Zhao et al. 2009).

Wildschut et al. (2013) treated wheat straw using ethanol for glucose conversion via enzymatic hydrolysis, and compared that with sulfuric acid catalytic treatment. They reported that the enzymatic digestibility significantly increased from 37.2% to 89.4% when 30 mM sulfuric acid was added with 60% aqueous ethanol. Moreover, furfural production increased from 19% to 34.4%, which was consistent with the higher removal of xylan during acid catalytic treatment. Figure 4.8 illustrates the variation of enzymatic digestibility of treated agricultural biomass for different solvent treatments. Amiri et al. (2014) also observed that enzymatic digestibility increased after the treatment of rice straw with 1% sulfuric acid with 75% aqueous ethanol. Sun et al. (2015) achieved about 90% of cellulose and hemicellulose conversion into reducing sugars via enzymatic hydrolysis of 70% glycerol-treated wheat straw. The treatment was performed at 220°C for a time duration of 3 hours. Similarly, 90% glucose yield was achieved after the treatment of

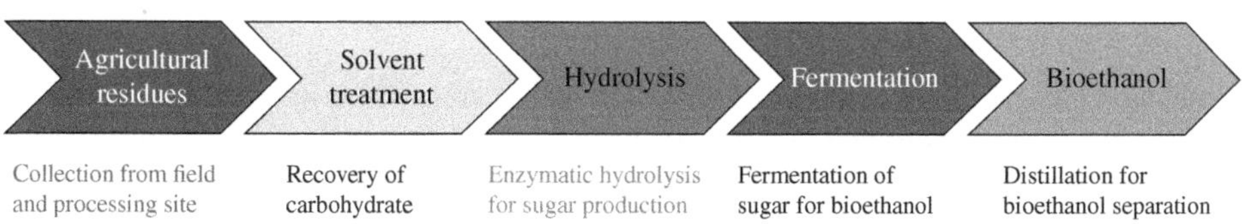

Figure 4.7 *Process steps for lignocellulosic bioethanol production.*

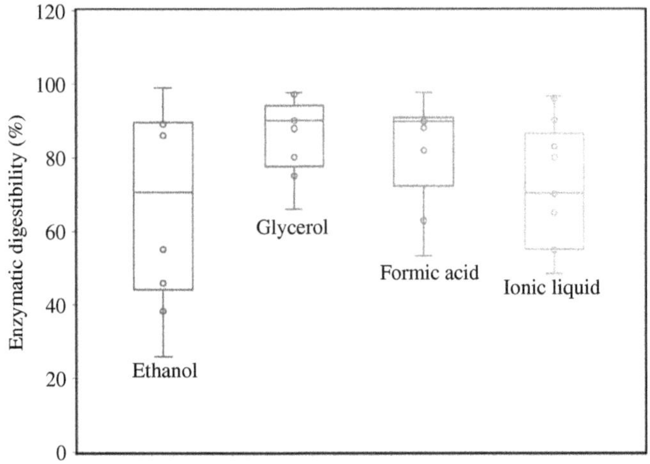

Figure 4.8 *Effects of different solvent pretreatment on enzymatic digestibility (Zhao et al. 2009; Wildschut et al. 2013; Amiri et al. 2014; Sun et al. 2015; Zhang et al. 2013; Li et al. 2011).*

sugarcane bagasse with 1.2% hydrochloric acid with 90% glycerol at 130°C for 1 hour (Zhang et al. 2013). Sugarcane bagasse treated with 78% formic acid at 107°C for 1 hour exhibited approximately 88% enzymatic digestibility, with 80.7% lignin removal and 89.2% hemicellulose removal (Zhang et al. 2013).

The characteristics of the pretreated materials as well as the catalytic conversion efficiency are associated with the severity of the pretreatment, which depends on the operating parameters such as concentration of catalysts, residence time, and reaction temperature. The severity factor (SF) of solvent treatment is expressed by the following equation (Obama et al. 2012):

$$SF = \log(R_o \cdot C^n) \tag{4.1}$$

where R_o, C, and n represent the reaction ordinate, concentration of catalysts, and an arbitrary constant, respectively. R_o is expressed as a function of treatment temperature (T in °C) and residence time (R in s) as follows (Sannigrahi et al. 2010):

$$R_o = R \cdot \exp\left(\frac{T(R) - 100}{14.75}\right) \tag{4.2}$$

The value of n is variable with catalyst type, and the effects of variation of biomass composition on the value of n has been little investigated. Mesa et al. (2010) proposed a single factor called the combined severity factor (CSF) considering all parameters for any type of catalyst as follows:

$$CSF = \log R_o + |pH - 7| \tag{4.3}$$

Salapa et al. (2018) investigated the acid-catalyzed acetone treatment of barley straw followed by enzymatic hydrolysis. They correlated the sugar yield with the treatment severity and found that cellulose-to-glucose conversion improved with an increase in CSF from −0.5 to 0.5. However, further increase in CSF impacted negatively on the sugar yield. Teramoto et al. (2009) also reported an increase in enzymatic digestibility efficiency for ethanol-treated bagasse with an increase in CSF up to 2. So far, only a few large-scale organic solvent-based treatment plants are in operation, as most of these traditional solvent treatments are associated with significant environmental, health, and safety issues. Some of the solvent treatment-based, industrial-scale, agricultural residue-processing technologies/plants are as follows.

- Chempolis Ltd. established a demonstration-scale bioethanol plant in Finland in 2008 under the "Chempolis biorefining plant" project. The company developed the Formicobio™ technology, which consisted of a novel biosolvent-based treatment system named Formicodeli™ for the fractionation of wheat straw, bagasse, empty fruit bunch and corn stalks, followed by enzymatic hydrolysis. The plant produces about 5000 tons per year of cellulosic ethanol and other chemicals, including furfural, from 25 000 tons of agricultural residues per year.
- In 2012, Borregaard AS built a demonstration-scale cellulose-based ethanol plant in Norway. The plant uses approximately 365 tons of sugarcane bagasse, straw, energy crops, and other lignocellulosic residues for producing 110 tons ethanol per year, and other products including lignin and various chemicals.
- Leaf Resources developed an industrial-scale solvent treatment technology, which includes a glycerol-based treatment process named Glycell™, which uses enzymatic hydrolysis to produce sugars from cellulose and hemicellulose.
- CIMV has developed acetic acid and formic acid-based treatment technology for lignocellulosic biomass processing, and constructed a pilot-scale plant based on this treatment technology in 2006.

Recently, ionic liquid and gamma valerolactone has been employed for lignocellulosic biomass processing because of their greenness compared with organic solvents (Zhang et al. 2017). The gamma valerolactone/water system has proven its efficacy on agricultural biomass treatment, as more than 92% glucose conversion has resulted from the enzymatic saccharification of treated cotton stalk (Wu et al. 2016). Recently, research on ionic liquid treatment of agricultural biomass has increased significantly. The ionic liquids, particularly imidazolium cation-based ionic liquids, dissolve cellulose, and transform the crystalline structure into an amorphous structure. Ionic liquid treatment also increases the cellulose specific surface area, leading to an increase in sugar yield from enzymatic hydrolysis (C. Li et al. 2011b; Halder et al. 2019b). For instance, C. Li et al. 2011b found that the surface area of 1-ethyl-3-methylimidazolium acetate treated corn stover increased from 0.7 to 15.1 m^2/g and 90% sugar yield with 76% of hemicellulose removal. Additionally, Brønsted acidic ionic liquid treatment of agricultural biomass can produce furfural, hydroxymethyl furfural, and levulinic acid, which can open up new prospects for biorefinery (Halder et al. 2019c). However, further research is required on the ionic liquid-based treatment process in order to overcome shortcomings and make the process feasible for industrial-scale development.

4.3.3 Gasification

Gasification involves the conversion of feedstock via a thermochemical process into fuel gases through partial oxidation between 600 and 900°C in the presence of a gasifying agent (Hosseini et al. 2015; Shahbaz et al. 2016). As already mentioned, agricultural waste consists of mainly cellulose, hemicelluloses, lignin, and extractive parts such that gasification technology is a sustainable option for transformation into sustainable energy. A typical process for gasification of biomass includes the following steps:

(i) Pre-treatment, such as drying, standardizing of particle size, or grinding.
(ii) Gasification – the thermal decomposition of feedstock to produce fuel gases. During gasification, various reactions occur, which convert biomass into product gases (Hossain et al. 2016). The fuel gas consists of CO, H_2, CH_4, CO_2, N_2, and light hydrocarbons, which is known as syngas (Mohammed et al. 2011; Hosseini et al. 2012).
(iii) Gas cleaning – reduction of tar content, NO_x, SO_x, and other particles from fuel gas to produce syngas, comprising a mixture of CO, H_2, CH_4, CO_2, N_2, and light hydrocarbons.
(iv) Utilization for heat/power/fuel – the fuel gas is used in a turbine or gas engine for production of electricity, as well as heat as a byproduct.

The syngas can be converted into chemicals by catalytic processes (Ibarra-Gonzalez and Rong 2018). Fuel gas production depends on the size of feedstock particles, moisture present in the feedstock, gasifying agent, types of reactors, equivalence ratio (see below), gasification process temperature, and catalyst.

- Feed particle size: small feed particle size improves energy efficiency but increases the cost of gasifying plant. A coarse particle size, however, lowers the milling costs but increases the residence time and the associated fixed costs (Alauddin et al. 2010).
- Moisture: water content should be as low as possible, but drying the feedstocks consumes a lot of energy that reduces the plant's net energy production. For each kg of moisture, around 2300 kJ is needed for vaporization, with a requirement of an additional 1500 kJ to increase the temperature of feedstock to 700°C (Basu 2010).
- Gasifying agent: agricultural wastes can be gasified by steam, air, oxygen-enriched air, a mixture of steam and air, steam–oxygen mixture, and oxygen–air–steam mixture (Campoy et al. 2009). The advantages and disadvantages of gasifying agents in agricultural waste are discussed in Table 4.4.
- Equivalence ratio (ER): this is a critical operating parameter. The ER is calculated by the division of the actual molar ratio of oxygen (air) to feedstock biomass from stoichiometrically required molar ratio of oxygen (air) to feedstock biomass. Increased ERs also help to increase gas yield and decrease tar in raw gas (Gai and Dong 2012). High equivalence ratios result in maximum feedstock and fuel gas combustion in the reactor, leading to low exit gas heating value. Upon reaching a peak of 0.26, the heating value of gas declines as ER rises (van den Enden and Lora 2004).
- Gasification temperature: the temperature of gasification is a result of various process characteristics such as gasification agent (air/oxygen), contents of waste (ash, moisture, and heating value), ER, and heat loss during the gasification process. The temperature of gasification significantly affects gas fuel composition, as the temperature controls the reaction thermodynamically (Alauddin et al. 2010; Basu 2010).
- Gasification technologies: The industrial reactors (gasifiers) vary fundamentally for (i) the way the feedstock and the gasifying agent interact with each other (cross-flow or counter-current flow); (ii) heat transfer rate; for example, heat can be supplied either externally or by a combustion agent directly into the reactor; (iii) feedstock residence

Table 4.4 *Pros and cons of gasifying agents in agricultural waste gasification.*

Gasification agent	Advantages	Disadvantages	References
Air	1. Moderate tar content 2. Inexpensive for small and laboratory scale	1. Low calorific value producer gas (4–7 MJ/Nm^3) 2. High N_2 content 3. Not applicable for methane production 4. Expensive for large scale	Hosseini and Wahid (2014), Moilanen and Nasrullah (2011), and Sikarwar et al. (2017)
Steam	1. Medium calorific value producer gas (10–15 MJ/Nm^3) 2. Useful for production of H_2-rich and N_2-free producer gas 3. Easy to remove excess steam 4. Inexpensive for both small and large scale	1. Process is very endothermic	Mohammed et al. (2011), Alauddin et al. (2010), Moilanen and Nasrullah (2011), Sikarwar et al. (2017), Li et al. (2009)
O_2	1. N_2-free producer gas 2. High calorific value producer gas (12–28 MJ/Nm^3) 3. Inexpensive for large scale	1. Expensive for small scale 2. In the case of excess oxygen (more than stoichiometric amount), the process become combustion, and produces "flue gas" rather producing "fuel gas" 3. Low H_2 content 4. High CO and CO_2	Basu (2010), Sikarwar et al. (2017), Florin and Harris (2008)

time. The three main reactors used in agricultural waste gasification process are: entrained flow; fixed; or fluidized bed reactors. Brief descriptions of the numerous reactors and effect of residence time and their main benefits and disadvantages in the gasification process are given in Table 4.5.

- Catalyst: the gasifying process challenges include lower heating and quality for syngas, removal of tar and carbon dioxide, and impurities in char (Herman et al. 2016). Different catalysts, adsorption materials and adsorption mechanisms were studied in order to deal with the above problems associated with the gasification process. In general, three types of catalysts, including dolomite, alkali or alkaline earth or nickel-basis, are used during the gasification process, as shown in Table 4.6 (Baker et al. 1987; Aznar et al. 1993; Lee et al. 2000; Devi et al. 2005; Hu et al. 2006; Göransson et al. 2011; Kaewpanha et al. 2014)

4.3.4 Pyrolysis

Pyrolysis has also been particularly interesting, among other endothermic thermochemical methods. Pyrolysis can decompose solid biomass directly into liquid, gas, and biochar in

Table 4.5 Pros and cons of reactors used in agricultural waste gasification.

Reactor type	Advantages	Disadvantages	References
Fixed bed (two types; down and updraft)	1. Simple design and economically viable at small scale 2. Applicable for different types of agricultural waste 3. Suitable for high moisture, less volatile matter, and high ash content feedstock 4. Low tar yield in downdraft gasifier	1. Not suitable for large applications because of non-uniform temperature 2. High char yield 3. High tar yield in updraft gasifier 4. Higher residence time	Basu (2018), Sikarwar et al. (2017), Widjaya et al. (2018)
Fluidized bed (two types: bubbling and circulating beds)	1. Economically viable for large scale because of identical temperature distribution 2. Short residence time 3. Low tar yields 4. Homogenous particle mixing and high gas–solid contact	1. Particulate matter high in syngas	Reed and Das (1988), Puig-Arnavat et al. (2010)
Entrained flow	1. Low tar yield 2. Economically viable for large scale 3. Producer gas has high quality	1. Not appropriate for agricultural waste due to fibrous nature and high moisture content	Sikarwar et al. (2017), Basu (2018)

Table 4.6 Pros and cons of catalyst used in agricultural waste gasification.

Catalyst	Advantages	Disadvantages	References
Nickel	1. Potential for tar reduction and to improve the water gas shift reaction 2. Improve yield of desired product gas and quality 3. Suitable on a commercial scale	1. High attrition in fluidized bed environments compared with fixed bed systems 2. Form nickel carbides due to coking	Aznar et al. (1993), Baker et al. (1987)
Olivine and limestone	1. Low cost and freely available in nature 2. Efficient for tar removal	1. Higher attrition in fluidized bed process	Hu et al. (2006), Devi et al. (2005)
Alkali and alkaline metal	1. Enhanced generation of cleaner gas 2. Enhanced generation of permanent gases when a nickel catalyst is added with alkali metal (sodium carbonate)	1. High cost 2. Agglomeration at high temperature 3. High ash yield	Lee et al. (2000), Göransson et al. (2011), Kaewpanha et al. (2014)

an inert environment. The fractions of these products depend on the biomass composition and type, heating rate of pyrolysis, temperature, and residence time. The biomass-derived pyrolysis oil can be used to produce hydrocarbons or for power generation (Kim et al. 2013). The gas can be used to synthesize liquid fuels as well as synthetic natural gas. The biochar produced from pyrolysis can act as an adsorbent or fertilizer, as well as to produce activated carbon (Antal et al. 1990, 2000; Yaman 2004; Mohan et al. 2007, 2011, 2014; De et al. 2013; K*ılıc* et al. 2013). The pyrolysis reaction processes are complicated, but mainly consist of three key steps as shown below:

$$\text{biomass} \rightarrow \text{untreated/reacted biomass residue} + \text{water} \tag{4.4}$$

$$\text{untreated/reacted biomass residue} \rightarrow (\text{biochar})_1 + (\text{volatiles} + \text{gases})_1 \tag{4.5}$$

$$(\text{biochar})_1 \rightarrow (\text{volatiles} + \text{gases})_2 + (\text{biochar})_2 \tag{4.6}$$

The first step is the removal of water and low molecular weight volatile components (Equation (4.4)); production of primary biochar as well as decomposition of large hydrocarbon molecules into smaller molecules takes place in the second step (Equation (4.5)); the biochar slowly decomposes and forms a carbon-rich residual which is called less reactive char (Equation (4.6)) in the third step.

The product distribution is affected by particle size, reaction time, pressure, and temperature. Operating conditions influence product yield and quality. The aim of pyrolysis is generally to increase the yield of the component, so the effects of these process conditions must be addressed.

- Effect of residence time: Mohamed et al. (2013) as well as Tsai et al. (2007) found that while residence time affects the liquid and gaseous yield and composition, it minimally impacts the char yield. The longer the vapor residence time, the more time is available for the components of biomass to react and repolymerize. Short residence is useful in the reduction of higher molecular weight components through secondary decomposition by the sweeping gas (Zanzi et al. 2002; Luo et al. 2004; Park et al. 2008). Pütün (2010) found that biochar yield decreases due to short residence time in pyrolysis of cotton seed. Residence time also affects the physical characteristics and quality of biochar through the formation of various-sized pores (micro and macro) (Tsai et al. 1997). The heating rate, temperature, and other pyrolysis parameters often dominate over the impact of residence time, making it very difficult to offer a clear idea on the function of residence time in biochar production (Encinar et al. 1996; Fassinou et al. 2009).
- Effect of particle size: The size of particles must be taken into consideration as it controls heat transfer rate into the feedstock. Demirbas (2004) found that when particle size changes from 0.5 to 2.2 mm, biochar production increased for olive husk and for corncob when pyrolysis was carried out at 677°C. The biochar yield increased with particle size on pyrolysis of various agricultural waste, as reported by Ateş et al. (2004) and Zhang et al. (2009). Mani et al. (2010) observed similar increasing trend in biochar yield for the pyrolysis of wheat straw with particle size from 0.25 to 0.475 mm, but little change between 0.475 and 1.35 mm. Such differences suggest that the particle size role in pyrolysis biochar production is not yet entirely understood, and further work is required in this field.

- Effect of heating rate: Heating rate affects the product formation in agricultural waste pyrolysis, and is typically higher in fast pyrolysis. High heating rates enhance rapid depolymerization of agricultural waste, which reduce the char yield and increase volatiles yield. Secondary reactions are more dominant at high heating rates, which help in the formation of gaseous product from decomposition of higher molecular weight components (Ateş et al. 2004). In slow pyrolysis, lower heating rates produce more biochar due to the slow decomposition of biomass. The biochar yield from safflower seed and *Charthamus tinctorius* L was reported to decrease when the heating rate was increased (Angın 2013; Sensoez and Angin 2008). Pütün et al. (2004) also studied the effect of heating rate on yields from rice straw pyrolysis.
- Temperature effect: Reaction temperature is the most important parameter as it affects both the composition of products and final yield. Very high and low temperatures stimulate the production of non-condensable gases and char, respectively. At higher temperatures, there is a decrease in the production of liquids and increase in the production of gases due to thermal cracking reactions of heavy hydrocarbons. Increasing the pyrolysis temperature by 300°C resulted in a 10% decrease in the yield of biochar from hazelnut shell (Pütün et al. 1999) and 17% decrease in the yield from sesame stalk (Ateş et al. 2004). In addition, Choi et al. (2012) recently reported a decrease in biochar production with increased pyrolysis temperatures. The pattern shows a clear decline in biochar yields when the pyrolysis temperature is increased. Xiao and Yang (2013) found that biochar quality and surface properties improve as a consequence of the formation of smaller pores by release of higher volatility as the pyrolysis temperature becomes higher, in the case of pyrolysis of rice straw. Similar conclusions were reached by Azargohar et al. (2013) in the case of pyrolysis of wheat straw, sawdust, and flax straw.
- Presence of catalyst: Catalysts reduce the amount of oxygen present, as well as oxygen-containing functional groups from the feedstock, and maximize the hydrogen-to-carbon ratio in liquid, gas, and solid products. Tar production was found to decrease when a small quantity of inorganic material was combined with biomass (Encinar et al. 1997). A few catalysts led to decreased liquid and gas yields and an increased biochar yield. However, few catalysts improve the liquid yield. Sometimes, ash present in the feedstock can also act as a catalyst (Samolada et al. 2000). Encinar et al. (1997) studied the effects of various catalysts such as KCl, NaCl, $AlCl_3{\cdot}6H_2O$, LiCl, $FeCl_3{\cdot}6H_2O$, and $ZnCl_2$ on the olive bagasse pyrolysis product yield, and compared the biochar yield. They found that catalysts used in the study increased the biochar yield extensively, but biochar yield only increased up to 44% in the presence of $ZnCl_2$, which is less compared with the others. Biochar yield decreased in the case of $Ca(OH)_2$ (Auta et al. 2014), activated alumina and sodium feldspar (Demiral and Şensöz 2008) catalysts. However, ZSM-5 catalyst increased biochar yield and quantity of gaseous products by decreasing of the concentration of liquid products (Samolada et al. 2000; Stefanidis et al. 2011; Smets et al. 2013). Each agricultural waste differs in composition distribution, moisture, and ash content. Therefore, a general theory of the catalyst effect on product distribution is difficult to form.
- Moisture content: The moisture/water content of agricultural waste impacts upon pyrolysis and influences product properties and consistency. In agricultural waste this moisture can be present water vapor, chemically bound water (adsorbed into biomass pores), or

free water (Bryden and Hagge 2003). The larger portion of the heat energy during pyrolysis is used to remove moisture from the biomass, whereas the remainder is used to increase pyrolysis temperature (Demirbas 2004). Low moisture in the feedstock improves the biochar production rate. Low moisture content is economically viable compared with high moisture, due to the lower energy required for pyrolysis as well as reduced time for pyrolysis (Asadullah et al. 2008).

Even though a suite of conventional techniques exists, there is a need for the development of novel techniques that can produce new products, to reduce agro-waste going to landfill. Next we summarize the possible novel uses of cellulosic products, without resorting to destruction of plant material by conventional techniques.

4.4 Novel Techniques and Envisioned Product Streams: A New Perspective

There is a dire need for the development of innovative solutions as well as novel tools for assessing and evaluating agro-waste valorisation strategies in a circular bioeconomy. A significant effort has been dedicated to developing successful lignocellulose-based technologies from agro-waste, to produce platform chemicals as well as various fuels. As mentioned earlier, agro-waste biomass primarily contains cellulose, hemi-cellulose, and lignin. Thus the primary residues of agricultural waste can be converted to various resources depending on the selected conversion processes, to produce environmentally sustainable and green bioproducts such as biofertilizers, bio-energy, biomaterials for renewable energy such as hydrogen and other gas, platform chemicals, and so on (Figure 4.9). For example, integrated processing of C5 and C6 sugars could be achieved by fractionation of lignocellulosic biomass.

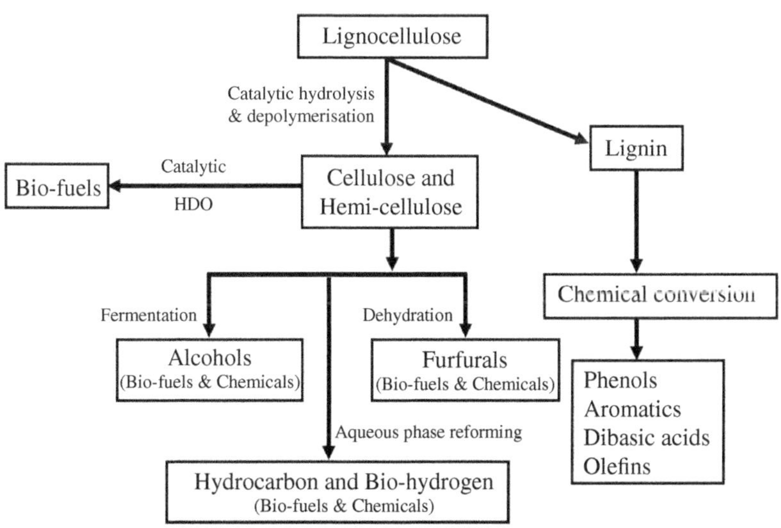

Figure 4.9 *Potential scheme for chemo-catalytic biorefinery.* Source: Kohli et al. (2019), https://www.mdpi.com/1996-1073/12/2/233. Licensed under CCBY 4.0.

Lignin can be removed from lignocellulosic materials by physicochemical or biochemical treatments. The properties of extracted lignin depend on the source and extraction procedures. Each method has its own balance in using chemical reagents and net energy consumption, which can affect molecular weight and lignin purity, other physical and chemical properties, as well as different possible chemical structures. Kraft lignin, the largest byproduct of the pulping process, is highly hydrophobic with aliphatic thiol groups. Lignin, resulting from the sulfite pulping process, has ligno-sulfonic acids and lignosulfonate along with phenolic or aliphatic hydroxyl groups. Soda-processed lignin has high purity and alkali solubility. Organosolvent-derived lignin is generally free of sulfur, along with having lower MW compared with lignin derived from other treatment methods. Steam-burst lignin has low MW and improved solubility in organic solvents. Dilute acid lignin contains sugar-derivative impurities. However, improved production yields of chemicals such as HMF (hydroxymethylfurfural), furfural, and levulinic acid can be achieved using acidic treatment in conjunction with organic solvents to deconstruct lignocellulosic biomass. Thus, a wide variety of properties of derived lignin could be obtained, depending on the treatment method used.

Ethanol, produced from lignocellulosics, can also be converted to fuel hydrogen. There is significant interest in hydrogen research both internationally and domestically, as industry and governments worldwide investigate and execute on decarbonization strategies. Thus, biorefinery approaches offer potential to recover low-cost, green, and value-added products from abundant and natural resources that can be fermented by microorganisms in an energy-efficient manner. Biohydrogen is an important product that can be obtained from agro-waste through various fermentation techniques. Biohydrogen thus produced is a sustainable energy form compared with other hydrogen production methods. Current hydrogen production techniques are mainly based on electrolysis of water as well as processing of hydrogen-rich compounds through thermocatalytic reforming, but these processes are energy-intensive. Although biohydrogen production and its utilization appears an attractive, environmentally friendly and feasible alternative to fossil-based fuels, the leap from the present fuel economy to a biohydrogen economy is still in its infancy, and will not be covered here.

In order to have a sustainable future, it is required that the society slowly moves away from the current and widely practiced linear, "take, make, dispose (waste-creation)" model to a "reduce, reuse, recycle, regenerate" model (Rhodes 2017). The production of carbohydrates and lignin, which is mainly derived from agro-waste, is estimated to exceed is estimated to exceed 2×10^{11} tons per year (Tuck et al. 2012). One of the innovative applications for recycling lignocellulosic biomass residues is to prepare important lignin-derived products for platform chemicals, resin, adhesive and biobased composites, and also cellulosics such as cellulose nanofibers (CNFs), cellulose nano/microcrystals (CNC/CMC), and cellulose derivatives (Ahmed et al. 2014).

Lignin, without modification, is not suitable for use as a chemical in resin production. A pulping-based process involves hydrolytic cleavages of esters and ethers linkages, which improves its reactivity for advanced resin-based composites. Another functionalization process is demethylation, which can modify lignin to phenolic lignin. Demethylation thus results in a catechol formation, providing extra binding sites on lignin that are required for the polymerization. To demethylate lignin, small aliphatic thiols can be used via a nucleophilic substitution reaction with sodium methoxide. Demethylation can remove one or

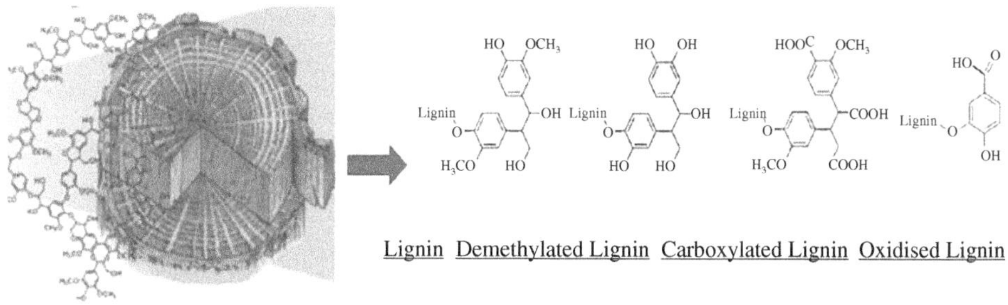

Figure 4.10 *Various modified lignins.*

both methyl groups to the phenolic hydroxyl group (Figure 4.10), and the subsequent material is expected to have high reactivity. Carboxylation is another modification technique to increase the reactivity by converting lignin hydroxyl groups into carboxyl groups. This conversion is possible by reacting lignin with maleic anhydride or succinic anhydride in the presence of ethylene glycol. Lignin oxidation can result in the formation of several functional products and fine chemicals. The most useful oxidants that preserve the lignin aromatic rings are nitrobenzene, air, and oxygen, as well as some metallic oxides (Tarabanko and Tarabanko 2017). The oxidation results in aldehydes and acids. The production rate depends on the nature of the oxidant and whether the oxidation is carried out in the presence or absence of catalysts (copper (II) and cobalt (II)).

Lignin can also be blended with existing resins to improve properties like adhesion, or can be functionalized with existing monomers such as vinyl actetate by aqueous phase graft polymerization (Panesar et al. 2013). Nemoto et al. (2010) developed lignin as a transparent resin, and successfully prepared various lignophenols (LPs) with hydroxyl groups, which were alkoxylated with various alkyl halides. These LPs had good film-forming ability along with comparable optical properties to that of bisphenol A-based polycarbonate (PC); hence they can be used as transparent materials, in place of other synthetic plastics such as PC (Nemoto et al. 2010). Chemically modified lignin could also be used for preparation of epoxy resin. The epoxy rings will undergo crosslinking/curing with the carboxylic acid groups of modified lignin. Blended polymers with lignin, other chemicals (epoxies) and curing agents (amine based) have been extensively explored by many researchers. Lignin can be used in phenol epoxy resins to reduce the cost of production. However, their reactive functionality must be improved by adding reagents.

Cellulose has been the topic of extensive investigations in macromolecular chemistry over the past three decades, due to its biodegradable, renewable, tough, water-insoluble and fibrous nature. Cellulosic materials are benign and non-toxic, hence causing no harm to the natural habitat. Cellulosics decompose and return to the natural carbon cycle. Numerous new functional materials from cellulose are being developed, mostly using nanotechnology, over a broad range of applications. Cellulose-based films have been extensively used for wrapping purposes (wound dressing). They also find application in renal failure treatment, controllable drug delivery systems, contact lenses, blood purification, biosensors (Czaja et al. 2007), as well as in new clinical applications ("scaffolds in tissue engineering, temporary skin substitute, a hemostatic agent, post-operative adhesion barrier, and as a culture": Fu et al. 2019). Cellulose derivatives are extensively used as a rheology modifier for 3D printing (Athukoralalage et al. 2019). Researchers have used various agricultural residues for cellulose production, such as tobacco stalks and leaves (Kulic and Radojicic 2011), wheat straw (Alemdar and Sain 2008a), potato tubers (Dufresne et al. 2000), sugar beet pulp (Habibi 2008), pea hull (Chen et al. 2009), wheat straw and soy hulls (Alemdar and Sain 2008b), empty oil palm fruit bunches (Jonoobi et al. 2011), banana rachis (Zuluaga et al. 2007), lotus leaf stalks (Chen et al. 2015), rice husk (Ludueña et al. 2011), and coconut husk (Rosa et al. 2010).

Bacterial cellulose (BC) is another form of cellulose synthesized by aerobic bacteria grown on food sources. The pure and crystalline nature of BC is gaining much interest in the research community due to its unique properties in comparison with plant-based cellulose (Moniri et al. 2017). In order to produce BC, the culture medium needs to be enriched with carbohydrates, nitrogen, proteins, vitamins, and inorganic salts (Almeida et al. 2013). But the production cost of BC using conventional techniques is very high, resulting in limited use of BC in commercial applications. However, the optimum conditions for cellulose production with microorganisms have not yet been established. Since coconut water (CW) is rich in carbohydrates, proteins, and trace elements, it can be used as a culture media to produce BC. The simplest example of this is *nata de coco*, an indigenous dessert of the Philippines made by culturing *Gluconacetobacter xylinus* (formerly *Acetobacter xylinum*) through fermentation with CW (Nugroho and Aji 2015). CW serves as a carbon source for *G. xylinus* and later converts to extracellular cellulose. After a period of time, a layer of BC forms on the surface of the fermented CW (Halib et al. 2012).

Incorporation of probiotics into food products is mostly problematic because the bacteria may not be viable by the time it reaches the stomach, or they might get destroyed during their transit of the gastrointestinal tract (López-Rubio et al. 2009). Probiotics immobilized on dietary fibers extracted from agro-wastes has shown a viable count and improved growth properties, indicating that agro-wastes can be used in the production of functional foods (Fung et al. 2011; Teh et al. 2009). However, there have not been many attempts to utilize agro-wastes for immobilizing probiotics.

Agricultural wastes are rich in a variety of biologically active species such as antioxidants. Oxidation destroys many food products, hence antioxidants derived from agro-waste are valuable products. They can be used as natural food additives that modify regular food products to make them more beneficial to health and wellbeing, thereby increasing their market value.

One of the most overlooked but major sources of plant material going to landfill is yard waste. There are tremendous opportunities to reduce the yard waste going to landfill in many developing and developed countries. Below we present a case study of yard waste management in Hong Kong.

4.5 Case Study: Yard Waste Management

4.5.1 Background of Yard Waste in Hong Kong

In 2018, the total amount of MSW disposal in Hong Kong was 4.17 million tonnes, or 11 428 tonnes per day (tpd). The disposal rate of MSW was 1.53 kg/person/day in 2018 (Environment Protection Department 2019). The Environment Bureau of Hong Kong SAR target is to reduce the MSW disposal rate to landfill to 40% by 2022 (Environment Bureau

Table 4.7 *Weight of yard waste disposed of by different Hong Kong government departments to landfill in 2018.*

Department	Weight (tonnes)
Agriculture, Fisheries, and Conservation Department	0
Architectural Services Department	3013
Civil Engineering and Development Department	3030
Drainage Services Department	575
Highways Department	2380
Leisure and Cultural Services Department	10 657
Water Supplies Department	1576
Lands Department	2300
Housing Department	598
Total	24 129

Source: Data from Environment Protection Department (2018), used with permission.

2014). This target included yard waste. Yard waste contains of all kinds of vegetation waste matters, and is also known as garden waste or green waste. Yard waste includes clippings from grass and plants, leaves, cut and cleaned bushes and shrubs, weeds, branches and twigs, trunks of trees, plants from festive decorations, plants in pots and vases, and cut flowers (Environment Bureau 2014).

In 2018, 341 tonnes of yard waste was disposed of in landfill each day due to super-typhoon Mangkhut. If yard waste caused by this event is excluded, only 171 tonnes of yard waste are disposed to landfill each day. In 2018, the weight of yard waste disposed of by different Hong Kong government departments to landfill was 24 129 tonnes (Table 4.7). The Leisure and Cultural Services department disposed of the highest quantity of yard waste compared with other departments, at 10 657 tonnes (Environment Protection Department 2018).

Recently, Hong Kong has started to focus on yard waste in terms of reduction and reuse. They have started to carry out data collection, campaigns for yard waste reduction, collection and separation, and finding treatment methods for those portions that are difficult to manage (Environment Bureau 2014).

4.5.2 Conventional Yard Waste Reduction and Treatment Strategy

Hong Kong has implemented different strategies to solve the yard waste issue. The strategies include yard waste reduction, yard waste separation, and collection and treatment.

Through yard waste reduction, the Environment Bureau of Hong Kong encourages people to minimize the use of display plants during festivals such as Lunar New Year or Christmas. The Food and Environmental Hygiene Department (FEDH) also replants unsold plants from the festival fair. In addition, minimization of yard waste generation during landscaping should also be taken into consideration (Environment Bureau 2014).

The Environment Bureau encourages householders to separate their yard waste from their food waste for collection purposes through major programs. The Environment Bureau suggests that commercial property management should introduce yard waste separation and collection requirements for cleansing or gardening contractors. In addition, the government is responsible for developing and implementing a suitable methodology for yard waste separation and collection, (Environment Bureau 2014).

Yard waste treatment that fits Hong Kong includes composting, anaerobic digestion, natural degradation, and reuse and recycling. Composting is one of the most environmentally benign methods to treat yard waste as it is economical and does not create extra pollution. Composters must be built near the major sources of yard waste such as large housing and commercial areas and sites managed by the Leisure and Cultural Services Department. Furthermore, organic waste treatment facilities (OWTFs) can also be used to solve the yard waste problem through anaerobic digestion. In total the digesters can be used to treat around 35 tonnes of yard waste every day (Environment Bureau 2014). There are two systems for composting: on-site composting using the 3-Bin Composting system, or off-site at the composting plants (Greening Landscape and Tree Management Section Development Bureau 2014). Natural degradation is another method that can be used to treat yard waste, but Hong Kong is only able to deal with limited capacities. For example, yard waste resulting from country parks is more suited for natural decomposition within country parks, because it is time-consuming and difficult to find another place to carry out the decomposition (Environment Bureau 2014). Yard waste can also be treated through reuse and recycling. For example, wood waste and festival decoration plants can be converted to wood fuel or pellets. EcoPark at Hong Kong treats 2 tonnes of wood waste into mulch every day (Environment Bureau 2014). Tree logs produced from vegetation maintenance can be transformed into site furniture and benches. It can also be used for sculptural or pruning work (Greening Landscape and Tree Management Section Development Bureau 2014). Shredded waste wood can be reused as mulch and returned back into the ecological cycle as nutrients (Greening Landscape and Tree Management Section Development Bureau 2014).

4.5.3 Novel Techniques and Strategies for Yard Waste Treatment

Yard waste is mainly plant cell wall, comprising cellulose, hemicellulose, and lignin. The components of yard waste are identical to agricultural waste. Hence, it is possible to apply the methods of the agricultural waste biorefinery to treat yard waste. For instance, sugarcane bagasse can be used as feedstock for succinic acid production through fermentation of *Yarrowia lipolytica* PSA02004 (Ong et al. 2019). The bagasse can be replaced by yard waste as a feedstock to perform fermentation. In another case, improvement of desert soils was accomplished using biochar produced from yard waste, cotton gin trash, pecan shells and pecan orchard prunings. Biochar production took place at 450°C under a nitrogen environment with a slow pyrolysis process (Y. Zhang et al. 2016). Organic yard waste also has been converted into ethanol through fermentation by *Rhodotorula mucilaginosa* (Schmitt et al. 2012).

4.6 Conclusions and Future Trends

Agricultural waste is one of the principal anthropogenic sources of methane present in our atmosphere. Widespread harvesting of crops provides substantial agricultural residues every year, with huge untapped potential to generate significant amounts of value-added products. This chapter aims to focus on scientific studies that are pushing the envelope for novel valorisation strategies for agro-based waste. The present chapter touches on different valorisation strategies, focusing on anaerobic fermentation, yard waste management, and novel products that can potentially be recovered through these strategies using different pre-treatment approaches. Agricultural waste, in general, can be tapped as a rich source of bioactive compounds through two different concepts, the biorefinery and the circular economy.

The chapter also takes a fresh look at the development of novel cellulosic nanomaterials such as CNF and CNC. These nanocellulosic materials are derivatives of lignocellulosic biomass. As discussed above, nanocellulosic materials and dietary fibers extracted from agro-wastes are reported to maintain higher viability of microbes, and hence have tremendous potential to be applied in future probiotic studies. In conclusion, based on scientific and logistical evidence, agricultural waste is a highly promising source of biogas if utilized appropriately. Valorisation through anaerobic digestion proves to be quite straightforward and effective compared to other approaches. However, the utilization of biomass boils down to the pretreatment method applied. Evidently, there is still a lot of ongoing research about the right combination of methods to provide better degradability and correspondingly higher methane yields. The fact that valorisation of agricultural waste is essential for attaining sustainable development goals is undeniable.

Acknowledgements

The work described in this chapter is a part of the European Horizon 2020 research and development project titled "NoAW – No Agro-Waste: Innovative approaches to turn agricultural waste into ecological and economic assets." We thank the Bill and Melinda Gates Foundation for supporting Dr Srinivas Mettu and Dr Carol Lin through the Grand Challenges Exploration (GCE) Initiative, for project OPP1211888 in the round "New Approaches for Manufacturing Gut Microbial Biotherapeutics" (Round 22).

References

Abdel-Shafy, H.I. and Mansour, M.S. (2018). Solid waste issue: sources, composition, disposal, recycling, and valorisation. *Egypt. J. Pet.* **27** (4): 1275–1290.

Ahmed, A.E.S., Hassan, M., El-Masry, A., and El-Gendy, A. (2014). Testing of medical tablets produced with microcrystalline cellulose prepared from agricultural wastes. *Polym. Compos.* **35** (7): 1343–1349.

Alauddin, Z.A.B.Z., Lahijani, P., Mohammadi, M., and Mohamed, A.R. (2010). Gasification of lignocellulosic biomass in fluidized beds for renewable energy development: a review. *Renewable Sustainable Energy Rev.* **14** (9): 2852–2862.

Alemdar, A. and Sain, M. (2008a). Biocomposites from wheat straw nanofibers: morphology, thermal and mechanical properties. *Compos. Sci. Technol.* **68**: 557–565.

Alemdar, A. and Sain, M. (2008b). Isolation and characterization of nanofibers from agricultural residues – wheat straw and soy hulls. *Bioresour. Technol.* **99** (6): 1664–1671.

Almeida, D.M., Prestes, R.A., Fonseca, A.F.D. et al. (2013). Minerals consumption by *Acetobacter xylinum* on cultivation medium on coconut water. *Braz. J. Microbiol.* **44** (1): 197–206.

Amiri, H., Karimi, K., and Zilouei, H. (2014). Organosolv pretreatment of rice straw for efficient acetone, butanol, and ethanol production. *Bioresour. Technol.* **152**: 450–456.

André, L., Pauss, A., and Ribeiro, T. (2018). Solid anaerobic digestion: state-of-art, scientific and technological hurdles. *Bioresour. Technol.* **247**: 1027–1037.

Angın, D. (2013). Effect of pyrolysis temperature and heating rate on biochar obtained from pyrolysis of safflower seed press cake. *Bioresour. Technol.* **128**: 593–597.

Antal, M.J. Jr., Mok, W.S., Varhegyi, G. et al. (1990). Review of methods for improving the yield of charcoal from biomass. *Energy Fuels* **4** (3): 221–225.

Antal, M.J., Allen, S.G., Dai, X. et al. (2000). Attainment of the theoretical yield of carbon from biomass. *Ind. Eng. Chem. Res.* **39** (11): 4024–4031.

Arvanitoyannis, I.S. and Tserkezou, P. (2008). Wheat, barley and oat waste: a comparative and critical presentation of methods and potential uses of treated waste. *Int. J. Food Sci. Technol.*: 694–725.

Asadullah, M., Rahman, M.A., Ali, M.M. et al. (2008). Jute stick pyrolysis for bio-oil production in fluidized bed reactor. *Bioresour. Technol.* **99** (1): 44–50.

Ateş, F., Pütün, E., and Pütün, A. (2004). Fast pyrolysis of sesame stalk: yields and structural analysis of bio-oil. *J. Anal. Appl. Pyrolysis* **71** (2): 779–790.

Athukoralalage, S.S., Balu, R., Dutta, N.K., and Roy Choudhury, N. (2019). 3D bioprinted nanocellulose-based hydrogels for tissue engineering applications: a brief review. *Polymers* **11** (5): 898.

Auta, M., Ern, L., and Hameed, B. (2014). Fixed-bed catalytic and non-catalytic empty fruit bunch biomass pyrolysis. *J. Anal. Appl. Pyrolysis* **107**: 67–72.

Azargohar, R., Jacobson, K., Powell, E.E. et al. (2013). Evaluation of properties of fast pyrolysis products obtained, from Canadian waste biomass. *J. Anal. Appl. Pyrolysis* **104**: 330–340.

Aznar, M.P., Corella, J., Delgado, J. et al. (1993). Improved steam gasification of lignocellulosic residues in a fluidized bed with commercial steam reforming catalysts. *Ind. Eng. Chem. Res.* **32** (1): 1–10.

Baker, E.G., Mudge, L.K., and Brown, M.D. (1987). Steam gasification of biomass with nickel secondary catalysts. *Ind. Eng. Chem. Res.* **26** (7): 1335–1339.

Basu, P. (2010). Gasification theory and modeling of gasifiers. Chapter 5. In: *Biomass Gasification and Pyrolysis*. Boston: Academic Press.

Basu, P. (2018). *Biomass Gasification, Pyrolysis and Torrefaction: Practical Design and Theory*. Academic Press.

Boucher, J. and Friot, D. (2017). *Primary Microplastics in the Oceans: A Global Evaluation of Sources*. Switzerland: IUCN Gland.

Boyden, A. (2001). *Reliable Data on Sustainable Wheat Straw Availability, Straw Production*. Agriculture Development Fund.

Bryden, K.M. and Hagge, M.J. (2003). Modeling the combined impact of moisture and char shrinkage on the pyrolysis of a biomass particle. *Fuel* **82** (13): 1633–1644.

Campoy, M., Gómez-Barea, A., Vidal, F.B. et al. (2009). Air–steam gasification of biomass in a fluidised bed: process optimisation by enriched air. *Fuel Process. Technol.* **90** (5): 677–685.

Chen, Y., Cheng, J.J., and Creamer, K.S. (2008). Inhibition of anaerobic digestion process: a review. *Bioresour. Technol.* **99** (10): 4044–4064.

Chen, Y., Liu, C., Chang, P.R. et al. (2009). Bionanocomposites based on pea starch and cellulose nanowhiskers hydrolyzed from pea hull fibre: effect of hydrolysis time. *Carbohydr. Polym.* **76**: 607–615.

Chen, Y., Wu, Q., Huang, B. et al. (2015). Isolation and characteristics of cellulose and nanocellulose from lotus leaf stalk agro-wastes. *BioResources* **10** (1): 2014.

Cherubini, F. and Ulgiati, S. (2010). Crop residues as raw materials for biorefinery systems – an LCA case study. *Appl. Energy* **87** (1): 47–57.

Choi, H.S., Choi, Y.S., and Park, H.C. (2012). Fast pyrolysis characteristics of lignocellulosic biomass with varying reaction conditions. *Renewable Energy* **42**: 131–135.

Czaja, W.K., Young, D.J., Kawecki, M., and Brown, R.M. (2007). The future prospects of microbial cellulose in biomedical applications. *Biomacromolecules* **8** (1): 1–12.

De Meester, M., Demeyer, J., Veighe, F. et al. (2012). The environmental sustainability of anaerobic digestion as a biomass valorisation technology. *Bioresour. Technol.* **121**: 396–403.

De, M., Azargohar, R., Dalai, A.K., and Shewchuk, S.R. (2013). Mercury removal by bio-char based modified activated carbons. *Fuel* **103**: 570–578.

Demiral, İ. and Şensöz, S. (2008). The effects of different catalysts on the pyrolysis of industrial wastes (olive and hazelnut bagasse). *Bioresour. Technol.* **99** (17): 8002–8007.

Demirbas, A. (2004). Effects of temperature and particle size on bio-char yield from pyrolysis of agricultural residues. *J. Anal. Appl. Pyrolysis* **72** (2): 243–248.

Demirbas, A. and Ozturk, T. (2005). Anaerobic digestion of agricultural solid residues. *Int. J. Green Energy* **1** (4): 483–494.

Devi, L., Ptasinski, K.J., Janssen, F.J. et al. (2005). Catalytic decomposition of biomass tars: use of dolomite and untreated olivine. *Renewable Energy* **30** (4): 565–587.

Dufresne, A., Dupeyre, D., and Vignon, M.R. (2000). Cellulose microfibrils from potato tuber cells: processing and characterization of starch–cellulose microfibril composites. **76** (14): 2080–2092.

Encinar, J., Beltran, F., Bernalte, A. et al. (1996). Pyrolysis of two agricultural residues: olive and grape bagasse. Influence of particle size and temperature. *Biomass Bioenergy* **11** (5): 397–409.

Encinar, J.M., Beltrán, F.J., Ramiro, A. et al. (1997). Catalyzed pyrolysis of grape and olive bagasse. Influence of catalyst type and chemical treatment. *Ind. Eng. Chem. Res.* **36** (10): 4176–4183.

van den Enden, P.J. and Lora, E.S. (2004). Design approach for a biomass fed fluidized bed gasifier using the simulation software CSFB. *Biomass Bioenergy* **26** (3): 281–287.

Environment Bureau. (2014). *A Food Waste & Yard Waste Plan for Hong Kong 2014–2022*. https://www.enb.gov.hk/sites/default/files/pdf/FoodWastePolicyEng.pdf (accessed 5 April 2020).

Environment Protection Department. (2018). *Quantities of yard waste collected and disposed of by various government departments*. https://gia.info.gov.hk/general/201812/05/P2018120500250_298935_1_1543982567444.pdf (accessed 5 April 2020).

Environment Protection Department. (2019). *Monitoring of Solid Waste in Hong Kong – Waste Statistics for 2018*. https://www.wastereduction.gov.hk/en/assistancewizard/waste_red_sat.htm (accessed 5 April 2020).

Fassinou, W.F., Van de Steene, L., Toure, S. et al. (2009). Pyrolysis of *Pinus pinaster* in a two-stage gasifier: influence of processing parameters and thermal cracking of tar. *Fuel Process. Technol.* **90** (1): 75–90.

Florin, N.H. and Harris, A.T. (2008). Enhanced hydrogen production from biomass with in situ carbon dioxide capture using calcium oxide sorbents. *Chem. Eng. Sci.* **63** (2): 287–316.

Forster-Carneiro, T., Fernández, L.A., Pérez, M. et al. (2004). Optimization of SEBAC start up phase of municipal solid waste anaerobic digestion. *Chem. Biochem. Eng. Q.* **18** (4): 429–439.

Forster-Carneiro, T., Riau, V., and Pérez, M. (2010). Mesophilic anaerobic digestion of sewage sludge to obtain class B biosolids. microbiological methods development. *Biomass Bioenergy* **34** (12): 1805–1812.

Forster-Carneiro, T., Berni, M.D., Dorileo, I.L., and Rostagno, M.A. (2013). Biorefinery study of availability of agriculture residues and wastes for integrated biorefineries in Brazil. *Resour. Conserv. Recycl.*: 78–88.

Fu, L.-H., Qi, C., Ma, M.-G., and Wan, P. (2019). Multifunctional cellulose-based hydrogels for biomedical applications. *J. Mater. Chem. B* **7** (10): 1541–1562.

Fung, W.-Y., Yuen, K.-H., and Liong, M.-T. (2011). Agrowaste-based nanofibers as a probiotic encapsulant: fabrication and characterization. *J. Agric. Food. Chem.* **59** (15): 8140–8147.

Gai, C. and Dong, Y. (2012). Experimental study on non-woody biomass gasification in a downdraft gasifier. *Int. J. Hydrogen Energy* **37** (6): 4935–4944.

Geyer, R., Jambeck, J.R., and Law, K.L. (2017). Production, use, and fate of all plastics ever made. *Sci. Adv.* **3** (7): e1700782.

Ghatak, H.R. (2002). Papermaking potential of congress grass: pulpability and fiber characteristics. *Tappi Journal* **1** (3): 24–27.

Ghatak, H.R. (2011). Biorefineries from the perspective of sustainability: feedstocks, products, and processes. *Renewable Sustainable Energy Rev.* **15** (8): 4042–4052.

Gontard, N., Sonesson, U., Birkved, M. et al. (2018). A research challenge vision regarding management of agricultural waste in a circular bio-based economy. *Crit. Rev. Env. Sci. Technol.* **48** (6): 614–654.

Göransson, K., Söderlind, U., He, J. et al. (2011). Review of syngas production via biomass DFBGs. *Renewable Sustainable Energy Rev.* **15** (1): 482–492.

Government of Prince Edward Island (1999). *Best Management Practices: Agricultural Waste Management* (ed. PEI Department of Fisheries, PEI Department of Agriculture and Forestry, Environment Canada). Canada: Government of Prince Edward Island.

Greening Landscape and Tree Management Section Development Bureau. (2014). *Guidelines on Yard Waste Reduction and Treatment.* https://www.greening.gov.hk/filemanager/content/pdf/faq/Guidelines_on_Yard_Waste_Reduction_and_Treatment_e.pdf (accessed 5 April 2020).

Guendouz, J., Buffiere, P., Cacho, J. et al. (2008). High-solids anaerobic digestion: comparison of three pilot scales. *Water Sci. Technol.* **58** (9): 1757–1763.

Habibi, Y. (2008). Optimization of cellouronic acid synthesis by TEMPO-mediated oxidation of cellulose III from sugar beet pulp. *Cellulose* **15** (1): 177–185.

Hai, H.T. (2010). *Benefits of the 3R Approach for Agricultural Waste Management (AWM) in Vietnam: Under the Framework of Joint Project on Asia Resource Circulation Research.* Hayama: Institute for Global Environment Stategies.

Halder, P., Paul, N., Joardder, M. et al. (2016). Feasibility analysis of implementing anaerobic digestion as a potential energy source in Bangladesh. *Renewable Sustainable Energy Rev.* **65**: 124–134.

Halder, P., Kundu, S., Patel, S. et al. (2019a). TGA-FTIR study on the slow pyrolysis of lignin and cellulose-rich fractions derived from imidazolium-based ionic liquid pre-treatment of sugarcane straw. *Energy Convers. Manag.* **200**: 112067.

Halder, P., Kundu, S., Patel, S. et al. (2019b). A comparison of ionic liquids and organic solvents on the separation of cellulose-rich material from river red gum. *Bioenergy Res.* **12** (2): 275–291.

Halder, P., Kundu, S., Patel, S. et al. (2019c). Progress on the pre-treatment of lignocellulosic biomass employing ionic liquids. *Renewable Sustainable Energy Rev.* **105**: 268–292.

Halib, N., Mohd Amin, M.C.I., and Ahmad, I. (2012). Physicochemical properties and characterization of nata de coco from local food industries as a source of cellulose. *Sains Malays.* **41**: 205–211.

Herman, A.P., Yusup, S., Shahbaz, M. et al. (2016). Bottom ash characterization and its catalytic potential in biomass gasification. *Procedia. Eng.* **148**: 432–436.

Hossain, M.A., Jewaratnam, J., and Ganesan, P. (2016). Prospect of hydrogen production from oil palm biomass by thermochemical process – a review. *Int. J. Hydrogen Energy* **41** (38): 16637–16655.

Hosseini, S.E. and Wahid, M.A. (2014). Utilization of palm solid residue as a source of renewable and sustainable energy in Malaysia. *Renewable Sustainable Energy Rev.* **40**: 621–632.

Hosseini, M., Dincer, I., and Rosen, M.A. (2012). Steam and air fed biomass gasification: comparisons based on energy and exergy. *Int. J. Hydrogen Energy* **37** (21): 16446–16452.

Hosseini, S.E., Wahid, M.A., and Ganjehkaviri, A. (2015). An overview of renewable hydrogen production from thermochemical process of oil palm solid waste in Malaysia. *Energy Convers. Manag.* **94**: 415–429.

Hu, Z.-H. and Yu, H.-Q. (2005). Application of rumen microorganisms for enhanced anaerobic fermentation of corn stover. *Process Biochem.* **40** (7): 2371–2377.

Hu, G., Xu, S., Li, S. et al. (2006). Steam gasification of apricot stones with olivine and dolomite as downstream catalysts. *Fuel Process. Technol.* **87** (5): 375–382.

Ibarra-Gonzalez, P. and Rong, B.-G. (2018). A review of the current state of biofuels production from lignocellulosic biomass using thermochemical conversion routes. *Chin. J. Chem. Eng.* **27** (7): 1523–1535.

Intergovernmental Panel on Climate Change (2014). *Climate Change 2014: Mitigation of Climate Change. Contribution of Working Group III to the Fifth Assessment Report of the Intergovernmental Panel on Climate Change*. IPCC https://www.ipcc.ch/report/ar5/wg3/ (accessed 5 April 2020).

Jonoobi, M., Khazaeian, A., Tahir, P.M. et al. (2011). Characteristics of cellulose nanofibers isolated from rubberwood and empty fruit bunches of oil palm using chemo-mechanical process. *Cellulose* **18**: 1085–1095.

Kaewpanha, M., Guan, G., Hao, X. et al. (2014). Steam co-gasification of brown seaweed and land-based biomass. *Fuel Process. Technol.* **120**: 106–112.

Kalamaras, S. and Kotsopoulos, T. (2014). Anaerobic co-digestion of cattle manure and alternative crops for the substitution of maize in South Europe. *Bioresour. Technol.* **172**: 68–75.

Karthikeyan, O.P. and Visvanathan, C. (2013). Bio-energy recovery from high-solid organic substrates by dry anaerobic bio-conversion processes: a review. *Rev. Environ. Sci. Biotechnol.* **12** (3): 257–284.

Kaur, G., Uisan, K., Ong, K.L. et al. (2018). Recent trends in green and sustainable chemistry & waste valorisation: rethinking plastics in a circular economy. *Curr. Opin. Green Sustain. Chemi.* **9**: 30–39.

Khoshnevisan, B., Tsapekos, P., Zhang, Y. et al. (2019). Urban biowaste valorisation by coupling anaerobic digestion and single cell protein production. *Bioresour. Technol.* **290**: 121743.

Khurshid, M. (2009). Biogas development in rural areas of Pakistan: a sustainable option for domestic energy. *Sci. Vision* **15** (2): 57–61.

Kılıc, M., Kırbıyık, Ç., Çepelioğullar, Ö. et al. (2013). Adsorption of heavy metal ions from aqueous solutions by bio-char, a by-product of pyrolysis. *Appl. Surf. Sci.* **283**: 856–862.

Kim, S.W., Park, D.K., and Kim, S.D. (2013). Pyrolytic characteristics of Jatropha seedshell cake in thermobalance and fluidized bed reactors. *Korean J. Chem. Eng.* **30** (5): 1162–1170.

Kohli, K., Prajapati, R., and Sharma, B.K. (2019). Bio-based chemicals from renewable biomass for integrated biorefineries. *Energies* **12** (2): 233.

Kral, I., Piringer, G., Saylor, M.K. et al. (2016). Environmental effects of steam explosion pretreatment on biogas from maize – case study of a 500-kW Austrian biogas facility. *Bioenergy Res.* **9** (1): 198–207.

Kulic, G. and Radojicic, V. (2011). Analysis of cellulose content in stalks and leaves of large leaf tobacco. *J. Agric. Sci.* **56**: 207–215.

Lal, R. (2005). World crop residues production and implications of its use as a biofuel. *Environ. Int.* **31** (4): 575–584.

Lal, R. (2008). Crop residues as soil amendments and feedstock for bioethanol production. *Waste Manag.* **28** (4): 747–758.

Lee, S.W., Nam, S.S., Kim, S.B. et al. (2000). The effect of Na_2CO_3 on the catalytic gasification of rice straw over nickel catalysts supported on kieselguhr. *Korean J. Chem. Eng.* **17** (2): 174–178.

Letcher, T.M. (2011). *Waste*. Elsevier.

Li, J., Yin, Y., Zhang, X. et al. (2009). Hydrogen-rich gas production by steam gasification of palm oil wastes over supported tri-metallic catalyst. *Int. J. Hydrogen Energy* **34** (22): 9108–9115.

Li, C., Cheng, G., Balan, V. et al. (2011). Influence of physico-chemical changes on enzymatic digestibility of ionic liquid and AFEX pretreated corn stover. *Bioresour. Technol.* **102** (13): 6928–6936.

Li, Y., Zhu, J., Wan, C. et al. (2011). Solid-state anaerobic digestion of corn stover for biogas production. *Trans. ASABE* **54** (4): 1415–1421.

Li, M.-F., Sun, S.-N., Xu, F. et al. (2012). Organosolv fractionation of lignocelluloses for fuels, chemicals and materials: a biorefinery processing perspective. In: *Biomass Conversion* (eds. C. Baskar, S. Baskar and R. Dhillon), 341–379. Springer.

Li, R., Duan, N., Zhang, Y. et al. (2017). Co-digestion of chicken manure and microalgae Chlorella 1067 grown in the recycled digestate: nutrients reuse and biogas enhancement. *Waste Manag.* **70**: 247–254.

Lin, J., Zuo, J., Gan, L. et al. (2011). Effects of mixture ratio on anaerobic co-digestion with fruit and vegetable waste and food waste of China. *J. Environ. Sci.* **23** (8): 1403–1408.

Liu, C., Wachemo, A.C., Tong, H. et al. (2018). Biogas production and microbial community properties during anaerobic digestion of corn stover at different temperatures. *Bioresour. Technol.* **261**: 93–103.

López-Rubio, A., Sanchez, E., Sanz, Y., and Lagaron, J.M. (2009). Encapsulation of living bifidobacteria in ultrathin PVOH electrospun fibers. *Biomacromolecules* **10** (10): 2823–2829.

Ludueña, L., Fasce, D.P., Alvarez, V.A., and Stefani, P.M. (2011). Nanocellulose from rice husk following alkaline treatment to remove silica. *Bioresources* **62** (2): 1440–1453.

Luo, Z., Wang, S., Liao, Y. et al. (2004). Research on biomass fast pyrolysis for liquid fuel. *Biomass Bioenergy* **26** (5): 455–462.

Mani, T., Murugan, P., Abedi, J. et al. (2010). Pyrolysis of wheat straw in a thermogravimetric analyzer: effect of particle size and heating rate on devolatilization and estimation of global kinetics. *Chem. Eng. Res. Des.* **88** (8): 952–958.

Matsumura, Y., Minowa, T., and Yamamoto, H. (2005). Amount, availability, and potential use of rice straw (agricultural residue) biomass as an energy resource in Japan. *Biomass Bioenergy* **29** (5): 347–354.

Mesa, L., González, E., Cara, C. et al. (2010). An approach to optimization of enzymatic hydrolysis from sugarcane bagasse based on organosolv pretreatment. *J. Chem. Technol. Biotechnol.* **85** (8): 1092–1098.

Mohamed, A.R., Hamzah, Z., Daud, M.Z.M. et al. (2013). The effects of holding time and the sweeping nitrogen gas Flowrates on the pyrolysis of EFB using a fixed–bed reactor. *Procedia Eng.* **53**: 185–191.

Mohammed, M., Salmiaton, A., Azlina, W.W. et al. (2011). Hydrogen rich gas from oil palm biomass as a potential source of renewable energy in Malaysia. *Renewable Sustainable Energy Rev.* **15** (2): 1258–1270.

Mohan, D., Pittman Jr, C.U., Bricka, M. et al. (2007). Sorption of arsenic, cadmium, and lead by chars produced from fast pyrolysis of wood and bark during bio-oil production. *J. Colloid Interface Sci.* **310** (1): 57–73.

Mohan, D., Rajput, S., Singh, V.K. et al. (2011). Modeling and evaluation of chromium remediation from water using low cost bio-char, a green adsorbent. *J. Hazard. Mater.* **188** (1–3): 319–333.

Mohan, D., Kumar, H., Sarswat, A. et al. (2014). Cadmium and lead remediation using magnetic oak wood and oak bark fast pyrolysis bio-chars. *Chem. Eng. J.* **236**: 513–528.

Moilanen, A. and Nasrullah, M. (2011). *Gasification Reactivity and Ash Sintering Behaviour of Biomass Feedstocks*. Espoo 2011. VTT Publications 769. 39 p. + app. 96 p. https://www.vttresearch.com/sites/default/files/pdf/publications/2011/P769.pdf.

Moniri, M., Boroumand Moghaddam, A., Azizi, S. et al. (2017). Production and status of bacterial cellulose in biomedical engineering. *Nanomaterials* **7** (9): 257.

Moore, C.J. (2008). Synthetic polymers in the marine environment: a rapidly increasing, long-term threat. *Environ. Res.* **108** (2): 131–139.

Müller, C., Townsend, K., and Matschullat, J. (2012). Experimental degradation of polymer shopping bags (standard and degradable plastic, and biodegradable) in the gastrointestinal fluids of sea turtles. *Sci. Total Environ.* **416**: 464–467.

Nemoto, T., Konishi, G.-I., Tojo, Y., and Funaoka, M. (2010). Development of lignin as a transparent resin: evaluation of thermal and optical properties of alkoxylated lignophenols. *Polym. J.* **42** (11): 896–900.

Nizami, A.S., Saville, B.A., and Maclean, H.L. (2013). Anaerobic digesters. Bioenergy Production by Anaerobic Digestion: Using Agricultural Biomass and Organic Wastes, p.139 https://books.google.com.au/books?hl=en&lr=&id=0S2YAAAAQBAJ&oi=fnd&pg=PA139&ots=yug2psDuwV&sig=VUIWOYsmgOsBNsWvrOkNDXFRS9s&redir_esc=y#v=onepage&q&f=false.

Nugroho, D.A. and Aji, P. (2015). Characterization of Nata de coco produced by fermentation of immobilized *Acetobacter xylinum*. *Agric. Agric. Sci. Procedia* **3**: 278–282.

Obama, P., Ricochon, G., Muniglia, L. et al. (2012). Combination of enzymatic hydrolysis and ethanol organosolv pretreatments: effect on lignin structures, delignification yields and cellulose-to-glucose conversion. *Bioresour. Technol.* **112**: 156–163.

Obi, F.O., Ugwuishiwu, B.O., and Nwakaire, J.N. (2016). Agricultural waste concept, generation, utilization and management. *Niger. J. Technol.*: 957–964.

Ong, K.L., Li, C., Li, X. et al. (2019). Co-fermentation of glucose and xylose from sugarcane bagasse into succinic acid by *Yarrowia lipolytica*. *Biochem. Eng. J.* **148**: 108–115.

Panesar, S.S., Jacob, S., Misra, M., and Mohanty, A.K. (2013). Functionalization of lignin: fundamental studies on aqueous graft copolymerization with vinyl acetate. *Ind. Crops Prod.* **46**: 191–196.

Park, H.J., Park, Y.-K., and Kim, J.S. (2008). Influence of reaction conditions and the char separation system on the production of bio-oil from radiata pine sawdust by fast pyrolysis. *Fuel Process. Technol.* **89** (8): 797–802.

Phonbumrung, T. and Khemsawas, C. (1998). Agricultural crop residue. *Proceedings of Sixth Meeting of Regional Working Group on Grazing and Feed Resources for Southeast Asia*, Legaspi, Philippines, 5–9 October 1998, 183–187.

Puig-Arnavat, M., Bruno, J.C., and Coronas, A. (2010). Review and analysis of biomass gasification models. *Renewable Sustainable Energy Rev.* **14** (9): 2841–2851.

Pütün, E. (2010). Catalytic pyrolysis of biomass: effects of pyrolysis temperature, sweeping gas flow rate and MgO catalyst. *Energy* **35** (7): 2761–2766.

Pütün, A., Özcan, A., and Pütün, E. (1999). Pyrolysis of hazelnut shells in a fixed-bed tubular reactor: yields and structural analysis of bio-oil. *J. Anal. Appl. Pyrolysis* **52** (1): 33–49.

Pütün, A.E., Apaydın, E., and Pütün, E. (2004). Rice straw as a bio-oil source via pyrolysis and steam pyrolysis. *Energy* **29** (12–15): 2171–2180.

Rajendran, K., Aslanzadeh, S., and Taherzadeh, M.J. (2012). Household biogas digesters—a review. *Energies* **5** (8): 2911–2942.

Reed, T.B. and Das, A. (1988). Handbook of biomass downdraft gasifier engine systems. Biomass Energy Foundation. https://books.google.com.au/books?hl=en&lr=&id=_zJHmz2EA3IC&oi=fnd&pg=PA84&ots=CvojdP328q&sig=ErlQuZWEn1mCGAd8sCLxv7HNITU&redir_esc=y#v=onepage&q&f=false.

Rhodes, C.J. (2017). The imperative for regenerative agriculture. *Sci. Prog.* **100** (1): 80–129.

Rickert, D. (1993). *Water Quality Assessment to Determine the Nature and Extent of Water Pollution by Agriculture and Related Activities. Prevention of Water Pollution by Agriculture and Related Activities.* Roma: Water Reports (FAO).

Rosa, M., Medeiros, E., Malmonge, J. et al. (2010). Cellulose nanowhiskers from coconut husk fibers: effect of preparation conditions on their thermal and morphological behavior. *Carbohydr. Polym.* **81**: 83–92.

Salapa, I., Topakas, E., and Sidiras, D. (2018). Simulation and optimization of barley straw organosolv pretreatment. *Ind. Crops Prod.* **113**: 80–88.

Samolada, M., Papafotica, A., and Vasalos, I. (2000). Catalyst evaluation for catalytic biomass pyrolysis. *Energy Fuels* **14** (6): 1161–1167.

Sannigrahi, P., Miller, S.J., and Ragauskas, A.J. (2010). Effects of organosolv pretreatment and enzymatic hydrolysis on cellulose structure and crystallinity in Loblolly pine. *Carbohydr. Res.* **345** (7): 965–970.

Scarlat, N., Dallemand, J.-F., and Fahl, F. (2018). Biogas: developments and perspectives in Europe. *Renewable Energy* **129**: 457–472.

Schmitt, E., Bura, R., Gustafson, R. et al. (2012). Converting lignocellulosic solid waste into ethanol for the State of Washington: an investigation of treatment technologies and environmental impacts. *Bioresour. Technol.* **104**: 400–409.

Sensoez, S. and Angin, D. (2008). Pyrolysis of safflower (*Charthamus tinctorius* L.) seed press cake: part 1. The effects of pyrolysis parameters on the product yields. *Bioresour. Technol.* **99** (13): 5492–5497.

Shahbaz, M., Yusup, S., Inayat, A. et al. (2016). Application of response surface methodology to investigate the effect of different variables on conversion of palm kernel shell in steam gasification using coal bottom ash. *Appl. Energy* **184**: 1306–1315.

Sikarwar, V.S., Zhao, M., Fennell, P.S. et al. (2017). Progress in biofuel production from gasification. *Prog. Energy Combust. Sci.* **61**: 189–248.

Smets, K., Roukaerts, A., Czech, J. et al. (2013). Slow catalytic pyrolysis of rapeseed cake: product yield and characterization of the pyrolysis liquid. *Biomass Bioenergy* **57**: 180–190.

Stefanidis, S.D., Kalogiannis, K.G., Iliopoulou, E. et al. (2011). In-situ upgrading of biomass pyrolysis vapors: catalyst screening on a fixed bed reactor. *Bioresour. Technol.* **102** (17): 8261–8267.

Stephen, J.D., Sokhansanj, S., Bi, X. et al. (2010). The impact of agricultural residue yield range on the delivered cost to a biorefinery in the Peace River region of Alberta, Canada. *Biosystems Eng.* **105** (3): 298–305.

Sukhesh, M.J. and Rao, P.V. (2018). Anaerobic digestion of crop residues: technological developments and environmental impact in the Indian context. *Biocatal. Agric. Biotechnol.* **16** (4): 513–528.

Suksong, W., Jehlee, A., Singkhala, A. et al. (2017). Thermophilic solid-state anaerobic digestion of solid waste residues from palm oil mill industry for biogas production. *Ind. Crops Prod.* **95**: 502–511.

Sun, F.F., Wang, L., Hong, J. et al. (2015). The impact of glycerol organosolv pretreatment on the chemistry and enzymatic hydrolyzability of wheat straw. *Bioresour. Technol.* **187**: 354–361.

Taherzadeh, M.J. and Karimi, K. (2007). Acid-based hydrolysis processes for ethanol from lignocellulosic materials: a review. *BioResources*: 472–499.

Tarabanko, V. and Tarabanko, N. (2017). Catalytic oxidation of lignins into the aromatic aldehydes: general process trends and development prospects. *Int. J. Mol. Sci.* **18** (11): 2421.

Teghammar, A., Karimi, K., Horváth, I.S. et al. (2012). Enhanced biogas production from rice straw, triticale straw and softwood spruce by NMMO pretreatment. *Biomass Bioenergy* **36**: 116–120.

Teh, S.-S., Ahmad, R., Wan-Abdullah, W.-N., and Liong, M.-T. (2009). Evaluation of agrowastes as immobilizers for probiotics in soy milk. *J. Agric. Food. Chem.* **57** (21): 10187–10198.

Teramoto, Y., Lee, S.-H., and Endo, T. (2009). Cost reduction and feedstock diversity for sulfuric acid-free ethanol cooking of lignocellulosic biomass as a pretreatment to enzymatic saccharification. *Bioresour. Technol.* **100** (20): 4783–4789.

Torales, A. (2013). *Anaerobic Digestion: Types, Processes, and Environmental Impact*. Nova Science Publishers.

Tsai, W., Chang, C., and Lee, S. (1997). Preparation and characterization of activated carbons from corn cob. *Carbon* **35** (8): 1198–1200.

Tsai, W., Lee, M., and Chang, Y. (2007). Fast pyrolysis of rice husk: product yields and compositions. *Bioresour. Technol.* **98** (1): 22–28.

Tuck, C.O., Pérez, E., Horváth, I.T. et al. (2012). Valorisation of biomass: deriving more value from waste. *Science* **337** (6095): 695–699.

United Nations (1997). *Glossary of Environment Statistics, Studies in Methods*. United Nations New York.

Vaneeckhaute, C., Meers, E., Michels, E. et al. (2013). Ecological and economic benefits of the application of bio-based mineral fertilizers in modern agriculture. *Biomass Bioenergy* **49**: 239–248.

Vasco-Correa, J., Khanal, S., Manandhar, A. et al. (2018). Anaerobic digestion for bioenergy production: global status, environmental and techno-economic implications, and government policies. *Bioresour. Technol.* **247**: 1015–1026.

Vats, N., Khan, A.A., and Ahmad, K. (2019). Observation of biogas production by sugarcane bagasse and food waste in different composition combinations. *Energy* **185**: 1100–1105.

Vegter, A.C., Barletta, M., Beck, C. et al. (2014). Global research priorities to mitigate plastic pollution impacts on marine wildlife. *Endang. Species Res.* **25** (3): 225–247.

Veluchamy, C. and Kalamdhad, A.S. (2017). Influence of pretreatment techniques on anaerobic digestion of pulp and paper mill sludge: a review. *Bioresour. Technol.* **245**: 1206–1219.

Widjaya, E.R., Chen, G., Bowtell, L. et al. (2018). Gasification of non-woody biomass: a literature review. *Renewable Sustainable Energy Rev.* **89**: 184–193.

Wildschut, J., Smit, A.T., Reith, J.H. et al. (2013). Ethanol-based organosolv fractionation of wheat straw for the production of lignin and enzymatically digestible cellulose. *Bioresour. Technol.* **135**: 58–66.

Wolf, B. and Snyder, G. (2013). *Sustainable Soils: The Place of Organic Matter in Sustaining Soils and their Productivity*. CRC Press.

Wolf, B., Fleming, J., and Batchelor, J. (1980). *Liquid Fertilizer Manual*. Peoria: IL7 National Fertilizer Solutions Association.

Wu, M., Yan, Z.Y., Zhang, X.M. et al. (2016). Integration of mild acid hydrolysis in γ-valerolactone/water system for enhancement of enzymatic saccharification from cotton stalk. *Bioresour. Technol.* **200**: 23–28.

WWF-Australia (2020). *Statement from WWF-Australia on Australia's Bushfire Emergency*. Australia: WWF-Australia.

Xiao, R. and Yang, W. (2013). Influence of temperature on organic structure of biomass pyrolysis products. *Renewable Energy* **50**: 136–141.

Xu, F., Wang, Z.-W., Tang, L. et al. (2014). A mass diffusion-based interpretation of the effect of total solids content on solid-state anaerobic digestion of cellulosic biomass. *Bioresour. Technol.* **167**: 178–185.

Yaman, S. (2004). Pyrolysis of biomass to produce fuels and chemical feedstocks. *Energy Convers. Manage.* **45** (5): 651–671.

Yan, Z., Song, Z., Li, D. et al. (2015). The effects of initial substrate concentration, C/N ratio, and temperature on solid-state anaerobic digestion from composting rice straw. *Bioresour. Technol.* **177**: 266–273.

Zanzi, R., Sjöström, K., and Björnbom, E. (2002). Rapid pyrolysis of agricultural residues at high temperature. *Biomass Bioenergy* **23** (5): 357–366.

Zhang, H., Xiao, R., Huang, H. et al. (2009). Comparison of non-catalytic and catalytic fast pyrolysis of corncob in a fluidized bed reactor. *Bioresour. Technol.* **100** (3): 1428–1434.

Zhang, Z., Wong, H.H., Albertson, P.L. et al. (2013). Laboratory and pilot scale pretreatment of sugarcane bagasse by acidified aqueous glycerol solutions. *Bioresour. Technol.* **138**: 14–21.

Zhang, Y., Idowu, O.J., and Brewer, C.E. (2016). Using agricultural residue biochar to improve soil quality of desert soils. *Agriculture* **6** (1): 10.

Zhang, Z., Harrison, M.D., Rackemann, D.W. et al. (2016). Organosolv pretreatment of plant biomass for enhanced enzymatic saccharification. *Green Chem.* **18** (2): 360–381.

Zhang, Q., Hu, J., and Lee, D.-J. (2017). Pretreatment of biomass using ionic liquids: research updates. *Renewable Energy* **111**: 77–84.

Zhao, X., Cheng, K., and Liu, D. (2009). Organosolv pretreatment of lignocellulosic biomass for enzymatic hydrolysis. *Appl. Microbiol. Biotechnol.* **82** (5): 815.

Zheng, Y., Zhao, J., Xu, F. et al. (2014). Pretreatment of lignocellulosic biomass for enhanced biogas production. *Prog. Energy Combust. Sci.* **42**: 35–53.

Zhou, Y., Li, C., Nges, I.A. et al. (2017). The effects of pre-aeration and inoculation on solid-state anaerobic digestion of rice straw. *Bioresour. Technol.* **224**: 78–86.

Zuluaga, R., Putaux, J.-L., Restropo, A. et al. (2007). Cellulose microfibrils from banana farming residues: isolation and characterization. *Cellulose* **14** (6): 585–592.

5

Valorisation of Woody Biomass

Md Khairul Islam[1], Chengyu Dong[1], Hsien-Yi Hsu[2], Carol Sze Ki Lin[2] and Shao-Yuan Leu[1]

[1] *Department of Civil and Environmental Engineering, Hong Kong Polytechnic University, Hung Hom, Hong Kong*
[2] *School of Energy and Environment, City University of Hong Kong, Kowloon Tong, Hong Kong*

5.1 Generation of Woody Biomass

Woody biomass derived from forest and agricultural residues is an important source of renewable energy and chemicals. According to the Food and Agriculture Organization of the United States (FAO), the global production of wood residues between 1974 and 2016 has increased from 1.85×10^4 tons to 5.69×10^8 tons (Food and Agriculture Organization 2018). Most of the woody biomass has been generated in Asia (51%), America (20%), and Europe (27%) (Food and Agriculture Organization 2018), which is derived from different sources, such as woody product wastes (Figure 5.1a) and yard wastes (Figure 5.1b,c). It contains different parts of woody plants, such as the trunks, branches, and leaves.

Living woody plant cells are generated from mitosis division of the apical and lateral meristems (Evert 2006). The apical meristems are located in the tips of the stems, roots, and branches and are responsible for the primary growth, i.e. lengthening of the plants; while the lateral meristems present under the bark of roots, stem, and branches are responsible for the secondary growth, that is, thickening of the plant. The stem, or the trunk, is currently the most useful part of the woody plant. The tree trunk consists of four segments: bark, phloem, vascular cambium, and the secondary xylem (Harada and Côté Jr 1985). Bark protects the

Waste Valorisation: Waste Streams in a Circular Economy, First Edition.
Edited by Carol Sze Ki Lin, Guneet Kaur, Chong Li and Xiaofeng Yang.

Figure 5.1 Examples of woody biomass derived from different sources: (a) woody product wastes; (b) yard wastes; and (c) trimmings.

plant from mechanical and biological attacks. The thin layer of cells between the phloem and the xylem is the vascular cambium, which divides the wood trunk into xylem in the inner side and phloem in the outer. Phloem is the living tissue that stores and transports nutrients to different parts of the plants, and the xylem is produced by the multiplication of meristematic cells in the vascular cambium (Lachaud et al. 1999).

5.2 General Classification and Properties of Woods

Woody materials are most commonly classified based on the seed structure and reproduction system of the plants (Figure 5.2). Hardwood, known as angiosperm, originates from deciduous trees such as oak, maple, eucalyptus, and beech. Softwood, or gymnosperm, originates from coniferous trees, such as pine, spruce, Douglas fir, and redwood. Hardwood and softwood vary in appearance and have different functional plant cells. In softwood, approximately 90% of the wood volume comprises tracheids, which transport liquid and provide mechanical support to the tree. In hardwood, fibers provide the mechanical support while vessels transport liquid throughout the plant. The transformation and storage of nutrients is performed by parenchyma cells in both softwood and hardwood (Daniel 2009).

Wood consists of elongated cells (Fujita and Harada 2000) in the longitudinal direction of the stem, and grows radially as concentric rings. The inner part of the wood ring is heartwood and the outer parts are sapwood, of which the types and contents of basic chemicals are provided in Figure 5.2. Heartwood consists of entirely dead tissues and provides mechanical support to the trees. The dead cells in heartwood contain deposited extractives, which improve biological resistance and darken the color of the heartwood. Transportation and storage of water, mineral and nutrients take place in the sapwood, although most cells in the sapwood are dead. The living cells of sapwood provide protection against microbial attacks. When the living cells in sapwood die, the sapwood is gradually converted into heartwood with aging.

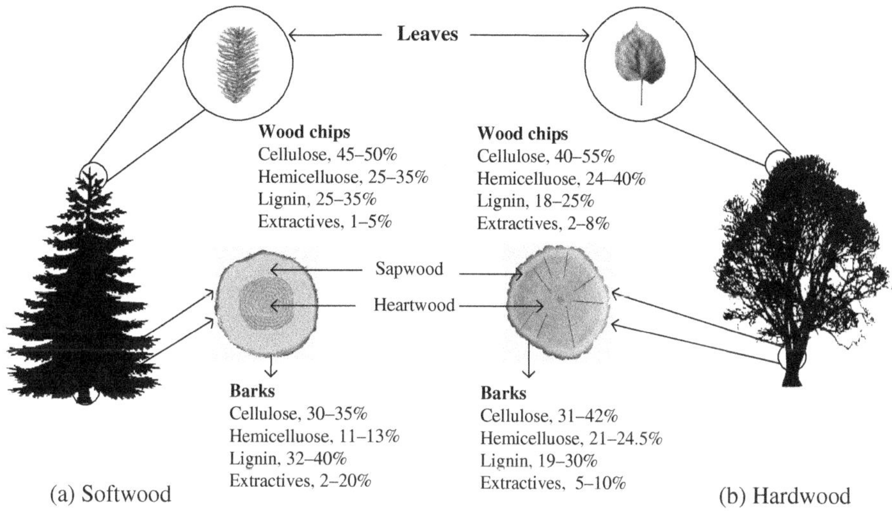

Figure 5.2 *Structure and chemical compositions of softwood and hardwood.* Source: Adapted from Kumar and Sharma (2017) and Popa (2015).

5.3 Wood Chemistry

Wood is composed of various heterogeneous chemical constituents. Cellulose, hemicellulose, and lignin are the main constituents. These building block chemicals are heterogeneously distributed throughout the wood cell walls (Figure 5.3), which vary significantly among different species. Although the cellulose contents in hardwood and softwood are almost same, the softwood contains more lignin and less hemicellulose than hardwood (Kibblewhite and Brookes 1976; Alén 2011). Low molecular weight compounds such as extractives and some inorganics are nonstructural constituents of wood.

5.3.1 Cellulose

Cellulose is the most abundant and widely applied biopolymer in nature. Glucose monosaccharide is the basic unit of cellulose, and units connect together with 1-4, β-β-glucosidic linkages to form a large polymer. The degree of polymerization of the wood cellulose molecule ranges from 800 to 10 000 (Klemm et al. 2005), which is about four to five times higher than artificial polyester and nylon 66 polymers (Papadopoulou et al. 2016). This high degree of polymerization provides cellulosic fiber with superior tensile strength and toughness for advance applications. Engineered nanofibrillated cellulose (NFC) and microfibrillated cellulose (MFC) have high strength and stiffness, and thus can be used to enhance the gas barrier properties of film and the mechanical strength of paper-based products (Osong et al. 2016). The linear cellulose molecules have a strong tendency to form intra- and intermolecular hydrogen bonds due to the presence of strongly electronegative oxygen atoms, making this hydrophilic biopolymer an outstanding candidate for cleaning and packaging. The bundles of cellulose molecules aggregate to form microfibrils, a basic unit of cellulose fibers. Cellulose fibers are formed by crystalline, highly ordered arrangements of microfibrils, or amorphous, less ordered arrangements of microfibrils

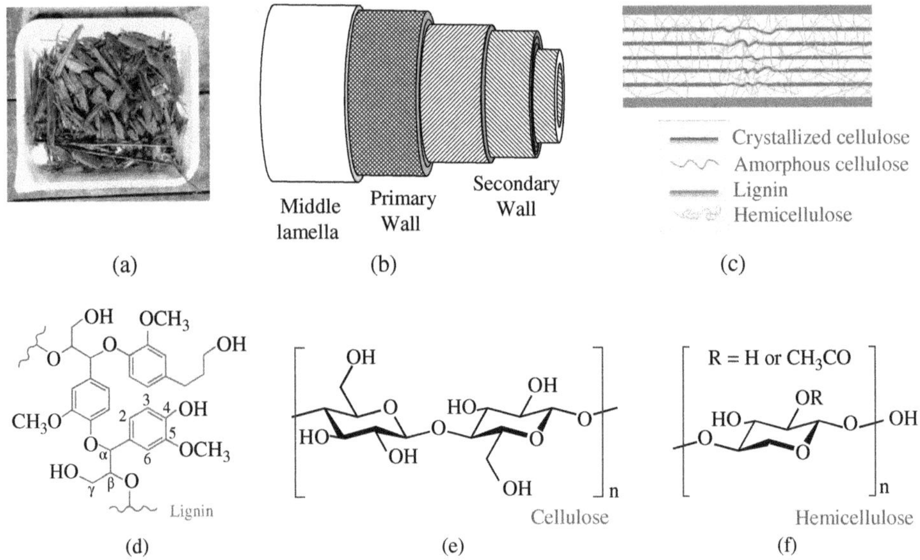

Figure 5.3 *Schematic illustration of the physicochemical structure of woody plant cell walls at various scales: (a) wood chips; (b) cell wall structure; (c) chemical composition of the secondary wall; (d) basic structures of lignin; (e) cellulose; and (f) hemicellulose.*

(Xiang et al. 2016). This characteristic makes cellulose a lightweight fiber with excellent tensile strength. The strong hydrogen bonding also imparts the cellulose molecules with high tensile strength and insolubility in most solvents. However, the cellulose is soluble in strong mineral acid solution, in which cellulose molecules are completely hydrolyzed to glucose.

5.3.2 Hemicelluloses

Hemicelluloses are hetero-polysaccharides of hexose and pentose sugar. Hemicelluloses are less ordered than cellulose and more easily dissolved in acids. The degree of polymerization of hemicelluloses is usually less than 300. This component easily hydrolyzes to monomer sugars in dilute mineral acids at low temperature (Zhang et al. 2006). The monomers are easily digestible and fermentable by various micro-organisms to produce desired products (Kumar et al. 2018). Lower energy consumption in the fractionation of hemicellulose sugars is a key factor for the production of platform chemicals like succinic acid, levulinic acid, xylitol, fumaric acid, malic acid, 2,5-furan dicarboxylic acid, 3-hydroxy propionic acid, aspartic acid, glucaric acid, glutamic acid, and arabinitol (Kumar et al. 2018). There are some basic differences between the hemicellulose structure in hardwood and softwood. Hardwood hemicellulose is mainly composed of glucuronoxylan and glucomannan, while softwood hemicelluloses are mainly made up of galactoglucomannan and arabinoglucuronoxylan. Combinations of different types and amounts of hexoses and pentoses in hemicellulose show different digestibility or fermentability characteristics during fermentation in a biorefinery (Wang et al. 2014).

Table 5.1 Lignin monomers and their composition (Pinkert et al. 2011).

Name	Structure	Softwood	Hardwood
p-hydroxyphenyl (H-lignin)	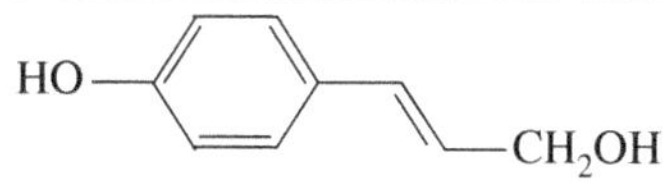	<5%	0–8%
Syringyl (S-lignin)	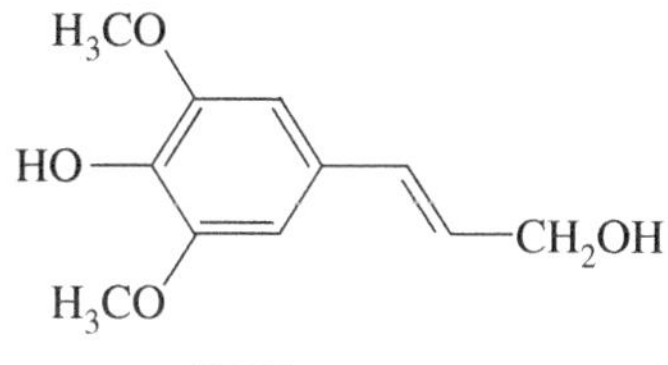	Trace amounts	46–75%
Guaiacyl (G-lignin)	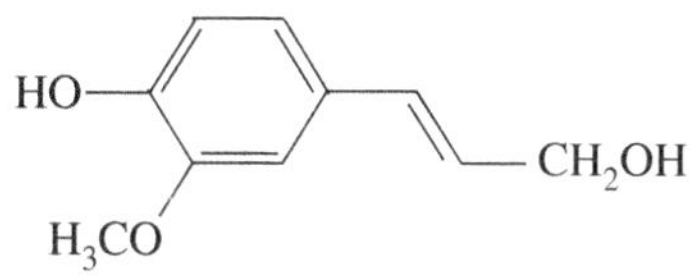	>95%	25–50%

5.3.3 Lignin

Lignin is an amorphous, hydrophobic biopolymer constructed by three monolignols (Table 5.1): p-hydroxyphenyl (H-lignin), syringyl (S-lignin), and guaiacyl (G-lignin), which is the main renewable source of aromatic chemicals (Sjostrom 2013). Lignin has many remarkable properties, like high resistance to chemical, biological, and thermal degradation. The presence of multiple functional groups in the matrix makes lignin a promising alternative for carbon-based materials (Ho-Yin et al. 2019). Lignin has exhibited superior functionality in UV absorbance, amphiphilicity, and anti-oxidation (Tian et al. 2017). The application of lignin and its derivatives in adhesives and green composites have been reported recently (Espinoza-Acosta et al. 2018; Faris et al. 2016). Owing to its low cytotoxicity and pH controllable properties, lignin has also been considered for bioimaging and drug delivery systems (Dai et al. 2017). The exact linkages between the building block components are very complex and have not yet been completely decoded (Özparpucu et al. 2017), but a few major linkages have been confirmed through investigating the dissociated monomers after digestion. β-aryl (β-o-4) and α-aryl (α-o-4) ether linkages are the representative chemical linkages in the plant cell wall, accounting for at least 43–50% and 6–8% (in number) in softwood, respectively (Li et al. 2015). The potential to cleave the two linkages in or outside the phenolic structures are process-specific.

Lignins are generally classified as hardwood lignin, softwood lignin, and grass lignin. Hardwood lignins are composed of syringyl and guaiacyl units, which originate from trans-sinapyl alcohols and transconiferyl alcohol respectively. Softwood lignin, on the other hand, is mainly composed of guaicyl, which comes from trans-coniferyl alcohol. The p-hydroxyphenyl unit of grass lignin is derived from trans-p-coumaryl alcohol and two foregoing precursors. Native lignins are insoluble in water and acids, but isolated lignins show the maximum solubility in different organic solvents such as dimethyl sulfoxide, acetone, dioxane, tetrahydrofuran, and so on. Isolated lignins can be classified as milled wood lignin, kraft lignin, alkali lignin, and enzymatically liberated lignin.

5.3.4 Extractives

Extractives are various low molecular weight organic compounds that are present in small fractions but with high functionality, and therefore with high potential for different uses. They have been used in recent years for pharmaceuticals (Nair et al. 2015; Constable et al. 2007), food additives (Moure et al. 2006), and chemical reagents (Eriksson et al. 2018; Miranda et al. 2017; Mattos et al. 2016). Extractives are bio-synthesized as secondary metabolites and protect plants from different insects and microbial attacks (Gierlinger et al. 2004; Ayvaz et al. 2010). Although wood contains less than 10% extractives, they greatly contribute to the physical properties of wood (e.g., color and odor). Extractive content determines the tensile strength, density, elasticity, permeability, flammability, and wettability of wood (Taylor et al. 2007; Jankowska et al. 2018). The major chemicals of wood extractives can be grouped as terpenoids, phenolics, alkaloids, fats, and waxes. The composition of extractives largely varies in different wood species and in the different parts (i.e., stem, leaves, barks, and roots) of the same wood species. The functions of extractives differ significantly, and different extractives can be discovered in different morphological areas of the plant. For instance, resin acids in the resin canals of plants protect wood from microbial attack; while fats and waxes exist in ray parenchyma tissue and act as a power source of woody plants. The majority of wood-based extractives are low boiling point organic compounds, while a small amount of inorganic extractives has also been reported.

5.3.4.1 Terpenoids

Terpenes and terpenoids are collectively known as isoprenoids because both are derived from isoprene units (C_5H_8). Terpenes are a pure hydrocarbon having low molecular weight, whereas terpenoids contain functional groups like carboxyl, carbonyl, and hydroxyl groups. To avoid complexity, terpenes and their derivatives are collectively known as terpenoids. Depending on the number of isoprene units in the structure, terpenes and terpenoids are classified into mono-, sesqui-, di-, tri-, and polyterpenoids.

Myrcene, limonene, β-phellandrene, α-pinene, β-pinene, 3-carene, camphene, borneol, and hinokitiol are dominant volatile monoterpenoids and can be easily recovered as turpentine by simple steam distillation. Monoterpenoids impart an odor to the wood and are mainly found in softwood oleoresin. Although monoterpenoids are not usually found in hardwood, tropical hardwood oleoresin contains a few of these chemicals as minor ingredients (Sjostrom 2013; Zhu et al. 2011).

Diterpenoids are the most important constituents of oleoresin canal and can be categorized into acyclic, bicyclic, tricyclic, tetracyclic, and macrocyclic terpenoids. Diterpenoids are found mainly in softwoods in the form of resin acids. Geranyl-linalool, β-epimanool, cis-abienol, manoyloxide, pimaral, cembrene, lambertianic acid, communic acid, and mercusic acid are the most common resin acids (Sjostrom 2013; Zhu et al. 2011).

Triterpenoids are oxygenated derivatives and are classified as tetracyclic lanostane, pentacyclic lupine, and pentacyclic oleanane derivatives. Triterpenoids and steroids are structurally similar and both synthesized from squalene. The difference in synthesis pathways result in different structures. The most prominent triterpenoids in wood are sitosterol, sitostanol, citrostadienol, botulin, and serratenediol; which are important raw materials for wood-based chemicals (Sjostrom 2013; Zhu et al. 2011).

5.3.4.2 *Fats and Waxes*

The lipophilic materials in hardwood parenchyma cells are predominantly fats and waxes, whereas the softwood parenchyma resin contains mainly fats. Wood contains more than 30 types of saturated and unsaturated fatty acids, alcohols, and alkanes. Palmitic, stearic, arachidic, behenic, lignoceric, oleic, linoleic, linolenic, pinolenic, and eicosatrienoic acids are the common fatty acids found in woody plants (Sjostrom 2013).

5.3.4.3 *Phenolic Compounds*

Heartwood and bark contain thousand types of phenolic and complex polyphenol compounds. Phenolics protect wood plants from fungal and microbial attack. Simple phenolics, stilbenes, lignans, flavonoids, and tannins are a major class of phenolics found in plants (Sjostrom 2013).

5.4 Chemical Composition Analysis

Chemical composition analysis is often conducted to evaluate the true value of the waste-derived biomass. The analytical procedure for structural and non-structural components of wood is demonstrated in Figure 5.4.

5.4.1 Structural Carbohydrates and Lignin

The analytical procedures for structural carbohydrates and lignin have been well-documented by the National Renewable Energy Laboratory (NREL) of the US Department of Energy (Sluiter et al. 2008). The basic concept of this process is to dissociate the polysaccharides in the plant cell wall into monomers without promoting the formation

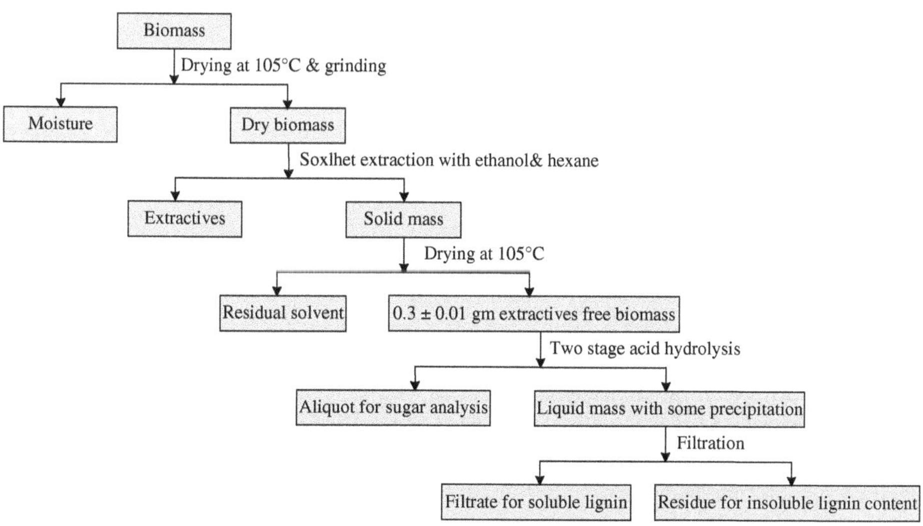

Figure 5.4 *Chemical composition analysis scheme of lignocellulosic biomass.*

of dehydration byproducts, such as furfural, hydromethoxy furfural, and levulinic acid. The dissociation of polysaccharides is carried out in two stages: (i) hydrolysis with strong sulfuric acid (72 wt%) at near-room temperature, and (ii) high-temperature secondary hydrolysis after diluting the primary hydrolysis slurry to 4 wt% of sulfuric acid. During primary hydrolysis, polysaccharides are completely separated from the lignin whereas secondary hydrolysis depolymerizes oligomers to monomer sugars (Bhagia et al. 2016). The fundamentals behind the NREL procedure were developed on the assumption of low proteins and lipids. This technique cannot be applied precisely for high lipid, protein, and pectin-containing biomass, such as grass, leaves, and vegetable samples.

5.4.2 Extractives

Extractives can be analyzed by soaking the biomass samples in polar and nonpolar organic solvents, such as methanol, ethanol, acetone, hexane, benzene, and diethyl ether, at a desired temperature with continuous solvent refluxing. Soxhlet extraction is the most effective method of extractive separation from plants, and is reported comprehensively (Ramluckan et al. 2014; Schwanninger and Hinterstoisser 2002).

5.5 Pretreatment

Conversion of woody biomass into useful products relies on a serious of complex processes, such as pretreatment, hydrolysis, fermentation, and separation. Pretreatment aims to alter and modify the basic structure of lignocellulosic biomass for functional application. The main purpose of pretreatment is to increase the biomass conversion yield by increasing the enzyme accessibility to the pretreated substrate. Lignocellulosic biomass is a complex matrix of the introduced building block chemicals, which provide strong recalcitrance to cellulose degradation (Himmel et al. 2007). Woody biomass contains more lignin and possesses greater recalcitrance than herbaceous biomass. To overcome the biomass recalcitrance, it is necessary to fractionate all the components at the very beginning of the bioconversion process. Different pretreatment techniques have been designed to remove the recalcitrance of lignocellulosic biomass (Park et al. 2017; Baker et al. 2017; Zhao et al. 2017; Brosse et al. 2009). Pretreatment removes lignin and hemicellulose, and increases the surface area and digestibility of cellulose. It can also prevent the degradation of pentose sugar and minimize the formation of inhibitory components for fermentation. Physical, biological, and chemical methods have been reported for the pretreatment of plant biomass, while biological pretreatment is less preferred industrially due to its low efficiency and processability (Abd-Aziz et al. 2018; Mussatto 2016). Physical pretreatment like steam explosion is usually used for woody biomass, but the process alone has not been extensively used commercially in comparison with chemical pretreatment approaches (Zhu and Pan 2010).

The most extensively studied chemical pretreatments utilize a wide range of chemicals at elevated temperature (Wang et al. 2018; Terán Hilares et al. 2017). Acid, alkali, organosolv, peroxide, sodium hypochlorite, sodium sulfide, ozone, and ammonia have all been used in the chemical pretreatments (Leu and Zhu 2013). We classify here the major chemical pretreatment concepts into three categories (i.e., organosolv, acid, and alkaline pretreatment), of which the key mechanisms are presented in Figure 5.5.

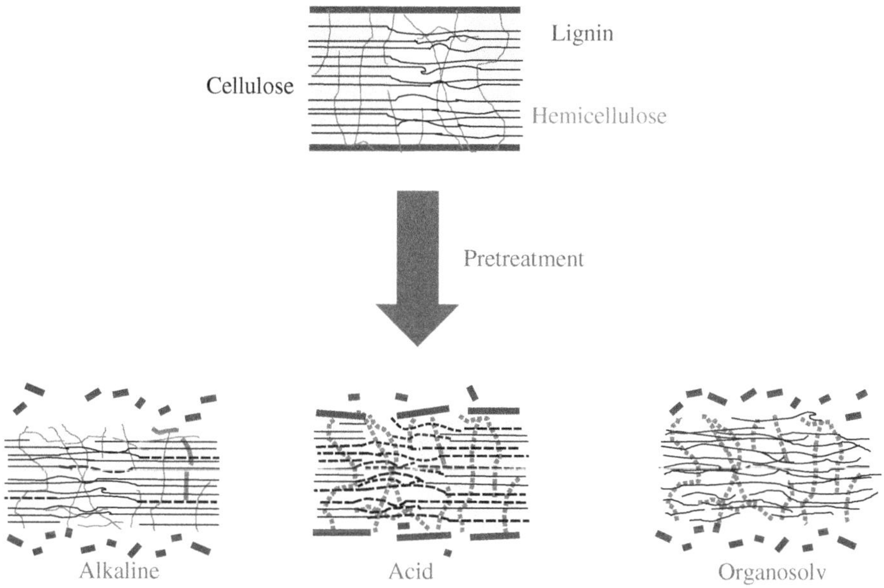

Figure 5.5 Mechanisms of different chemical pretreatments.

The chemical reactions in cleaving the two representative ether linkages under different types of biorefinery processes are illustrated in Figure 5.6. The presented model lignin structures include two G-lignins linked at the α-o-4 position. The monolignol at the bottom right contains a –OH group linking to its phenolic structure at the 4th position while its α- and β- positions both linked with other lignins. The monolignol at the upper left links with other lignins at the 4th and β- positions, but its α- position links only with a –OH group. The concept diagram demonstrates the major functions and limitations of applied reagents and/or catalysts to cell wall decomposition during thermochemical processes. Cleavage of the two linkages under an alkaline pretreatment process is shown in Figure 5.6a. At elevated pH, the cleavage reactions can initiate most rapidly at the –OH group at the 4th position of the phenolic structure, while it can be carried out only very slowly when the 4th position links to another lignin. During the pretreatment process, more –OH structures can be created to continue the cleavage of the linkages. Alkali can also cleave some hydrolysable linkages in lignin and polysaccharides, which result in a reduction in the degree of polymerization and crystallinity of the substrate (Kim et al. 2016). For acid pretreatment processes, bond cleavage can start randomly at either the α- or β- positions regardless of whether there is a hydroxyl group bonding with the aromatic ring at the 4th carbon. With different reagents the α- and β- linkages can be cleaved and the original functional groups can be replaced by hydrogen, ethyl, or sulfonate groups. In addition, acid-based biorefinery processes can effectively cleave glucosidic linkages in hemicelluloses and amorphous cellulose, which further improve the accessibility of enzymes to cellulose to promote enzymatic hydrolysis (Leu and Zhu 2013).

Acid pretreatments have been successfully applied to solubilize hemicellulose and efficiently used to remove the recalcitrance of biomass. Both dilute and concentrated

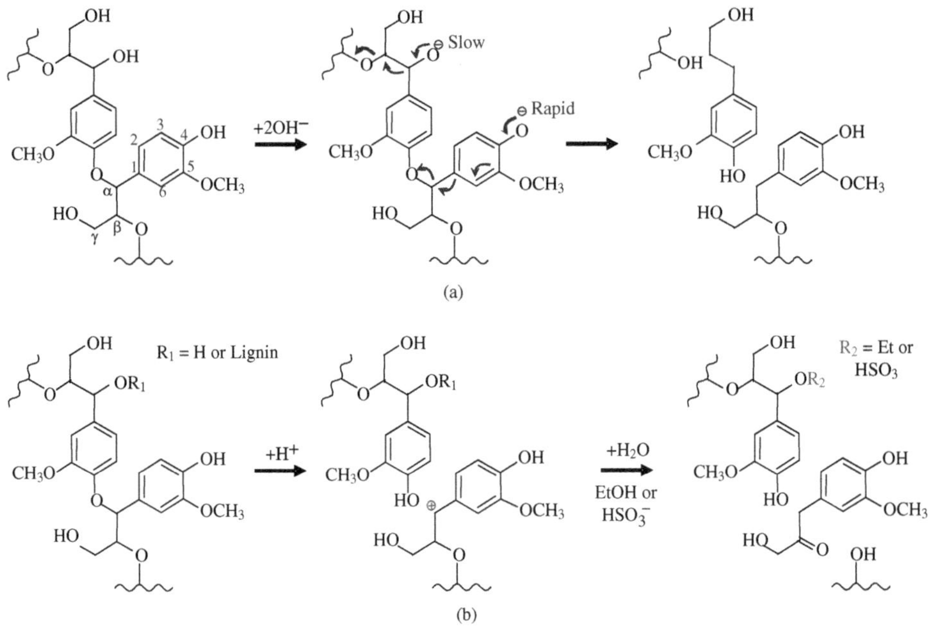

Figure 5.6 *(a) Cleavage of α-o-4 and β-o-4 bonds in an alkaline pretreatment process; (b) cleavage of α-o-4 and β-o-4 bonds in dilute acid, organosolv, and sulfite pretreatment processes.*

H_2SO_4 acids are most commonly used for pretreatment of lignocellulosic biomass (Mosier et al. 2005). Dilute acid pretreatment processes are more preferable than concentrated acid processes because concentrated acids mediate the generation of fermentation inhibitors (Taherzadeh and Karimi 2008). Acid pretreatment can be done at low temperatures (<120°C) for longer retention times or at high temperatures (>180°C) for shorter retention times. Other acids such as nitric acid, hydrochloric acid, phosphoric acid, acetic acid, and formic acid are also successfully studied in biomass pretreatment (Gámez et al. 2006; Rodrıguez-Chong et al. 2004).

Alkali preferentially removes lignin over hemicellulose from biomass (Leu and Zhu 2013). Alkali reacts with acetyl and uronic acid groups in the hemicellulose and thus increases the accessibility of enzymes (Chang and Holtzapple 2000). The hydrolysis of ester linkages more effectively removes lignin with minimal cellulose and hemicellulose solubility (Sun and Cheng 2002). NaOH is more effective than other alkalis in reducing cellulose crystallinity and degree of polymerization, and altering the structure of lignin (Kumar and Wyman 2009).

Organosolv techniques are mainly used for the extraction of high-quality lignin. Ethanol, methanol, propanol, butanol, butanediol, ethylene glycol, glycerol, acetic acid, acetone, and tetrahydrofuran are most commonly used in combination with inorganic acids, salts, or alkali catalysts (Zhao et al. 2009; Bajpai 2016). Organic solvents are most effective in lignin degradation than other chemical pretreatments. Lignin removal increases the accessible surface of substrate for enzyme saccharification (Agbor et al. 2011). Pretreatment temperature, reaction time, solvent concentration, and catalyst concentration greatly determine the crystallinity, fiber length, and surface area of pretreated biomass. The effects of the different catalysts such as H_2SO_4, NaOH, and $MgSO_4$ have been studied on ethanol organosolv of pine. The results showed that H_2SO_4 is the most effective catalyst for pine (J.Y. Park et al. 2010; N. Park et al. 2010). The combination of ethanol organosolv to ball milling increased the lignin separation and glucose yield and decreased the pretreatment severity of organosolv (Hideno et al. 2013). Ionic liquid and deep eutectic solvents are recyclable organic compounds, and have been widely exploited as solvents for pretreatment of lignocellulose before further processing to energy and chemicals (Moniruzzaman and Goto 2018; Kanbayashi and Miyafuji 2016). Presence of solvent in the substrate reduces the enzyme activity of cellulose hydrolysis (Mosier et al. 2005). Removal and recovery of solvent also reduces the operational cost of the process (Sun and Cheng 2002). Organic solvents should be carefully handled because they are inflammable and can cause explosions.

5.6 Saccharification and Fermentation

Saccharification is the conversion process to break down polysaccharides into monomeric sugars (Zhu et al. 2017). Acid hydrolysis and enzymatic hydrolysis are commonly applied procedures for saccharification of lignocellulosic biomass. Acid hydrolysis can directly depolymerize the polysaccharide chains in the lignocellulosic biomass without any pretreatment, which can effectively release more than 90% of monomer sugars from the feedstock (Lavarack et al. 2002; Frederick Jr et al. 2008). However, concentrated acid hydrolysis can further react with the dissolved substrates (i.e., carbohydrates), resulting in high concentrations of fermentation-inhibiting products, such as organic acids, furfural, and hydroxymethylfurfural (HMF) (Badger 2002).

Enzymatic hydrolysis is currently the most widely applied/discussed approach for biomass conversion. This process accounts for approximately 27% of the total operation cost of the overall bioconversion process, and therefore needs to be optimized with the pretreatment processes for cost reduction (Bbosa et al. 2018). Saccharification efficiency can be affected by enzyme-related (Gupta et al. 2016) and substrate-related factors (Leu and Zhu 2013). The mechanisms of the interactions between dissociated and residual components to enzyme activities are the functions of the biomass and pretreatment process. Figure 5.7a–d shows some images of the pretreated substrate before and after enzymatic hydrolysis. The surface of woody biomass before pretreatment (Figure 5.7a) is smooth, with the inner surface well protected from hydrolysis. The pretreatment by ethanol, alkaline, and SPORL opens the plant cell wall (red arrow) and creates different levels of dissociation (Figure 5.7b–d, respectively). With increased severity of pretreatment conditions and degree of lignin removal, the damage in the cell wall become visible, together with lignin precipitation on the substrate surface (Figure 5.7d). The cell wall structure is further decomposed after hydrolysis, while enzymes may be captured under a scanning electron microscope (SEM) within 24 hours of hydrolysis (arrow mark in Figure 5.7e). The cell wall can be completely decomposed after enzymatic hydrolysis (Figure 5.7f).

Enzyme hydrolysis of biomass suffers with a low reaction rate due to the high crystallinity of cellulose, and less accessible surface area for enzyme and non-productive binding to cellulase enzyme (Dadi et al. 2006). Enzymatic hydrolysis requires pretreatment to break the regular structure of cellulose and to increase surface area for enzyme access (Badger 2002).

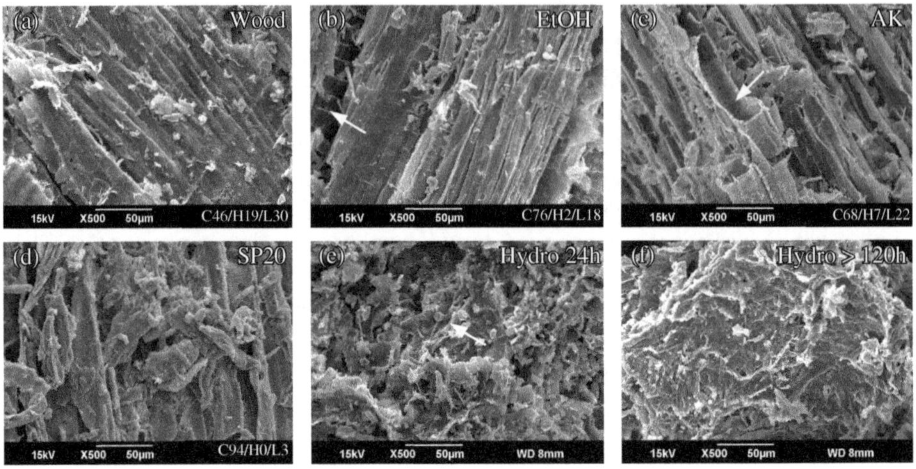

Figure 5.7 *SEM images of substrate at different stages of pretreatment or hydrolysis (see text).*

The yield of enzymatic hydrolysis increased from 20% to 90% after proper pretreatment of biomass (Table 5.2). The efficiency of both acid and enzyme hydrolysis depends on the crystallinity, molecular structure, degree of polymerization, surface area, swelling, lignin content, and hemicellulose content (Dadi et al. 2006). The extent of hydrolysis to be done depends on the targeted chemicals. Complete hydrolysis is necessary to produce ethanol. On the other hand, complete hydrolysis reduces the efficiency of production of butanol and lactic acid from the biomass.

Fermentation is a metabolic process that consumes sugars and produces alcohols or other chemicals depending upon the strains used in the fermentation. Fermentable sugars from the biomass can be converted into different chemicals by different enzymes. In the United States, less than 5% of total chemical consumption is obtained from the fermentation of lignocellulosic biomass. Common fermentation products in the recent literatures include ethanol, butanol, and lactic acid (Macedo and Brigham 2014; Zheng et al. 2015; Wang et al. 2015). The US Department of Energy and Department of Agriculture are committed to replace at least 30% of petroleum consumption with biofuel by 2030 (Perlack 2005). The US Department of Energy (Zhu and Zhuang 2012) reported that the United States currently uses 140 billion gallons of petroleum transportation fuel per year. To replace 30% of petroleum with bioethanol will require about 60 billion gallons of ethanol (US Congress 2005) The US alone can sustainably produce annually 1.366 billion tons of dry biomass, of which 998 million tons is from agricultural residues and 368 million tons from forest wood residues (Somerville 2006). This total biomass can theoretically produce 130 billion gallons of ethanol, which is about 60% of the current petroleum energy used in the US. Moreover, the use of bioenergy alters energy dynamics and reduces greenhouse gas emissions. China has also promised to contribute heavily to biofuel production and utilization (Nguyen et al. 2017). China is expected to replace 30% of total oil imports with cellulosic ethanol by 2050 (Zhao 2015).

Table 5.2 *Pretreatment conditions and saccharification yield for different biomass feedstocks.*

Feedstocks	Solvents (%)	Catalyst (%)	Solid/ liquid	Composition of raw substrate Cellulose	Hemi.	Lignin	Temp. (°C)	Time (min)	Composition of pretreated substrate Cellulose	Hemi.	Lignin	Enzyme load (FPU)	SED (%)	Ref.
Lodgepole pine	65 EtOH	1.1 H_2SO_4	0.14	49.0	21.7	26.4	170	60	74.8	3.58	17.59	20	100	Del Rio et al. (2010)
Lodgepole pine	65 BuOH	1.1 H_2SO_4	0.14	49.0	21.7	26.4	170	60	74.64	2.35	18.25	20	90	
Pitch pine	50 EtOH	1 $MgCl_2$	0.1	44	30	26	210	10	46	0	25		61.1	N. Park et al. (2010)
Pitch pine	50 EtOH	2 NaOH	0.1	44	30	26	210	20	34	0	12		85.4	
Hemp hurd	45 MeOH	3 H_2SO_4	0.04	43	21	23	165	20	48	18	22		67.9	Gandolfi et al. (2014)
Eucalyptus	25w EtOH	1 CH_3COOH	0.2	42.2	34.1	28.1	200	60	83	3.5	13.5	9.5	~100	Teramoto et al. (2009)
Japanese cypress	50 EtOH	0.4 HCl	0.18	41.5	25	32	170	45	75	0	25	22.8	~70	Hideno et al. (2013)
Liriodendron tulipifera	50w EtOH	1 NaOH	—	41.1	21.2	21.4	140	50	60	22.3	20		67	Koo et al. (2012)
Miscanthus	80 EtOH	1 H_2SO_4		49.09	24.18	25.11	170	60	48.09	8.5	22.01		45	Obama et al. (2012)
Norway spruce	65 EtOH	—	0.07	38.8	22.2	26.4	210	15	35.7	2.9	3.9	15	87	Agnihotri et al. (2015)
Empty palm fruit branch	78 FA	—	0.1	37.01	15	17.4	107	90	82.44	5.6	4.91	15	99	Cui et al. (2014)
Willow wood	55w EtOH	0.1 H_2SO_4	0.10	32.9	14.2	28.5	187	180	46.6	12.2	26.6	33	87	Huijgen et al. (2011)
Bamboo	75 EtOH	2 H_2SO_4	0.2	41.3	24	22.8	180	60	89.7	6.4	3.3	15	83.4	Li et al. (2012)
Alfalfa stem	9w PAA	—	0.16	30.2	18	15.2	100	120	30.1	16.3	14.9	60	47	Xu and Tschirner (2011)
Crofton wood	40w PAA	—	0.25	37.6	22.4	16.4	90	90	73	33.27	1	20	38	Zhao et al. (2008)
Siam wood	50w PAA	—	0.25	41	22.3	20.7	90	90	64.6	26.7	6.1	10	70	Zhao et al. (2010)
Oil palm trunk	7 H_2SO_4	4 Na_2SO_3	0.2	38.1	24.4	21.4	170	30	68.8	4.9	30.1	15	66	Noparat et al. (2017)

BuOH, butanol; EtOH, ethanol; FA, formaldehyde; MeOH, methanol; PAA, peracetic acid; SED, substrate enzyme digestibility.

5.7 New Functions of Wood Residues

Wood has been found in many ancient and historical structures like Egyptian pyramids, Chinese temples, and tombs, as construction materials. Beside the more traditional uses in construction, furniture, pulp paper, and bioethanol production, wood has been used recently for the manufacture of advanced materials like wood–plastic composites (WPCs) and cellulose nanomaterials. Some modern applications of wood materials are given in Table 5.3. These composite materials have good mechanical properties to replace traditional materials for different construction purposes. Wood extractives are widely used in the pharmaceutical industry because of their medicinal properties.

5.7.1 Wood–Plastic Composite for Construction Purposes

WPCs are prepared by combining two phases of polymeric matrices: thermosetting or thermoplastic polymer, and wood flour. WPC of desired mechanical properties can be prepared by varying the composition of the matrix phase, reinforcing fiber, coupling agent and lubricant (Leu et al. 2012; Javier et al. 2015). Aras et al. (2015) showed that the mechanical properties of the green materials produced from plastic waste and wood are almost similar to those of commercial wood and formaldehyde-based plastic.

For sustainable building, and renewable and ecofriendly construction, civil and construction engineers are focusing more on the utilization of woody biomass to produce low-cost and energy-efficient materials for construction. Wide varieties of wood and wood-based composites are also used for both residual and non-residual construction purposes.

Plywood, composite lumber, wafer board, and flake board are the traditional WPC materials used as building and construction materials. These traditional materials are produced by laminating thin wood sheets or strips of veneers glued together with adjacent layers, and are widely used in furniture and underlayment of floors and roofs. Wood is broken down to small particles and fibers, and then glued together with adhesive for the preparation of particleboard and fibreboard. Some fibreboards can be prepared without any adhesive because one of the major wood components can serve as a binding material.

5.7.2 Cellulose Nanomaterials

Cellulose nanocomposites consist of any reinforcing materials having dimensions less than 100 nm in any direction (Mariano et al. 2014). Cellulose nanocrystals (CNCs) and nanofibrillated cellulose (NFC) are the reinforcing agents in cellulose nanocomposites. The CNC is prepared by treating pulp fiber to break the linkages between cellulose and hemicellulose, and to release crystalline cellulose (Siró and Plackett 2010). The NFC, on the other hand, is produced by the application of shear stress perpendicular to the pulp fiber to break internal linkages of nanofibres (Xu et al. 2013). Because of the low cost, large aspect ratio and fiber entanglement in the final product (Xu et al. 2013), NFC is widely used for the production of cellulose composites (Isogai 2013), such as UV curable acrylic resin, poly-lactide, polyethylene, poly-polypropylene, polystyrene, glass fiber, carbon nanotubes, phenol-formaldehyde resin, latex, nano-clay and paper.

Table 5.3 *Current uses of wood products.*

Wood products	Fraction of wood	Market price, USD/ton	Main use	Byproducts		References
				Type	Amount (%)	
Timber or lumber	~ 100	250–400	Construction	Wood powder/barks	8–10	Fu et al. (2014); Borri and Corradi (2011)
Plywood	80–90	300–500	Furniture	Wood powder/barks	12–15	Safin et al. (2015); Pizzi et al. (1993)
Wood–plastic composite	50–60	300–1000	Decoration	—		Ashori (2008); Nourbakhsh and Ashori (2010)
Fiberboard	40–50	350–800	Furniture	—		Popovska et al. (2016); Reddy and Yang (2005)
Paper	30–50	600–1000	Printing	Lignin	20–25	M'hamdi et al. (2017); Liao (2005)
Nanomaterials	10–20	500–3000	Nanotechnology	Lignin	20–25	Isogai (2013); Sanchez et al. (2011)
Chemicals	—	~	Natural feedstock	—		Cherubini and Strømman (2011); Czernik and Bridgwater (2004)
Biofuels	—	400–600	Fuels	Lignin	20–25	Kalyani et al. (2017); Agarwal (2007)

5.7.3 Wood Extractives

Plant extractives are now being used for the production of flavors, fragrances, oil, and pharmaceutical preparations (Gandini 2008). Paclitaxel is a plant extractive used for the treatment of different types of cancer (Tanaka et al. 2017; Foglietta et al. 2018; Bernabeu et al. 2016). Significant anti-cancer activity of keratin nanoparticles-based paclitaxel in breast cancer models has been reported recently (Bernabeu et al. 2016). Paclitaxel loading on keratin nanoparticles increases the solubility and drug delivery of paclitaxel. The traditional use of turpentines as a solvent is substituted by cheap petrochemicals. Terpene and terpenoids can be used as a feedstock to produce rose alcohols, linalool, and nerol. Linalool is a key intermediate in the production of vitamin E (Langenheim 2003). Turpentines form a low molecular weight polymer having better flexibility, and this polymer can be used in papermaking, adhesives, printing inks, and coating (Chen 1992). Yumrutaş et al. (2008) reported that turpentines can be used as an additive for gasoline. Plant tannins contain molecules of complex phenolic structures obtained mainly from the bark and heartwood of the trees. Tannins have been used in the leather industry for the vegetable tanning process; but more convenient synthetic tannin has replaced the use of vegetable tannins (Pizzi 1983).

5.8 Conclusions and Future Trends

Recent development of technology has resulted in dramatic changes in consumer behavior and views of woody products and the forestry industry. While the demand for structural timber remains, the need of paper and related products has continuously decreased, lignocellulosic biomass has been seen as an emerging renewable feedstock to replace petroleum refinery products. With the depletion of fossil fuels, bioenergy and biomaterials will eventually substitute plastics, resins, additives, solvents, carbon fibers, medicine, and/or many other materials. Currently, however, technical and economic challenges still remain, preventing wide acceptance of the second-generation biorefinery. As the feedstock of the conventional refinery is still sufficient, there is no immediate demand to develop and/or use the related products. Developing technology should focus on assisting municipal waste treatment or handling the byproducts of related industry (e.g., kraft lignin). Accumulation of experience of this application should create useful information for the future development of the biomass biorefinery, and therefore needs to be emphasized and practiced.

Acknowledgement

The authors thank the Hong Kong General Research Fund (15212317), Environment and Conservation Fund (ECF 85/2017), and Sinopec Chemical Commercial Holding Company Limited for financial support. The authors also thank the University Research Facility in Chemical and Environmental Analysis (UCEA) of The Hong Kong Polytechnic University for sample analyses.

References

Abd-Aziz, S., Ibrahim, M.F., and Jenol, M.A. (2018). Biological pretreatment of lignocellulosic biomass for volatile fatty acid production. In: *Emerging Areas in Bioengineering*, vol. **1** (ed. H.N. Chang), 191–201. Wiley-VCH Verlag.

Agarwal, A.K. (2007). Biofuels (alcohols and biodiesel) applications as fuels for internal combustion engines. *Prog. Energy Combust. Sci.* **33** (3): 233–271.

Agbor, V.B., Cicek, N., Sparling, R. et al. (2011). Biomass pretreatment: fundamentals toward application. *Biotechnol. Adv.* **29** (6): 675–685.

Agnihotri, S., Johnsen, I.A., Bøe, M.S. et al. (2015). Ethanol organosolv pretreatment of softwood (*Picea abies*) and sugarcane bagasse for biofuel and biorefinery applications. *Wood Sci. Technol.* **49** (5): 881–896.

Alén, R. (2011). *Biorefining of Forest Resources*. Paperi ja Puu Oy.

Aras, U., Kalaycıoğlu, H., Yel, H., and Bitek, G. (2015). Effects of ammonium nitrate on physico-mechanical properties and formaldehyde contents of particleboard. *Procedia Social Behav. Sci.* **195**: 2130–2134.

Ashori, A. (2008). Wood–plastic composites as promising green-composites for automotive industries! *Bioresour. Technol.* **99** (11): 4661–4667.

Ayvaz, A., Sagdic, O., Karaborklu, S., and Ozturk, I. (2010). Insecticidal activity of the essential oils from different plants against three stored-product insects. *J. Insect Sci.* **10** (1): 21.

Badger, P. (2002). Ethanol from cellulose: a general review. In: *Trends in New Crops and New Uses*, vol. **1** (eds. J. Janick and A. Whipkey), 17–21. ASHS Press.

Bajpai, P. (2016). *Pretreatment of Lignocellulosic Biomass for Biofuel Production*. Springer.

Baker, P., Charlton, A., and Hale, M. (2017). Fungal pre-treatment of forestry biomass with a focus on biorefining: a comparison of biomass degradation and enzyme activities by wood rot fungi across three tree species. *Biomass Bioenergy* **107**: 20–28.

Bbosa, D., Mba-Wright, M., and Brown, R.C. (2018). More than ethanol: a techno-economic analysis of a corn stover-ethanol biorefinery integrated with a hydrothermal liquefaction process to convert lignin into biochemicals. *Biofuels, Bioprod. Biorefin.* **12** (3): 497–509.

Bernabeu, E., Gonzalez, L., Legaspi, M.J. et al. (2016). Paclitaxel-loaded TPGS-b-PCL nanoparticles: in vitro cytotoxicity and cellular uptake in MCF-7 and MDA-MB-231 cells versus mPEG-b-PCL nanoparticles and Abraxane®. *J. Nanosci. Nanotechnol.* **16** (1): 160–170.

Bhagia, S., Nunez, A., Wyman, C.E., and Kumar, R. (2016). Robustness of two-step acid hydrolysis procedure for composition analysis of poplar. *Bioresour. Technol.* **216**: 1077–1082.

Borri, A. and Corradi, M. (2011). Strengthening of timber beams with high strength steel cords. *Composites Part B* **42** (6): 1480–1491.

Brosse, N., Sannigrahi, P., and Ragauskas, A. (2009). Pretreatment of *Miscanthus* × *giganteus* using the ethanol organosolv process for ethanol production. *Ind. Eng. Chem. Res.* **48** (18): 8328–8334.

Chang, V.S. and Holtzapple, M.T. (2000). Fundamental factors affecting biomass enzymatic reactivity. *Appl. Biochem. Biotechnol.* **84** (1): 5–37.

Chen, G.-F. (1992). Developments in the field of rosin chemistry and its implications in coatings. *Prog. Org. Coat.* **20** (2): 139–167.

Cherubini, F. and Strømman, A.H. (2011). Chemicals from lignocellulosic biomass: opportunities, perspectives, and potential of biorefinery systems. *Biofuels, Bioprod. Biorefin.* **5** (5): 548–561.

Congress, U.S. (2005). *Energy Policy Act of 2005. Public Law*: 109–158. https://www.govinfo.gov/app/details/PLAW-109publ58 (accessed 5 April 2020).

Constable, D.J., Jimenez-Gonzalez, C., and Henderson, R.K. (2007). Perspective on solvent use in the pharmaceutical industry. *Org. Process Res. Dev.* **11** (1): 133–137.

Cui, X., Zhao, X., Zeng, J. et al. (2014). Robust enzymatic hydrolysis of formiline-pretreated oil palm empty fruit bunches (EFB) for efficient conversion of polysaccharide to sugars and ethanol. *Bioresour. Technol.* **166**: 584–591.

Czernik, S. and Bridgwater, A. (2004). Overview of applications of biomass fast pyrolysis oil. *Energy Fuels* **18** (2): 590–598.

Dadi, A.P., Varanasi, S., and Schall, C.A. (2006). Enhancement of cellulose saccharification kinetics using an ionic liquid pretreatment step. *Biotechnol. Bioeng.* **95** (5): 904–910.

Dai, L., Liu, R., Hu, L.-Q. et al. (2017). Lignin nanoparticle as a novel green carrier for the efficient delivery of resveratrol. *ACS Sustainable Chem. Eng.* **5** (9): 8241–8249.

Daniel, G. (2009). Wood and fibre morphology. In: *Pulp and Paper Chemistry and Technology*, vol. **1** (eds. M. Ek et al.), 49–71. de Gruyter.

Del Rio, L.F., Chandra, R.P., and Saddler, J.N. (2010). The effect of varying organosolv pretreatment chemicals on the physicochemical properties and cellulolytic hydrolysis of mountain pine beetle-killed lodgepole pine. *Appl. Biochem. Biotechnol.* **161** (1–8): 1–21.

Eriksson, D., Arshadi, M., Kataria, R., and Bergsten, U. (2018). Lipophilic extractives in different tree fractions and forestry assortments of *Pinus sylvestris* due for thinning or final cutting. *Scand. J. For. Res.*: 1–9.

Espinoza-Acosta, J.L., Torres-Chávez, P.I., Olmedo-Martínez, J.L. et al. (2018). Lignin in storage and renewable energy applications: a review. *J. Energy Chem.* **27** (5): 1422–1438.

Evert, R.F. (2006). *Esau's Plant Anatomy: Meristems, Cells, and Tissues of the Plant Body: Their Structure, Function, and Development*. Wiley.

Faris, A.H., Rahim, A.A., Ibrahim, M.N.M. et al. (2016). Combination of lignin polyol–tannin adhesives and polyethylenimine for the preparation of green water-resistant adhesives. *J. Appl. Polym. Sci.* **133** (20): 43437.

Foglietta, F., Spagnoli, G.C., Muraro, M.G. et al. (2018). Anticancer activity of paclitaxel-loaded keratin nanoparticles in two-dimensional and perfused three-dimensional breast cancer models. *Int. J. Nanomed.* **13**: 4847.

Food and Agriculture Organization. 2018. *FAOSTAT, Forestry Production and Trade*. http://www.fao.org/faostat/en/#data/FO (accessed 22 April 2020).

Frederick, W. Jr., Lien, S., Courchene, C. et al. (2008). Production of ethanol from carbohydrates from loblolly pine: a technical and economic assessment. *Bioresour. Technol.* **99** (11): 5051–5057.

Fu, M., Liu, Y., Li, N. et al. (2014). Application of modern timber structure in short and medium span bridges in China. *J. Traffic Transp. Eng.* **1** (1): 72–80.

Fujita, M. and Harada, H. (2000). Ultrastructure and formation of wood cell wall. In: *Wood and Cellulosic Chemistry* (eds. D.N.-S. Hon and N. Shiraishi), 1–49. CRC Press.

Gámez, S., González-Cabriales, J.J., Ramírez, J.A. et al. (2006). Study of the hydrolysis of sugar cane bagasse using phosphoric acid. *J. Food Eng.* **74** (1): 78–88.

Gandini, A. (2008). Polymers from renewable resources: a challenge for the future of macromolecular materials. *Macromolecules* **41** (24): 9491–9504.

Gandolfi, S., Ottolina, G., Consonni, R. et al. (2014). Fractionation of hemp hurds by organosolv pretreatment and its effect on production of lignin and sugars. *ChemSusChem* **7** (7): 1991–1999.

Gierlinger, N., Jacques, D., Schwanninger, M. et al. (2004). Heartwood extractives and lignin content of different larch species (*Larix* sp.) and relationships to brown-rot decay-resistance. *Trees* **18** (2): 230–236.

Gupta, V.K., Kubicek, C.P., Berrin, J.-G. et al. (2016). Fungal enzymes for bio-products from sustainable and waste biomass. *Trends Biochem. Sci* **41** (7): 633–645.

Harada, H. and Côté, W. Jr. (1985). Structure of wood. In: *Biosynthesis and Biodegradation of Wood Components* (ed. T. Higuchi), 1–42. New York: Academic Press.

Hideno, A., Kawashima, A., Endo, T. et al. (2013). Ethanol-based organosolv treatment with trace hydrochloric acid improves the enzymatic digestibility of Japanese cypress (*Chamaecyparis obtusa*) by exposing nanofibers on the surface. *Bioresour. Technol.* **132**: 64–70.

Himmel, M.E., Ding, S.-Y., Johnson, D.K. et al. (2007). Biomass recalcitrance: engineering plants and enzymes for biofuels production. *Science* **315** (5813): 804–807.

Ho-Yin, T., Cheng, S.-C., Yeung, C.S. et al. (2019). Development of a waste-derived lignin-porphyrin bio-polymer with enhanced photoluminescence at high water fraction with wide pH range and heavy metal sensitivity investigations. *Green Chem.* **21** (6): 1319–1329.

Huijgen, W.J., Smit, A.T., Reith, J.H., and Uil, H.D. (2011). Catalytic organosolv fractionation of willow wood and wheat straw as pretreatment for enzymatic cellulose hydrolysis. *J. Chem. Technol. Biotechnol.* **86** (11): 1428–1438.

Isogai, A. (2013). Wood nanocelluloses: fundamentals and applications as new bio-based nanomaterials. *J. Wood Sci.* **59** (6): 449–459.

Jankowska, A., Boruszewski, P., Drozdzek, M. et al. (2018). The role of extractives and wood anatomy in the wettability and free surface energy of hardwoods. *BioResources* **13** (2): 3082–3097.

Javier, C.-S., Jorge, D.-D., Sergio, A.-R., and Roberto, Z.-G. (2015). Optimization of the tensile and flexural strength of a wood-PET composite. *Ing., Invest. Tecnol.* **16** (1): 105–112.

Kalyani, D.C., Zamanzadeh, M., Müller, G., and Horn, S.J. (2017). Biofuel production from birch wood by combining high solid loading simultaneous saccharification and fermentation and anaerobic digestion. *Appl. Energy* **193**: 210–219.

Kanbayashi, T. and Miyafuji, H. (2016). Effect of ionic liquid treatment on the ultrastructural and topochemical features of compression wood in Japanese cedar (*Cryptomeria japonica*). *Sci. Rep.* **6**: 30147.

Kibblewhite, R.P. and Brookes, D. (1976). Distribution of chemical components in the walls of kraft and bisulphite pulp fibres. *Wood Sci. Technol.* **10** (1): 39–46.

Kim, J.S., Lee, Y., and Kim, T.H. (2016). A review on alkaline pretreatment technology for bioconversion of lignocellulosic biomass. *Bioresour. Technol.* **199**: 42–48.

Klemm, D., Heublein, B., Fink, H.P., and Bohn, A. (2005). Cellulose: fascinating biopolymer and sustainable raw material. *Angew. Chem. Int. Ed.* **44** (22): 3358–3393.

Koo, B.-W., Min, B.-C., Gwak, K.-S. et al. (2012). Structural changes in lignin during organosolv pretreatment of *Liriodendron tulipifera* and the effect on enzymatic hydrolysis. *Biomass Bioenergy* **42**: 24–32.

Kumar, A.K. and Sharma, S. (2017). Recent updates on different methods of pretreatment of lignocellulosic feedstocks: a review. *Bioresour. Bioprocess.* **4** (1): 7.

Kumar, R. and Wyman, C.E. (2009). Effects of cellulase and xylanase enzymes on the deconstruction of solids from pretreatment of poplar by leading technologies. *Biotechnol. Progr.* **25** (2): 302–314.

Kumar, V., Binod, P., Sindhu, R. et al. (2018). Bioconversion of pentose sugars to value added chemicals and fuels: recent trends, challenges and possibilities. *Bioresour. Technol.* **269**: 443–451.

Lachaud, S., Catesson, A.-M., and Bonnemain, J.-L. (1999). Structure and functions of the vascular cambium. *C.R. Acad. Sci., Ser. III* **322** (8): 633–650.

Langenheim, J.H. (2003). *Plant Resins*. Oregon: Timber Press.

Lavarack, B., Griffin, G., and Rodman, D. (2002). The acid hydrolysis of sugarcane bagasse hemicellulose to produce xylose, arabinose, glucose and other products. *Biomass Bioenergy* **23** (5): 367–380.

Leu, S.-Y. and Zhu, J. (2013). Substrate-related factors affecting enzymatic saccharification of lignocelluloses: our recent understanding. *Bioenergy Res.* **6** (2): 405–415.

Leu, S.-Y., Yang, T.-H., Lo, S.-F. et al. (2012). Optimized material composition to improve the physical and mechanical properties of extruded wood–plastic composites (WPCs). *Constr. Build. Mater.* **29**: 120–127.

Li, Z., Jiang, Z., Fei, B. et al. (2012). Ethanol organosolv pretreatment of bamboo for efficient enzymatic saccharification. *BioResources* **7** (3): 3452–3462.

Li, C., Zhao, X., Wang, A. et al. (2015). Catalytic transformation of lignin for the production of chemicals and fuels. *Chem. Rev.* **115** (21): 11559–11624.

Liao, S.-H. (2005). Expert system methodologies and applications—a decade review from 1995 to 2004. *Expert Syst. Appl.* **28** (1): 93–103.

Macedo, N. and Brigham, C.J. (2014). From beverages to biofuels: the journeys of ethanol-producing microorganisms. *Int. J. Biotechnol. Wellness Ind.* **3** (3): 79–87.

Mariano, M., El Kissi, N., and Dufresne, A. (2014). Cellulose nanocrystals and related nanocomposites: review of some properties and challenges. *J. Polym. Sci., Part B: Polym. Phys.* **52** (12): 791–806.

Mattos, B.D., Lourençon, T.V., Gatto, D.A. et al. (2016). Chemical characterization of wood and extractives of fast-growing *Schizolobium parahyba* and *Pinus taeda*. *Wood Mater. Sci. Eng.* **11** (4): 209–216.

M'hamdi, A.I., Kandri, N.I., Zerouale, A. et al. (2017). Life cycle assessment of paper production from treated wood. *Energy Procedia* **128**: 461–468.

Miranda, I., Sousa, V., Ferreira, J., and Pereira, H. (2017). Chemical characterization and extractives composition of heartwood and sapwood from *Quercus faginea*. *PLoS One* **12** (6): e0179268.

Moniruzzaman, M. and Goto, M. (2018). Ionic liquid pretreatment of lignocellulosic biomass for enhanced enzymatic delignification. *Adv. Biochem. Eng./Biotech.* **168**.

Mosier, N., Wyman, C., Dale, B. et al. (2005). Features of promising technologies for pretreatment of lignocellulosic biomass. *Bioresour. Technol.* **96** (6): 673–686.

Moure, A., Gullón, P., Domínguez, H., and Parajó, J.C. (2006). Advances in the manufacture, purification and applications of xylo-oligosaccharides as food additives and nutraceuticals. *Process Biochem.* **41** (9): 1913–1923.

Mussatto, S.I. (2016). *Biomass Fractionation Technologies for a Lignocellulosic Feedstock Based Biorefinery*. Elsevier.

Nair, S.G., Shah, J.V., Shah, P.A. et al. (2015). Extractive spectrophotometric determination of five selected drugs by ion-pair complex formation with bromothymol blue in pure form and pharmaceutical preparations. *Cogent Chem.* **1** (1): 1075852.

Nguyen, Q., Bowyer, J., Howe, J., et al. (2017). *Global Production of Second Generation Biofuels: Trends and Influences*. Minneapolis: Dovetail Partners, Inc. https://dovetailinc.org/upload/tmp/1579558792.pdf (accessed 8 April 2020).

Noparat, P., Prasertsan, P., Sompong, O., and Pan, X. (2017). Sulfite pretreatment to overcome recalcitrance of lignocellulose for enzymatic hydrolysis of oil palm trunk. *Energy Procedia* **138**: 1122–1127.

Nourbakhsh, A. and Ashori, A. (2010). Wood plastic composites from agro-waste materials: analysis of mechanical properties. *Bioresour. Technol.* **101** (7): 2525–2528.

Obama, P., Ricochon, G., Muniglia, L., and Brosse, N. (2012). Combination of enzymatic hydrolysis and ethanol organosolv pretreatments: effect on lignin structures, delignification yields and cellulose-to-glucose conversion. *Bioresour. Technol.* **112**: 156–163.

Osong, S.H., Norgren, S., and Engstrand, P. (2016). Processing of wood-based microfibrillated cellulose and nanofibrillated cellulose, and applications relating to papermaking: a review. *Cellulose* **23** (1): 93–123.

Özparpucu, M., Rüggeberg, M., Gierlinger, N. et al. (2017). Unravelling the impact of lignin on cell wall mechanics: a comprehensive study on young poplar trees downregulated for cinnamyl alcohol dehydrogenase (CAD). *Plant J.* **91** (3): 480–490.

Papadopoulou, E.L., Pignatelli, F., Marras, S. et al. (2016). Nylon 6, 6/graphene nanoplatelet composite films obtained from a new solvent. *RSC Adv.* **6** (8): 6823–6831.

Park, J.-Y., Shiroma, R., Al-Haq, M.I. et al. (2010). A novel lime pretreatment for subsequent bioethanol production from rice straw–calcium capturing by carbonation (CaCCO) process. *Bioresour. Technol.* **101** (17): 6805–6811.

Park, N., Kim, H.-Y., Koo, B.-W. et al. (2010). Organosolv pretreatment with various catalysts for enhancing enzymatic hydrolysis of pitch pine (*Pinus rigida*). *Bioresour. Technol.* **101** (18): 7046–7053.

Park, Y.C., Kim, T.H., and Kim, J.S. (2017). Effect of organosolv pretreatment on mechanically pretreated biomass by use of concentrated ethanol as the solvent. *Biotechnol. Bioprocess Eng.* **22** (4): 431–439.

Perlack, R.D. (2005). *Biomass as Feedstock for a Bioenergy and Bioproducts Industry: The Technical Feasibility of a Billion-Ton Annual Supply*. Oak Ridge National Laboratory.

Pinkert, A., Goeke, D.F., Marsh, K.N., and Pang, S. (2011). Extracting wood lignin without dissolving or degrading cellulose: investigations on the use of food additive-derived ionic liquids. *Green Chem.* **13** (11): 3124–3136.

Pizzi, A. (1983). Tannin-based wood adhesives, , Chapter 4. In: *Wood Adhesives Chemistry and Technology* (ed. A. Pizzi), 177–246. New York: Marcel Dekkar.

Pizzi, A., Valenzuela, J., and Westermeyer, C. (1993). Non-emulsifiable, water-based, mixed diisocyanate adhesive systems for exterior plywood. Part II. Theory application and industrial results. *Holzforschung Wood Res. Technol.* **47** (1): 68–71.

Popa, V.I. (2015). Wood bark as valuable raw material for compounds with biological activity. *Celuloză şi Hârtie* **64**: 5 17.

Popovska, V.J., Iliev, B., and Spiroski, I. (2016). Characteristics of medium density fiberboards for furniture production and interior application. *South East Eur. J. Archit. Des.* **2016**: 1–5.

Ramluckan, K., Moodley, K.G., and Bux, F. (2014). An evaluation of the efficacy of using selected solvents for the extraction of lipids from algal biomass by the soxhlet extraction method. *Fuel* **116**: 103–108.

Reddy, N. and Yang, Y. (2005). Biofibers from agricultural byproducts for industrial applications. *Trends Biotechnol.* **23** (1): 22–27.

Rodrıguez-Chong, A., Ramı´ rez, J.A., Garrote, G., and Vázquez, M. (2004). Hydrolysis of sugar cane bagasse using nitric acid: a kinetic assessment. *J. Food Eng.* **61** (2): 143–152.

Safin, R., Khasanshin, R.R., Shaikhutdinova, A.R., Ziatdinov, R.R. The technology for creating of decorative plywood with low formaldehyde emission. In *IOP Conference Series: Materials Science and Engineering*. 2015. IOP Publishing.

Sanchez, C., Belleville, P., Popall, M., and Nicole, L. (2011). Applications of advanced hybrid organic–inorganic nanomaterials: from laboratory to market. *Chem. Soc. Rev.* **40** (2): 696–753.

Schwanninger, M. and Hinterstoisser, B. (2002). Comparison of the classical wood extraction method using a Soxhlet apparatus with an advanced extraction method. *Eur. J. Wood Wood Prod.* **60** (5): 343–346.

Siró, I. and Plackett, D. (2010). Microfibrillated cellulose and new nanocomposite materials: a review. *Cellulose* **17** (3): 459–494.

Sjostrom, E. (2013). *Wood Chemistry: Fundamentals and Applications*. Elsevier.

Sluiter, A., Hames, B., Ruiz, R. et al. (2008). *Determination of Structural Carbohydrates and Lignin in Biomass: Laboratory Analytical Procedure*. Golden, CO: National Renewable Energy Laboratory.

Somerville, C. (2006). The billion-ton biofuels vision. *Science* **312** (5778): 1277.

Sun, Y. and Cheng, J. (2002). Hydrolysis of lignocellulosic materials for ethanol production: a review. *Bioresour. Technol.* **83** (1): 1–11.

Taherzadeh, M.J. and Karimi, K. (2008). Pretreatment of lignocellulosic wastes to improve ethanol and biogas production: a review. *Int. J. Mol. Sci.* **9** (9): 1621–1651.

Tanaka, S., Miyazaki, H., Shiozaki, A. et al. (2017). Cytosolic Cl^- affects the anticancer activity of paclitaxel in the gastric cancer cell line, MKN28 cell. *Cell. Physiol. Biochem.* **42** (1): 68–80.

Taylor, A.M., Gartner, B.L., and Morrell, J.J. (2007). Heartwood formation and natural durability—a review. *Wood Fiber Sci.* **34** (4): 587–611.

Teramoto, Y., Lee, S.-H., and Endo, T. (2009). Cost reduction and feedstock diversity for sulfuric acid-free ethanol cooking of lignocellulosic biomass as a pretreatment to enzymatic saccharification. *Bioresour. Technol.* **100** (20): 4783–4789.

Terán Hilares, R., Swerts, M.P., Ahmed, M.A. et al. (2017). Organosolv pretreatment of sugar cane bagasse for bioethanol production. *Ind. Eng. Chem. Res.* **56** (14): 3833–3838.

Tian, D., Hu, J., Bao, J. et al. (2017). Lignin valorisation: lignin nanoparticles as high-value bio-additive for multifunctional nanocomposites. *Biotechnol. Biofuels* **10** (1): 192.

Wang, Y., Abdel-Rahman, M.A., Tashiro, Y. et al. (2014). l-(+)-lactic acid production by co-fermentation of cellobiose and xylose without carbon catabolite repression using *Enterococcus mundtii* QU 25. *RSC Adv.* **4** (42): 22013–22021.

Wang, Y., Tashiro, Y., and Sonomoto, K. (2015). Fermentative production of lactic acid from renewable materials: recent achievements, prospects, and limits. *J. Biosci. Bioeng.* **119** (1): 10–18.

Wang, S., Zhao, W., Lee, T.S. et al. (2018). Dimethyl sulfoxide assisted ionic liquid pretreatment of switchgrass for isoprenol production. *ACS Sustainable Chem. Eng.* **6** (3): 4354–4361.

Xiang, Z., Gao, W., Chen, L. et al. (2016). A comparison of cellulose nanofibrils produced from *Cladophora glomerata* algae and bleached eucalyptus pulp. *Cellulose* **23** (1): 493–503.

Xu, L. and Tschirner, U.W. (2011). Peracetic acid pretreatment of alfalfa stem and aspen biomass. *BioResources* **7** (1): 203–216.

Xu, X., Liu, F., Jiang, L. et al. (2013). Cellulose nanocrystals vs. cellulose nanofibrils: a comparative study on their microstructures and effects as polymer reinforcing agents. *ACS Appl. Mater. Interfaces* **5** (8): 2999–3009.

Yumrutaş, R., Alma, M.H., Özcan, H., and Kaşka, Ö. (2008). Investigation of purified sulfate turpentine on engine performance and exhaust emission. *Fuel* **87** (2): 252–259.

Zhang, P., Hu, H., and Shi, S. (2006). Application of hemicellulose. *Tianjin Pap. Mak.* **2**: 16–18.

Zhao, J. (2015). Development of China's biofuel industry and policy making in comparison with international practices. *Sci. Bull.* **60** (11): 1049–1054.

Zhao, X.B., Wang, L., and Liu, D.H. (2008). Peracetic acid pretreatment of sugarcane bagasse for enzymatic hydrolysis: a continued work. *J. Chem. Technol. Biotechnol.* **83** (6): 950–956.

Zhao, H., Jones, C.L., Baker, G.A. et al. (2009). Regenerating cellulose from ionic liquids for an accelerated enzymatic hydrolysis. *J. Biotechnol.* **139** (1): 47–54.

Zhao, X., Zhang, L., and Liu, D. (2010). Pretreatment of Siam weed stem by several chemical methods for increasing the enzymatic digestibility. *Biotechnol. J.* **5** (5): 493–504.

Zhao, X., Li, S., Wu, R. et al. (2017). Organosolv fractionating pre-treatment of lignocellulosic biomass for efficient enzymatic saccharification: chemistry, kinetics, and substrate structures. *Biofuels, Bioprod. Biorefin.* **11** (3): 567–590.

Zheng, J., Tashiro, Y., Wang, Q., and Sonomoto, K. (2015). Recent advances to improve fermentative butanol production: genetic engineering and fermentation technology. *J. Biosci. Bioeng.* **119** (1): 1–9.

Zhu, J. and Pan, X. (2010). Woody biomass pretreatment for cellulosic ethanol production: technology and energy consumption evaluation. *Bioresour. Technol.* **101** (13): 4992–5002.

Zhu, J. and Zhuang, X. (2012). Conceptual net energy output for biofuel production from lignocellulosic biomass through biorefining. *Prog. Energy Combust. Sci.* **38** (4): 583–598.

Zhu, J., Zhang, X., and Pan, X. (2011). *Sustainable Production of Fuels, Chemicals, and Fibers from Forest Biomass*. American Chemical Society.

Zhu, X., Tong, S., Li, X. et al. (2017). Conversion of biomass into high-quality bio-oils by degradative solvent extraction combined with subsequent pyrolysis. *Energy Fuels* **31** (4): 3987–3994.

6

Recovery of Nutrients and Transformations of Municipal/Domestic Food Waste

Divyani Panwar[1], Parmjit S. Panesar[1], Gisha Singla[2], Meena Krishania[2] and Avinash Thakur[3]

[1] *Food Biotechnology Research Laboratory, Department of Food Engineering and Technology, Sant Longowal Institute of Engineering and Technology, Longowal - 148 106, Punjab, India*
[2] *Center of Innovative and Applied Bioprocessing, Sector-81, S.A.S Nagar, Mohali, Punjab 140306, India*
[3] *Department of Chemical Engineering, Sant Longowal Institute of Engineering and Technology, Longowal - 148 106, Punjab, India*

6.1 Introduction

Continual increase in human population (1.1% per year) has led to an increased demand for packaged and processed foods. This increase in demand for food production has resulted in an intensive increase in a variety of agriculture practices, transportation, and storage. Various modern technologies have also been developed which have resulted in an intensification of food production. However, this technological advancement in the field of food processing generates a surplus volume of food byproducts and wastes. Food processing industries such as the fruits and vegetable industry, beverage industry, dairy industry, bakery industry, sugar industry, and meat industry are involved in the production of processed food commodities, and thereby generate various types of food wastes (Joshi and Sharma 2011).

Waste Valorisation: Waste Streams in a Circular Economy, First Edition.
Edited by Carol Sze Ki Lin, Guneet Kaur, Chong Li and Xiaofeng Yang.

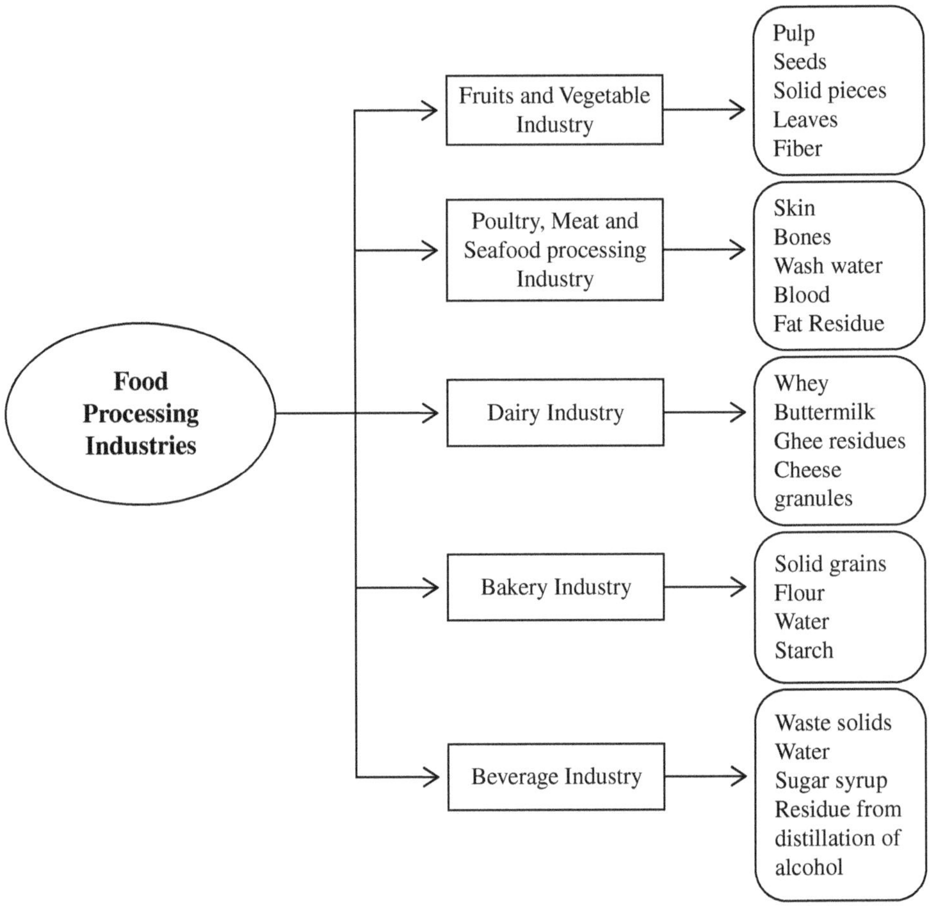

Figure 6.1 *Food processing industries and types of waste generated.* Source: Adapted from Panesar et al. (2014), reproduced with permission of Taylor & Francis.

Waste is defined as any substance or material which is unwanted, undesirable, and is of no use. This includes municipal solid wastes (commercial wastes and household wastes), hazardous wastes, agricultural wastes, clinical wastes and food processing wastes (Joshi and Sharma 2011). Food waste is generated throughout the food supply chain, which includes the stages of postharvest handling, processing, distribution, storage, sale, cooking, and final consumption (Food Waste Reduction Alliance 2016). Examples of food wastes include whey, bones, peels, skin, flour, and so on. The main categories of food industries and types of wastes generated through them are described in Figure 6.1.

According to the Food and Agriculture Organization (FAO) (2011), approximately a third of food produced for human consumption is lost or wasted, which accounts for about 1.3 billion tonnes/yr. This includes 45% of fruits and vegetables, 20% of dairy products, 20% of meat, 30% of cereals, and 35% of fish and seafood (Figure 6.2).

Food waste is generated during different stages of food processing, starting from initial production to final consumption, and approximately 25–50% of the product is wasted throughout the food chain. Significant amounts of food are lost during the initial and middle phases of food production in low-income countries, which may be due to financial issues,

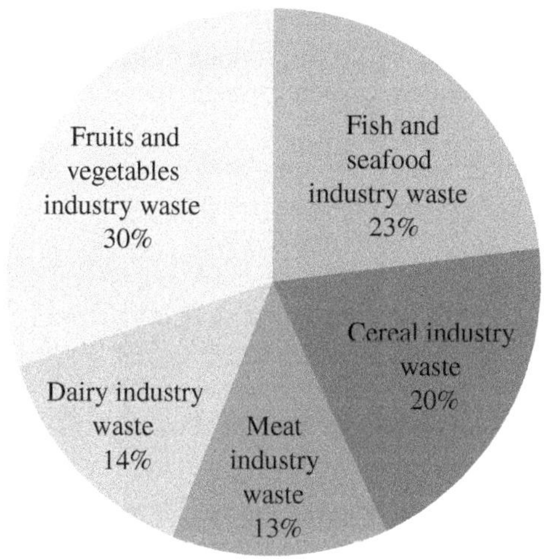

Figure 6.2 *Composition of food waste generated throughout the supply chain (FAO 2011).*

management issues, or technical reasons during different stages of the food supply chain. However, in medium and high-income countries, food is wasted during the final consumption stage, which is mainly due to poor consumer behavior, poor quality standards of final product, insufficient planning of purchases and storage of products past their expiry date (Kosseva 2013 a,b).

Food waste and its intensification is becoming a major issue around the globe. The proliferation of waste poses a serious threat to health and the environment, as well as the global economy. Therefore, there is an urgent need to take pertinent measures and to develop innovative technologies for treatment, minimization, reuse and valorisation of food waste (Paritosh et al. 2017). "Waste valorisation" is the process of recovering valuable compounds from wastes and using them for developing new products, such as chemicals, materials and fuels, via different biotechnological processes and chemical transformations (Otles et al. 2015).

Keeping the above aspects in mind, this chapter provides comprehensive information on different aspects of characteristics of food waste, its supply chain and recovery of valuable products from anaerobic digestion (AD) of food waste, and novel approaches for the transformation of food waste into valuable products through chemical and biotechnological routes.

6.2 Characteristics of Food Waste and its Supply Chain

Food waste is generated throughout the food supply chain and is produced when the desired product is separated from undesirable products (Hegde et al. 2018). The raw material enters the food production process and the final product exists either as the product desired, product-specific waste, or as non-product-specific waste (Russ and Schnappinger 2007). Non-product-specific waste does not affect the quality of product. However, product quality is directly affected by product-specific waste. Generation of product-specific waste cannot be avoided and it is highly organic in nature. The amount and type of product-specific waste can only be tailored using technical methods as it directly affects the quality of the final product. Therefore, food wastes are basically characterized by a high ratio of product-specific waste (Jayathilakan et al. 2012).

Various food processing industries, such as meat, dairy, fruits and vegetables, bakery, sugar processing, and beverage industries, can contribute to the generation of a variety of food wastes. Food processing wastes can be solid or liquid in nature. Solid wastes include both packaging and organic wastes, and are a result of various processing operations. Examples of solid wastes may include tomato waste, apple pomace, rinds, skin, bones, fruit and vegetable trimmings, inedible dough, solid grains, and packaging materials such as plastic, glass, and metals. Solid wastes comprise rich sources of starch, lignin, cellulose, dietary fibers, lipids, and monosaccharides (especially glucose and fructose). Solid wastes are mainly characterized in terms of density, calorific value, moisture content, pH, volatile matter, and optical properties (Chandrappa and Das 2012). Various characteristics of solid wastes include high moisture content, high nutrient status, lower content of inert contaminants such as stones, and varying amounts of suspended solids depending upon the source (Ammar 2013).

Liquid wastes are generated when a large quantity of water is used during various applications in the food industry, including cleaning, cooking, sanitation, and transportation (Hegde et al. 2018; Ammar 2013). Examples of liquid wastes generated may include whey generated from the cheese industry, potato processing wastewater, apple pomace sludge, brewery effluent, and oil mill effluent. Liquid effluents from food industries consist of organic matter, nitrogen in various forms, fats, oils and grease, total suspended solids (TSS), and other inorganic materials (Hegde et al. 2018). The strength of the waste is usually estimated in terms of biological oxygen demand (BOD) and chemical oxygen demand (COD) (Panesar et al. 2014). Wastewater from food industries is usually nontoxic in nature since the levels of hazardous and non-biodegradable compounds are considerably low. It is usually organic and colored, has high levels of BOD and COD, and has a characteristic odor which is possibly due to the generation of various gases during the biodegradation in water. Liquid wastes may contain different types of microorganisms which may be pathogenic in nature. Quantities of solid and liquid wastes generated from different processes is depicted in Table 6.1.

Table 6.1 *Solid and liquid wastes generated from different sources.*

Food waste	Total solids (g/kg)	Liquid volume (m^3/kg)
Eggs	0.111	—
Poultry	0.270	50.000
Red meat	0.440	25.000
Ice cream	—	0.08
Cheese whey	—	9.00
Skimmed milk	—	0.07
Spinach	20	—
Potatoes	66	0.012
Turnip greens	15	—
Tomatoes (caustic peeling)	14	0.010

Source: From Mishra et al. (2006) and Gorsuch (1986).

6.2.1 Characteristics of Waste Generated from Food Industries

Characteristics of various type of wastes generated from major food processing industries are explained below.

6.2.1.1 Fruits and Vegetables Industry

Surplus amounts of waste in the form of solids and liquid effluents are generated from the fruits and vegetables industry during different processing stages. These wastes have diverse characteristics which mainly depend upon the type of commodity processed, condition (ripeness and damage) of raw product, type of unit operation used, variations in season and multiplicity of the product, rate of water used during processing, and condition and type of equipment used (Shah et al. 2011; Williams 2005).

These effluents contain high organic loads, high BOD, COD, salts, suspended solids such as soil particles, organic sugars, pesticide residues, starches, and cleansing agents (Shah et al. 2011). High levels of nutrients such as nitrogen and phosphorus and disinfectants such as chlorine are also present in wastewater. The BOD, COD, and pH value of wastes from the fruit and vegetable industry ranges from 43 to 41 000 mg/l; 78 to 65 000 mg/l; and 3.1 to 11.9 respectively (Thakur et al. 2011).

6.2.1.2 Dairy Industry

The dairy industry is a crucial part of the food industry responsible for the production of a variety of milk products and byproducts. With such diversification of products and byproducts, a surplus amount of liquid waste is generated from the dairy processing industry. Processing of dairy products requires various unit operations such as centrifugation, homogenization, and pasteurization at every stage, including cleaning and sanitization, milk receiving, storing of raw materials, processing of raw materials into finished products, packaging, storing, and distribution, which are responsible for the generation of huge amounts of wastes (Shah et al. 2011).

The dairy industry generates between 3.74 and 11.22 million m^3 of waste per year (Tikariha and Sahu 2014). Wastewater from dairy industries is characterized by high organic load, high BOD and COD values, considerable variations in pH, as well as a high content of TSS, and oil/grease (Kosseva 2013a,b). Large amounts of milk solids, detergents, proteins, minerals (Na, Cl, K, Ca, Mg), fats, phosphorus and nitrogen are also present in wastewater (Demirel et al. 2005). The BOD, COD, and pH levels of dairy industry wastes range from 100 to 6000 mg/l; 250 to 6000 mg/l and 4.5 to 11.3 respectively (Britz et al. 2006).

6.2.1.3 Meat, Poultry and Seafood Processing Industry

With the increase in meat consumption and production, an increased quantity of non-consumable byproducts and wastes is generated from the meat, poultry and seafood industry (Liang and Hultin 2003). Massive quantities of liquid and solid wastes are generated at every stage during the processing of meat, poultry and seafood (Rathinaraj and Sachindra 2013). Most waste in the meat industry is discarded from slaughterhouses where carcasses are produced, meat is prepared and is cut into pieces for consumption (Panesar et al. 2014). Wastes from the meat, poultry, and fish industry may include skin, entrails, blood, bones, and so on, and these byproducts comprise nearly 60–70% of the slaughtered animal, of which approximately 40% is edible; 20% cannot be sold or consumed as meat or meat products (Bhaskar et al. 2007).

Effluents from these industries are highly polluted and are composed of blood, fat, hair, bits of meat and bone, undigested food, manure, urine, grease, tendons, and a variety of organic compounds (Arvanitoyannis and Ladas 2007).

Wastes from the meat, poultry and seafood industry are characterized by high BOD, COD, TSS, total nitrogen and phosphorus content (Lin et al. 2013). Wastewater has high turbidity and is odorous. It also contains large amounts of grease, fat, protein, and has a warm temperature. It has a high density of pathogenic microorganisms such as coliform bacteria, fecal coliform and fecal streptococcus groups of bacteria (Irshad et al. 2016). The BOD, COD, TSS and FOG (fats, oils, and grease) values of some wastes from various meat, poultry, and seafood industries range between 400 and 3000 mg/l; 500 and 19 000 mg/l; 200 and 18 768 mg/l; and 10 and 550 mg/l respectively (Carawan et al. 1979; Banks and Wang 2006).

6.2.1.4 *Cereals and Grains Processing Industry*

Cereals and grains are considered as the most important part of the human diet due to their high nutritional and energy values. To meet their demand, they are cultivated worldwide in large quantities. In principle, cereals and grains may include rice, wheat, barley, oats, sorghum, and so on. A variety of wastes and byproducts are generated from the cereals industry during different stages of processing, including harvesting, postharvest and postproduction stages. A surplus amount of waste such as husk, germ, bran, and grits is generated during cereal processing. The nature of these wastes depends upon the type of cereal used and the processing conditions (Pratape et al. 2011).

Grain dust is one of the most serious problems in the grain processing industries, and is hazardous to health, as well as the environment due to the presence of silica, jute particles, carbon, molds, and spores. The dust concentrations in major grain mills including rice, lentils, and wheat ranges from 10 to 780 mg/m^3. The air standard for suspended particulate matter (SPM) at ambient temperatures in industrial premises is approximately 0.5–1.0 mg/m^3 in 24 hours. The waste from the cereals and grains industry is characterized by high BOD and COD levels (Pratape et al. 2011). The BOD and COD levels of these wastes range between 35 000 and 50 000 mg/l and 100 000 and 150 000 mg/l respectively (Nandy et al. 2002).

6.2.2 Food Waste Supply Chain

An important part of food is lost or discarded as "food waste" that was initially produced for human consumption. According to the FAO (2014), "Food supply chain is defined as the connected series of activities for production, processing, distribution, and final consumption of food." Food is wasted at every stage of the food supply chain; however, the amounts of food wasted may vary from one stage to another. The food supply chain begins from raw material production, proceeds with industrial manufacturing and processing, and ends at retailing and final consumption (Otles et al. 2015). Food is wasted throughout this complete food supply cycle due to a variety of technological, economic or social reasons, and therefore the term *food supply chain waste* (FSCW) is being used by various researchers (Lin et al. 2013).

The food supply chain is composed of five stages, and food waste is generated in accumulative quantities at each segment.

Losses at the production stage usually occur during or immediately after harvesting from the farm (Lipinski et al. 2013). It includes losses that occur due to mechanical damage, damage during various harvesting operations such as crop sorting or fruit picking, or due to contamination by animals or insects (Otles et al. 2015). Losses during handling and storage occur after the product has left the farm for further handling, storage and transport (Lipinski et al. 2013). Otles et al. (2015) stated that these losses occurring between the farm and the distribution stage may be due to lack of storage facilities, transportation method, or due to degradation by pests and microorganisms. Food losses during processing and packaging occur due to spillage or degradation during industrial or domestic processing (for example canning, bread baking, etc.) and packaging. Food waste during this phase includes waste in the form of spilled milk, damaged fish, livestock trimming during slaughtering, and damaged fruits due to careless washing, peeling, slicing or inefficient industrial processes (Lipinski et al. 2013; Otles et al. 2015). Next, food wastes occur during the distribution of food product and in the marketing system, including wholesale and retail markets. During this phase, food wasted includes edible products which are discarded due to quality standards, food spilled or damaged in the market, or the edible product which is not sold before its expiry date (Lipinski et al. 2013). At this stage, food is wasted due to the lack of proper transportation, poor supplier–buyer relationships, poor infrastructure, inappropriate packaging, or a conscious decision to throw away food (Otles et al. 2015). Losses during consumption include food wasted at household or business level (restaurants/caterers). Food is wasted in the form of food purchased or cooked but not eaten, or edible products are discarded due to poor quality (Lipinski et al. 2013). Food waste at the consumer level is generated due to poor shopping habits of an individual, lack of knowledge and awareness to use food efficiently, poor packaging, or cultural issues (Marangon et al. 2014).

The food waste generated between field and fork is categorized as pre-consumption and post-consumption wastes (Karthikeyan et al. 2017). The generation of food waste starts from the farm even before the product enters the marketing system, and ends at the stage where the product is either consumed or is discarded as waste. Food waste predominantly occurs at the retail and consumption stages of the food chain as a result of carelessness or spoilage and expiry of food (Lipinski et al. 2013). Food loss and waste vary significantly between developed and developing countries (Figure 6.3). In medium- and high-income countries, food waste is usually generated at the consumption stage, and in low-income countries, food is generally discarded during the early and middle stages of the food supply chain (FAO 2011).

The rate of waste generation from this supply chain, and the organic nature of this waste, represents a serious problem for the industry concerned (Schieber et al. 2001). A surplus amount of waste is produced from various food industries due to the inefficiency of the food supply chain (Pfaltzgraff et al. 2013). Because of the magnitude and complex nature of food waste, there is an urgent need to adapt various strategies for reducing food loss in the world.

In developing countries, food supply chains can be strengthened by encouraging small-scale industries and farmers to modify and expand their production and marketing of products or by investing in infrastructure, transportation and food industries. The food supply chain in developed countries can be reduced extensively by raising awareness

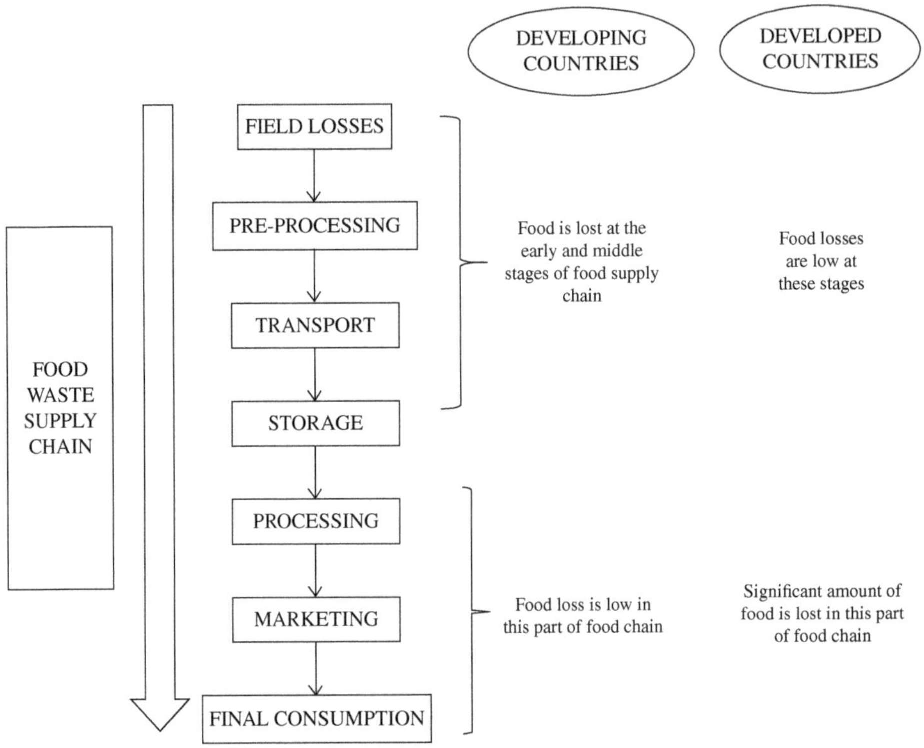

Figure 6.3 *Food waste in developing and developed countries throughout the food supply chain.* Source: Adapted from Tobiassen (2016).

among buyers and sellers and by developing proper waste management strategies and appropriate technologies (FAO 2011).

A high amount of organic food waste with high BOD and COD contents is generated from food industries, posing a great threat to the environment. Therefore, strict regulations have been enforced for proper handling, treatment and disposal of such wastes. However, these wastes are also found to be high in biodegradable materials such as fats, lipids, proteins, and various other valuable constituents. Various new strategies and methods have been developed for the recovery and utilization of these valuable constituents (Panesar et al. 2014).

6.3 Recovery of Valuable Products from Anaerobic Digestion of Food Waste

A considerable amount of food waste is generated from various food industries, and is responsible for polluting land, air, and water because of its high BOD/COD levels and its organic nature. Due to a rapid increase in population and urbanization, increasing food wastage and its accumulation is becoming a global issue. A substantial amount of environmental pollution is caused by improper disposal and inappropriate treatment methods of this large amount of food waste, which can further cause various health issues and high financial costs. Therefore, there is an urgent need for developing food waste treatment methods and legislative frameworks for proper treatment and utilization of food waste (Paritosh et al. 2017).

With an increase in attention towards sustainability, the concept of "recovery" is gaining popularity. As discussed in Section 6.2.2, food waste is considered as a biodegradable source of nutrients and organic material that can be utilized for the recovery of various valuable products (Otles et al. 2015). Plentiful efforts are being made for the proper utilization of food waste to recover and produce various useful products. Recovery of valuable components from food waste not only offers significant economic benefits, but also promotes proper sustainable development of society (Xu et al. 2018). Various chemical and biological processes are used for the treatment of food industry wastes, which include composting, aerobic digestion, pyrolysis, anaerobic digestion, incineration, and wet oxidation. Among these processes, anaerobic digestion (AD) is considered as the most efficient and cost-effective technology for the treatment and recovery of valuable products from food waste (Speece 1983).

Anaerobic digestion is one of the oldest technologies used for the treatment of food waste (Arsova 2010). AD is defined as a biological process that includes a series of steps involving microbial decomposition of biodegradable material in the absence of oxygen (Nayono 2009). During this process, the organic food waste is degraded by anaerobic microorganisms, resulting in the formation of biogas, which mainly consists of methane and carbon dioxide, and a nutrient-rich residue called digestate (Sheets et al. 2015; Yang et al. 2015). AD is a highly complex process carried out by the symbiotic association of anaerobic bacteria (Pavlostathis and Giraldo-Gomez 1991). The AD process is divided into four stages which include hydrolysis, fermentation (acidogenesis), acetogenesis, and methanogenesis (Nayono 2009).

During hydrolysis, complex organic polymers such as proteins, polysaccharides, fats, and lipids are hydrolyzed by hydrolase enzyme excreted by hydrolytic bacteria, thereby resulting in small soluble monomeric units (simple sugars, amino acids, and fatty acids) (Bajpai 2017; Paritosh et al. 2017). The next stage after hydrolysis is the acid-forming phase called acidogenesis. During this stage, the products formed during hydrolysis are further broken down into simple organic compounds such as short-chain fatty acids (propionic acid, acetic acid or formic acid), ketones, alcohols, ammonia, carbon dioxide, hydrogen and some sulfur compounds by the action of acidogenic or fermentative bacteria (Arsova 2010; Paritosh et al. 2017). The third stage of AD is acetogenesis, in which the products produced during acidogenesis are further converted into acetic acid, hydrogen and carbon dioxide by the action of acetogenic bacteria (Paritosh et al. 2017). The final stage of AD is methanogenesis. In this phase, the products formed during acetogenesis (acetic acid, carbon dioxide and hydrogen) are transformed into methane and carbon dioxide by methanogenic bacteria (Li et al. 2010). The overall process of anaerobic digestion is explained in Figure 6.4.

Anaerobic digestion is an efficient and vital technology used globally for the proper disposal and treatment of food waste (Li et al. 2009). This technology has proven to be a promising and environmentally friendly approach for the management of food waste as well as the production of a variety of valuable products including biogas and digestate. Biogas can be used as a source of renewable energy. Aside from being used as a fertilizer, digestate is also a great source for a variety of valuable nutrients such as potassium, nitrogen and, phosphorus. (Angelidaki et al. 2003).

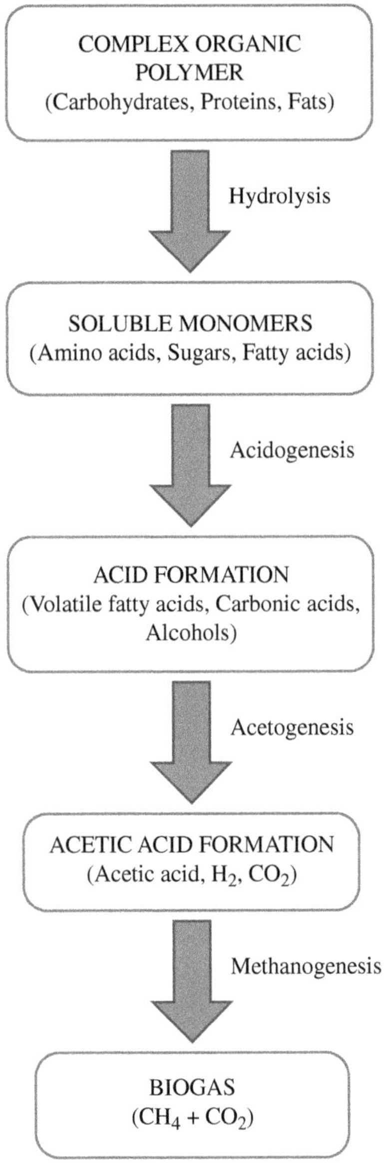

Figure 6.4 *Schematic diagram of the anaerobic digestion process.* Source: Adapted from Paritosh et al. (2017), https://www.hindawi.com/journals/bmri/2017/2370927/#copyright. Licensed under CCBY 4.0.

6.3.1 Biogas

Biogas is one of the valuable products obtained from anaerobic digestion, and is composed of methane (CH_4) and carbon dioxide (CO_2). Small amounts of hydrogen (H_2), ammonia (NH_3), hydrogen sulfide (H_2S), carbon monoxide (CO), nitrogen (N_2), and oxygen (O_2) are also present in biogas (Dahiya et al. 2017). Biogas is considered as a versatile source of energy which can be successfully used to replace the use of fossil fuels and natural gas used to produce a variety of chemicals. Biogas is produced in an oxygen-free environment where organic matter is decomposed by anaerobic bacteria. It can be produced from a variety of sources which may include landfills, animal manure, agricultural residues, food wastes, and so on (Wu et al. 2015). However, the quality and yield of biogas may vary with the type and source of raw material used (Taghizadeh-Alisaraei et al. 2017). Recovery of biogas helps in managing the organic wastes generated from agriculture and food industries, and manure generated from animals, thereby providing financial and environmental benefits. Production of biogas using food wastes is affected by various process parameters which include reactor design, inoculum type, volatile fatty acid, temperature, pH, and the carbon/nitrogen ratio (Dahiya et al. 2017). With the advancement in production of biogas, bioremediation technology with improved process efficiency has been developed (Dahiya and Joseph 2015).

Extensive research has been done on the production of biogas (CH_4) using food waste as a substrate (Dahiya et al. 2017). Waste from agriculture and food industries such as distillery, brewery, sugar, and dairy can be used as a substrate to produce biogas (Jain and Mishra 1989; Panesar et al. 1999; Marwaha et al. 2000). Wastewater from breweries has been successfully used for the generation of biogas with approximately 80% yield when treated by the upflow anaerobic sludge blanket (UASB) process (Hack 1985). Sugar mill wastewater has also been used for biogas production, wherein anaerobic digestion is followed by an aerobic oxidation pond for the stabilization of effluent (Bhaskaran and Chakrabarty 1966).

A hydrogen–methane two-stage fermentation method has been successfully developed for production of biogas using sugar-rich wastewaters, bread wastes, soyabean paste and brewery wastes as substrates. During the first stage, hydrogen is produced which can be used for generation of electricity, and during the second stage, methane is produced, which is used for generation of heat energy (Nishio and Nakashimada 2007).

Production of biogas is considered as the most significant advantage of AD, as approximately 70–80% of the energy of raw materials is preserved in methane (Mahony et al. 2002). Biogas production using food waste as a substrate is gaining popularity and has reached the commercial level globally.

6.3.2 Digestate

Digestate is defined as the product remaining after the removal of biogas from anaerobic digestion. It contains a partially degraded organic fraction, water, and macro- and micronutrients which have not been digested during the anaerobic process. The physicochemical characteristics of the digestate depend upon the composition and type of substrate used, as well as the working parameters of the AD process (Al Seadi et al. 2013). Compared with raw manures and slurries, digestate is found to have a low total solids, low organic carbon content, low carbon to nitrogen (C : N) ratio, low viscosity, but a high pH value, high phosphorus content, and high ammonia content (Möller et al. 2008; Tambone et al. 2010). These characteristics make the digestate suitable to be used directly as a fertilizer without any processing, which has successfully replaced the use of chemical fertilizers. However, in order to eliminate the restriction of using manure in some specific areas, recovering and recycling of nutrients from digestate is gaining popularity among farmers, researchers and industrialists (Drosg et al. 2015). This is expected to facilitate proper management of nutrients and overcome the issues of depleting natural reserves of phosphorus and potassium. Recycling of nutrients and organic matter from digestate is considered as the most sustainable method for the utilization of digestate as fertilizer (Al Seadi et al. 2013).

Digestate can either be processed partially to reduce its volume, or processed completely to separate the liquid fraction (pure water) and solid fraction (organic matter). Digestate processing results in various products that include:

(i) End-products with higher concentrations of nutrients compared with the untreated digestate;
(ii) Nutrients in the form of minerals;
(iii) End products that can be marketed and used as biofertilizers (Drosg et al. 2015).

In order to achieve efficient recycling of digestate, it should be of high quality and should be free from pathogens and other impurities (Al Seadi et al. 2013). A wide range of technologies and treatments can be applied for processing, concentration, and extraction of nutrients from digestate.

6.3.2.1 Solid–Liquid Recovery

Solid–liquid separation is considered to be the first step of digestate processing. The separated solid fraction can be easily transported due to the reduction of water content, and can either be directly utilized as a biofertilizer or further processed to produce various value-added products (Tambone et al. 2015; Drosg et al. 2015). The separated liquid can either be recycled or treated for the recovery of valuable nutrients such as phosphorus and nitrogen (Nges et al. 2015; Gong et al. 2013). Solid–liquid separation offers a variety of advantages, which include better management of digestate, reduced transportation costs for the solid fraction, and prevention of uncontrolled decomposition processes. A variety of techniques are used for the separation of solids and liquid, such as screw press, decanting centrifuge, belt press, and so on (Møller et al. 2000).

(i) *Screw press*: The screw press technique is one of the most commonly used techniques for separation of solids and liquids from the digestate. These separators are mainly used if the digestate is high in fiber content. Separation efficiency is directly influenced by the type of screen used, size of screen, amount of fiber content present in the digestate, and pressure applied (Weiland 2006). The screw press method is considered as an effective method of separation as it has been reported to produce a solid fraction having approximately 25–30% dry matter. However, the liquid fraction contains a large amount of small particles, which leads to poor recovery of nutrients (N, P, and K) (Møller et al. 2002).
(ii) *Decanter centrifuge*: A decanter centrifuge is a high-speed mechanical device that is frequently used for digestate processing. It uses centrifugal forces for the separation of liquids and solids from the digestate. Furthermore, they can also be used for the recovery of phosphorus contained in digestate (Møller 2001). In a decanter centrifuge, the separation of solid and liquid from digestate is caused by applying a centrifugal force which results in reduction of settling time of solid particles and gives two fractions, solid and liquid (Hjorth et al. 2011). The efficiency of a decanter centrifuge largely depends upon the flow rate of digestate, amount of dry matter in the input slurry,

as well as on the characteristics of the input slurry. Decanter centrifuges have been reported to have an efficiency of approximately 95%, but come with high functioning and energy costs (Cocolo 2013).

(iii) *Belt press*: The working principle of a belt press for the separation of solids and liquids from digestate is similar to that of screw press. Various precipitating and flocculating agents can be used to improve the separation efficiency. Characteristics of the input slurry, type and size of filter used, amount and type of flocculating and precipitating agents are some factors affecting the separation efficiency of a belt press. The belt press has been reported to have higher efficiency and low energy requirements compared with the screw press and decanter centrifuge, respectively (Cocolo 2013).

(iv) *Discontinuous centrifuge*: A discontinuous centrifuge is another type of centrifuge used for the processing of digestate. This is a batch-type centrifuge, where the digestate is fed continuously into the centrifuge in consecutive cycles at a particular time. The separation of solid–liquid fractions from the digestate occurs due to centrifugal forces. The discontinuous centrifuge was reported to have the same efficiency and energy demands when compared with a decanter centrifuge, but it produced solid fractions with slightly higher concentrations of total solids (Drosg et al. 2015).

(v) *Precipitation/flocculation*: Separation of solids–liquids from digestate can be effectively achieved via precipitation/flocculation. In this method, the small suspended particles present in digestate remain in solution due to their negative charge. Positively charged precipitating agents and flocculants are added and they tend to bind around the negatively charged particles, leading to formation of colloids. Consequently, these colloids are precipitated as solids and can be easily separated by settling in clarifiers.

Precipitating agents and flocculants such as aluminum sulfate ($Al_2(SO_4)_3$), ferric chloride ($FeCl_3$), ferric sulfate ($Fe_2(SO_4)_3$) and lime ($Ca(OH)_2$) are usually added in order to increase the separation efficiency and properly recover the nutrients (Drosg et al. 2015).

6.3.2.2 Processing of the Solid Fraction

The solid fraction obtained after the efficient processing of digestate contains approximately 20–30% dry matter. This solid fraction can be directly used as biofertilizer or soil conditioner. However, it still contains a considerable amount of biodegradable material which can lead to further microbial activity and odor emissions. Therefore, further processing of the solid fraction is required to obtain the desired biofertilizer (Al Seadi et al. 2013).

(i) *Composting*: This is the most widely used method for the processing of the solid fraction. This method involves the use of aerobic microorganisms for the degradation of remaining organic matter into CO_2 and water. Compost produced is a stabilized organic matter containing humic substances, which is an ideal biofertilizer as it enhances soil performance (Teglia et al. 2011). The solid fraction produced from digestate is too wet to be used directly for composting. Therefore, bulking material (such as woodchips) is added to successfully accomplish an efficient and stable composting process (Al Seadi et al. 2013). Vermicomposting is considered as an effective and sustainable approach, which utilizes earthworms for the proper stabilization of digestate (Monfet et al. 2018).

(ii) *Drying*: Drying of the solid fraction is another method of digestate processing which aims to reduce water content of the digestate. Drying stabilizes the organic solid waste at a faster rate, which makes the digestate easy to handle, store and transport, reduces the volume of the digestate, increases its nutrient concentration, and pasteurizes solids, which helps in reducing odors and ammonia emission (Monfet et al. 2018; Drosg et al. 2015). Various drying techniques can be applied in order to dry the digestate or the solid fraction, which may include: belt dryer, solar drying, drum dryer, biodrying, and fluidized bed dryer. Dried digestate can be marketed in the same form or can be converted into pellets, which improves its marketability and its utilization as a fertilizer or biofuel (Al Seadi et al. 2013).

6.3.2.3 Extraction of Phosphorus

Phosphorus extraction from the solid fraction has been successfully reported from dried sludge and from ashes produced from incineration of sludge. Combustion processes can be used to recover phosphorus from the ashes. Combustion of digestate produces ashes which contains variety of nutrients which includes 20–25% P_2O_5, K, Al, Mg, Si compounds, and some heavy metals such as Cu, Zn, and Cd. Numerous techniques have been designed by various companies for the recovery of phosphorus from combustion ashes (Schoumans et al. 2010).

Pyrolysis is another technique that has been successfully utilized for the recovery of phosphorus, in which the digestate is exposed to a high temperature of 150–900°C in the absence of oxygen, which produces syngas, bio-oil and biochar (Lemmens et al. 2007). Further experiments revealed that the amount of phosphorus recovered in biochar is much greater than that present in ashes (Schoumans et al. 2010). Some examples of processes developed for the recovery of phosphorus include PHOXNAN/LOPROX, Seaborne, LEACHPHOS, PASH process, and ECPHOS.

6.3.2.4 Processing and Nutrient Recovery Techniques for the Liquid Fraction

After the separation of solid and liquid fractions from the digestate, valorisation of the liquid fraction is of great interest as it contains considerable amounts of nutrients and suspended solids. This liquid fraction can be used for the dilution of high solid feedstock, can be either applied directly as a nitrogen-rich liquid fertilizer, or re-fed to the digestor (Romero-Güiza et al. 2014; Di Iaconi et al. 2010). If proper treatment of liquid fraction for the recovery of nutrients is not performed, the liquid fraction is sent to the wastewater treatment plants for the removal of nitrogen and phosphorus. Various techniques are available for proper treatment and efficient recovery of nutrients from the liquid fraction, which include membrane filtration, evaporation, ammonia stripping, and struvite precipitation (Monfet et al. 2018).

Ammonia stripping Steam or air stripping can be used for removal of ammonia from liquid effluents. The ammonia is released as ammonium sulfate using a sulfuric acid scrubber. This ammonium sulfate can be used as a valuable commercial-grade fertilizer (Capodaglio et al. 2015; Drosg et al. 2015). However, in steam stripping, ammonia-steam produced can be condensed directly to produce ammonia-water containing approximately 25–30% ammonia, thereby eliminating the requirement for a sulfuric acid scrubber (Drosg et al. 2015). For effective ammonia removal, pH of about 10.8–11.5 is maintained by the addition of lime. This method has an advantage over other methods as it leads to the recovery of pure nitrogen fertilizer and it does not require expensive chemicals. However, need for alkali to raise the pH, high-energy inputs, and the need for a stripping tower makes this an expensive process (Poletti et al. 2012; La et al. 2014).

Struvite precipitation Struvite precipitation is gaining popularity as a successful technique for recovering ammonia and phosphorus. In this method, magnesium salts are added to the digestate and a complex compound called magnesium ammonium phosphate (MAP) or struvite ($MgNH_4PO_4{\cdot}6H_2O$) is formed (Drosg et al. 2015; Gerardo et al. 2013; Uysal et al. 2010). This compound is highly nutritive and is insoluble, thereby forming a precipitate. Struvite precipitate is then separated from water and recovered as green fertilizer (Capodaglio et al. 2015; Poletti et al. 2012). Recovery of nutrients can be efficiently performed by adding magnesium in high amounts to maintain the nutrient concentration of approximately 1.3 : 1 : 0.9 for Mg:N:P and increase the pH level to 8.5–9.0. However, addition of $MgCl_2$ may lead to slower and incomplete reactions as well incomplete removal of chloride ions from the remaining solution. Therefore, this remaining solution can only be applied as a fertilizer for chloride-tolerant crops (Drosg et al. 2015).

A large amount of chemicals is required in this process, which may lead to high operational costs. The recovery cost of precipitate may be another disadvantage of the struvite precipitation process. Some examples of struvite-forming processes being commercially used are NuReSys (BE), ANPHOS, Colsen (NL), PRISA (GE), and Nutritec by Sustec (NL) (Vaneeckhaute et al. 2017).

Membrane separation technologies Membrane separation technologies are gaining popularity among industries for the processing of digestate and recovery of nutrients. It is a simple physical separation technology in which the liquid fraction is allowed to pass through a semi-permeable membrane. This technique does not require addition of any chemicals and its maintenance costs are lower when compared to other separation technologies (Monfet et al. 2018; Bamelis et al. 2015). However, the major problem reported for this technology is membrane fouling and clogging. Therefore, membranes must be cleaned and replaced at regular intervals to maintain the efficiency of the separation process (Guo et al. 2012). The end product of membrane filtration is a concentrate rich in ammonia and potassium which can be directly used as a soil conditioner and fertilizer (Drosg et al. 2015). Membrane processes are classified according to the pore size of membranes, such as microfiltration (>100 nm), ultrafiltration (>1 nm), reverse osmosis (no pores), and nanofiltration (<10 nm).

Advanced processes Advanced processes for the recovery of nutrients from the liquid fraction mainly consist of oxidation processes, which transform ammonia and nitrogen into nitrogen gas or nitrate. Various processes include ozonation and chemical oxidation, electrochemical oxidation, as well as microwave and hybrid processes.

In ozonation and chemical oxidation method, various strong oxidizers such as ozone and hydrogen peroxide are used for the oxidation of organic matter into simple molecules. This method has been reported as a successful technology to reduce sludge production and control odor emissions. This method is generally applied to digestate with low ammonia concentrations, where the ammonia is converted into nitrate (Monfet et al. 2018).

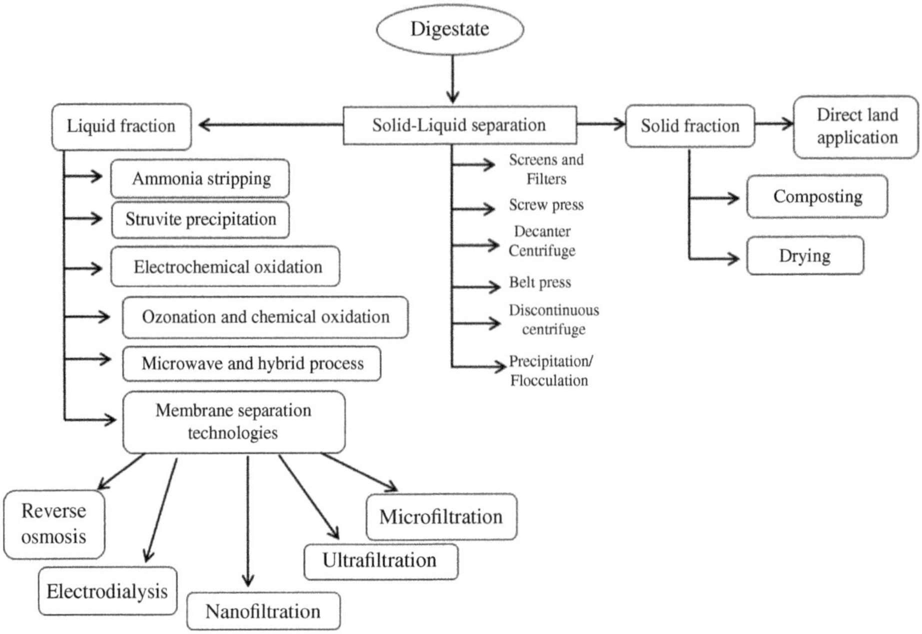

Figure 6.5 *Overview of digestate processing techniques.*

Electrochemical oxidation uses the process of either direct anodic oxidation or indirect oxidation for the conversion of ammonia into nitrogen. Two major disadvantages of the electrochemical oxidation process are loss of performance during the scaling-up process, and high costs due to the presence of electrodes (He et al. 2015).

In microwave and hybrid processes, electromagnetic radiations are used to induce polar molecules to rotate and produce thermal energy by dielectric heating. This process is used for evaporation of ammonia by utilizing microwave heat, thereby removing ammonia from liquids. An overview of digestate processing techniques is described in Figure 6.5.

6.4 Novel Approaches and Obtainable Products: Biotechnological Processes and Chemical Transformations

Major challenges faced by our society include the problems of waste accumulation and continuous depletion of fossil fuels and natural resources, which have led to rapid increases in the cost of raw materials, the threat of global warming, as well as development of expensive methods and recent legislations for proper disposal and treatment of food waste (Pfaltzgraff et al. 2013). Million of tonnes of residential waste (RW), commercial waste (CW), industrial waste (IW) and municipal solid waste (MSW) are produced around the world.

Of these, food waste has emerged as a potential substrate for production of various fuels, materials and chemicals due to its natural abundance, organic nature, low cost, and presence of various chemical components such as carbohydrates, proteins, and phenolics (Panesar et al. 2014). Increasing supply risks and rising prices of resources can best be countered by reusing our own outputs and reducing use of virgin materials (World Economic Forum and Ellen MacArthur Foundation 2014). In order to meet the increased demand for fuels, chemicals and energy, various approaches are being developed for the effective valorisation of food waste to produce variety of value-added products (Lin et al. 2013). Global consumerism and population growth are the main reasons for the increase in production of waste material. Therefore, advanced conversion methods with pollution control systems have been employed for the production of clean energy such as biogas (methane and CO_2), syngas (hydrogen and CO), liquid biofuel (ethanol and biodiesel), or pure hydrogen.

Two main approaches, biotechnological and chemical methods, are used for the production of various high-value products from food waste. With the chemical method, catalysts are used for the production of a variety of novel compounds, whereas with biotechnological methods, live organisms/microbes are used under specific conditions (Young 2010).

6.4.1 Chemical Transformations

Global consumerism and population growth are the main reasons for the increase in production of waste material which was traditionally disposed of by ineffective methods, such as landfilling, incineration, and composting (Young 2010). Landfilling has resulted in increased release of methane and greenhouse gases (GHGs). In remote areas, although incineration results in production of heat and energy, it also emits air pollutants which include dioxins and persistent organic pollutants (POPs), and it has a high operational cost (Uçkun et al. 2014). These types of waste disposal methods can cause severe environmental problems, such as release of methane which has 20–30 times more GHG potential than CO_2, leachate production, which can lead to soil and water contamination, and widespread growth of pathogenic microflora (Matsakas and Christakopoulos 2015; Matsakas et al. 2016). It can also result in the release of foul smells and odors, although it may be beneficial in soil for plant growth.

Therefore, advanced conversion methods with pollution control systems have been employed for the production of clean energy, such as biogas (methane and CO_2), syngas (hydrogen and CO), liquid biofuel (ethanol and biodiesel) or pure hydrogen. The current and outdated triple-R model (reduce, reuse, recycle) is being replaced by the cradle-to-cradle concept, in which waste is used as a raw material and recycled (McDonough and Braungart 2010).

Various thermochemical conversion processes such as gasification, pyrolysis and torrefaction have been performed at elevated temperatures, which result in faster conversion when compared with biochemical treatment processes (Figure 6.6). An old technology, gasification, has been used in oxygen-deprived conditions by using carbonaceous materials from organic wastes, such as household, municipal and agro industries, and resulting in production of syngas, which contains carbon monoxide and hydrogen. Using this technology, inorganic materials present in the waste are converted into slag or vitrified slag or ash (Young 2010). Some new approaches such as hydrothermal carbonization and plasma processing are also helpful to turn domestic food waste or high moisture waste into value-added products.

At this time, there is a great need for developed technology to maintain human health and welfare. Therefore, technologies have been developed for the conversion of domestic waste processes into green energy (Young 2010). Possible approaches for treating domestic food waste and giving valuable products are mentioned in the following sections.

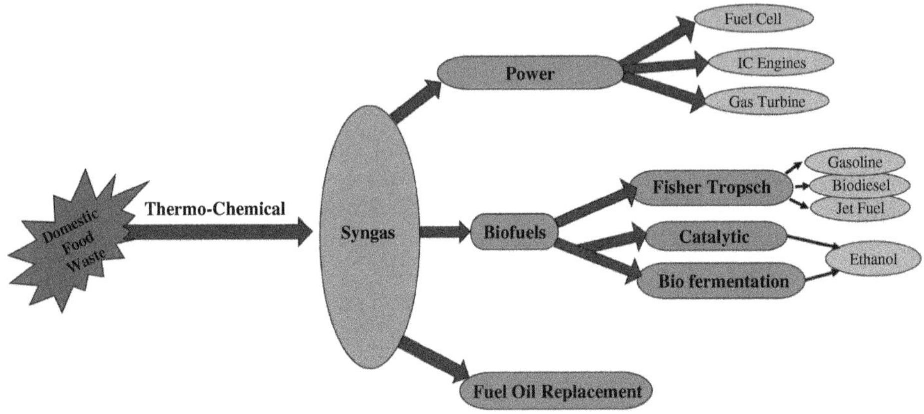

Figure 6.6 *Waste conversion and energy production from organic waste.*

6.4.1.1 Incineration

Mass burning or combustion is the oldest and most commonly practiced method for a variety of municipal or household waste conversion processes, with use of incinerators and hence requiring minimal processing conditions. Earlier models of incinerators produce high heat, toxic gases and other pollutants with little to no energy recovery, and they have been replaced by modern incinerators that have high heat recovery and air pollution control systems. Usually, a combustion temperature of 850°C is maintained, resulting in the production of CO_2, water, and non-combustible materials which remain as bottom ash containing residual carbon. However, a major problem with this combustion process is that ash and pollutant removal must take place while the incinerators are working (Ionescu et al. 2013). Furthermore, thermochemical methods for gasification and pyrolysis consist of product cooling and collection chambers, which can control organic and inorganic pollutant emissions.

6.4.1.2 Gasification

Gasification is a partial oxidation process that requires limited amounts of oxygen compared with complete stoichiometric combustion (incineration), where an oxygen-rich atmosphere is required (Basu 2013). Energy efficiency and environmental performance of municipal solid waste (MSW) gasification and incineration were compared by Panepinto et al. (2015), and it was suggested that the energy conversion efficiency can be improved by using combined-cycle gas turbines or by co-firing of syngas in large-capacity power plants. Partial oxidation of solid wastes takes place, resulting in conversion of wastes into combustible gases at high temperature (800–1200°C) to produce syngas, which consists of carbon monoxide, hydrogen and methane, or a mixture of gases known as gasifier agents. Certain undesirable components such as particulate matter, tar, alkali metals, chlorine and sulfide are also present (Lombardi et al. 2015). Gases emitted from this gasification method contain particulate matter, acidic gases and organic pollutants (dioxins). The particulate matters in these emissions can be controlled by electrostatic precipitators, bag filters, and by addition of slaked lime (Arena 2012). Various types of gasification reactor have been used, which include entrained flow, fluidized bed, cyclone and packed bed gasifiers (Klinghoffer and Castaldi 2013). Commercially available gasifiers are available with capacities of up to a few thousand tonnes per day (Higman 2014).

This method has various advantages over incineration, since the syngas produced can be used as clean fuel in conventional burners, coupled with a boiler or steam turbine. But this technology is now in the development stage, and the number of plants globally for producing syngas is limited (Arena 2012). In addition, there are some limitations of this method, such as formation of tar compounds which consist of mixtures of high molecular weight hydrocarbons, leading to blocking, fouling, and corrosion. Physical methods such as filters, scrubbers, and wet electrostatic precipitators can be used to remove tar (Vivanpatarakij and Assabumrungrat 2013), and chemical methods such as thermal and catalytic cracking have also been used (Campoy et al. 2014; Li et al. 2012). The promising method of syngas production by steam gasification reduces formation of tar and provides high yields of hydrogen without exothermic reactions (He et al. 2009).

Syngas yield is influenced by waste composition, particle size, temperature, and steam to feedstock ratio. Carbon conversion efficiency and tar decomposition were increased by using catalysts such as calcine and dolomite at elevated temperatures such as 950°C (Matsakas et al. 2017). Studies by Consonni and Viganò (2012) revealed that energy conversion efficiency can be improved if gasification is performed at pressures above atmospheric pressure.

In recent years, thermal plasma gasification has been an emerging technology; it has a rapid reaction rate, high flux density, and low oxidant amount. Studies have shown if the temperature of the plasma is increased, tar formation and other pollutants will be reduced in syngas production. Solid residues produced in the form of vitrified slag can be used in construction (Byun et al. 2010).

A gasification method using a mobile grille gasifier to process domestic waste without initial separation and grinding has been used in Brazil, and resulted in production of syngas (combination of H_2, CH_4, and CO) with lower amounts of emitted gases and metals than the permissible emissions according to the legislation in Brazil. Toxic metals are grouped as Class I (Cd, Hg, Tl). Therefore, the above method of gasification along with combustion of syngas leads to low pollution to the environment (Lopes et al. 2018).

Waste management and waste-to-energy conversion processes in Saudi Arabia nowadays are increasing at a rapid pace due to urbanization and a growing population. A model of gasification and anaerobic digestion conversion processes with financial indicators has been developed. It has been found that facility generation capacity, generated electricity revenue and capacity factor affects gasification (Hadidi and Omer 2017).

6.4.1.3 Pyrolysis

This process involves thermal degradation of organic waste like gasification; however, this process is performed in the absence of oxygen, which leads to the production of combustible gases. Solid waste consisting of coal, animal–human waste, scraps of food, paper, cardboard, plastics and rubber are suitable for pyrolysis. Pyrolytic oil produced from this process can be used as synthetic biodiesel for production of other useful products and fine chemicals. Degradation is conducted at temperatures between 300 and 650°C (Mohan et al. 2006) leading to the production of char and condensable gases (CO, CO_2, H_2, and CH_4). This technique can also be used for production of charcoal and coke from biomass and coal

(Chen et al. 2008; Maschio et al. 1992). The char produced has numerous applications, such as for energy production, soil amendment, and long-term carbon sequestration. Yield and chemical composition depend upon feedstock properties, temperature of reaction, and heating rate (Matsakas et al. 2017).

Pyrolysis can be divided into two types, slow pyrolysis and fast pyrolysis, based on the heating rate. Residence time in fast pyrolysis is only a few seconds, compared with several minutes in slow pyrolysis. In addition, the primary product in fast pyrolysis is bio-oil and gas, whereas it is char in slow pyrolysis (Basu 2013). Recently, a new concept of flash pyrolysis has been demonstrated for production of syngas (Williams and Barton 2011). Various models of integrated pyrolysis with microwave heating under vacuum conditions have been designed which have a shorter residence time, homogeneous heating, and high heat value (Yin 2012). The efficiency of this process depends upon structural arrangements, conductivity, and dielectric properties (Gedam and Regupathi 2012).

Pyrolysis plants on a commercial scale have been operated with a combination of gasification and combustion systems. Commonly used reactors are rotary kilns, tubular reactors, fixed bed and fluidized bed reactors (Lombardi et al. 2015). A recent study has shown that if the temperature is sufficiently increased, then it will have minor effects on the particle size (Luo et al. 2010). Liquid and char fractions consist of organic pollutants and heavy metals; however, with the use of catalysts, the amount of certain undesirable inorganic byproducts such as Cl, Br, and S can be reduced (Miskolczi et al. 2013).

An experimental pyrolysis plant for biofuel (bio-oil, biochar, and syngas) production from municipal green waste (MGW) feedstock using a short sealed rotary furnace was built in Rockhampton, central Queensland, Australia. It was found that 19.97% bio-oil, 40.83% biochar and 29.77% syngas was produced using MGW (Kabir et al. 2015).

Transport and industrial sorting of domestic waste resulted in an increase in costs, so an alternative method that can be more sustainable and produce less harmful byproducts was required. It has been found that, below 300°C, pyrolysis products contained biochar, bio-oil and syngas. Operating parameters, phase distribution, and chemical composition depends upon the feedstock material used. Analysis of components produced was performed by chromatography. High temperatures of pyrolysis resulted in an increased amount of valuable gases and bio-oils produced. Biochar produced can be used as a filter material for purification of water (Jouhara et al. 2018).

At an industrial level, pyrolysis is carried out in rotary kilns as it provides sufficient heat transfer and low energy consumption. Slow pyrolysis of wood, food, garden waste and paper is carried out at 400–500°C. Pyrolysis operates both with organic and inorganic wastes, and yield of gases and oil increases as temperature increases. Using this method, ready-to-use fuel can be produced easily and safely (Czajczyńska et al. 2017).

6.4.1.4 Torrefaction

The word "torrefaction" originated from a French word, which means roasting. It was initially used for treatment of coffee beans, in oxygen-rich environments (Basu 2013). This technique is described as a mild form (isothermal) of pyrolysis, which involves various step reactions including devolatilization, depolymerization, and carbonization performed at temperatures ranging between 200 and 300°C in an inert reduced environment. Types of material used as biomass include food waste, plastic, discarded tires and wood residues.

Yield of char decreases with increase in temperature (Matsakas et al. 2017). From this reaction, brown-black uniform solid residue, certain condensable compounds such as water, organics and lipids, and non-condensable gases like CO_2, CO, and CH_4 are produced (Tumuluru et al. 2012). Heavy metals such as Pb and Zn with high boiling points and dioxin-like compounds remain in char, but Hg with a low boiling point tends to enter the gas phase (Matsakas et al. 2017).

This technique can be classified as light (below 240°C) or severe (above 270°C). It leads to an improvement in physical, chemical and biochemical composition of biomass, such as in grindability, particle shape, size and distribution, palletability, proximate and ultimate composition, and calorific value. Char has been utilized as high-quality fuel, co-firing in power plants, entrained flow gasification, small-scale combustion facilities, water purification adsorbent, and for in-situ soil remediation purposes (Uslu et al. 2008; Basu 2013).

A torrefaction plant proposed by the Joint Institute for High Temperatures of the Russian Academy of Sciences (JIHT RAS) consisted of four zones: drying zone, preheating zone, torrefaction zone, and cooling zone, including a gas piston engine (GPE) (Kosov et al. 2014). Excess heat production was controlled by increasing the volume of combustion products through the torrefaction reactor (Zaichenko and Sytchev 2015). Russia accounts for 22% of total world's forests and 40% of the world reserves of peat. The torrefaction process is utilized there for forest sanitation, removal of sick and damaged trees, plants and wood waste from wood processing enterprises. Properties of the raw material and torrefaction temperature affect the properties of the final product formed. This work will help in replacement of fossil fuels, solid biofuels, or co-firing by using organic waste (Kuzmina et al. 2016).

Microwave-assisted torrefaction resulted in reduction of grinding energy consumption with an increase in microwave power (Iroba et al. 2017), which led to an improvement of about 29–70% and 30–68% in grinding energy savings for domestic waste such as municipal solid waste and grass clippings. Torrefaction results in a reduction in moisture absorption, which improves efficiency in processing, transportation, and fungal degradation. This method of torrefaction has become remarkable for converting waste to biofuel and electricity for homes in cold weather (Iroba et al. 2017).

Renewable bio-energy and value-added products have been developed using torrefaction. Lower emissions of polychlorinated di-benzo-p-dioxins (PCDD) and polychlorinated di-benzofurans (PCDF) was observed from processing solid waste in fuel blends. Char content has been reduced by chlorine migration to the gas phase. Trace metals catalytically remained active in char (Edo et al. 2017).

6.4.1.5 Hydrothermal Carbonization

This is a novel, wet, exothermic, low-temperature (180–350°C) conversion technique for treating domestic food waste which leads to the production of residual material with intrinsic value, while minimizing GHG production under autogenous pressure. Similar to pyrolysis and torrefaction, this reaction has been performed in the absence of oxygen (Libra et al. 2011). However, this technique has more advantages than pyrolysis and torrefaction, as it can be used for feedstocks containing moisture (Berge et al. 2011).

Different types of feedstocks with high moisture like domestic food waste can be used in this technique (Berge et al. 2011; Lin et al. 2016). Conversion of water into steam is avoided in this method as high pressure is maintained. Thus, the energy used to heat water is saved. Various reactions occur in the hydrothermal carbonization processes, such as hydrolysis, condensation, decarboxylation, and dehydration, resulting in the production of hydro-char.

Although hydro-char contains less aromatic carbon, it has more surface functional groups of hydrogen and oxygen, which increases its cation exchange capacity (Wiedner et al. 2013). This hydro-char can be used as an adsorbent of organic pollutants in water purification processes. At high temperature and pressure, water-insoluble organic chlorine is reduced in hydro-char due to conversion of organic chlorine to water-soluble chlorine (Prawisudha et al. 2012). Liquid formed in this reaction can easily be separated by mechanical or physical methods (Matsakas et al. 2017). If this hydrothermal carbonization technique is combined with pyrolysis or incineration, it can be effective in practice for waste remediation and energy recovery (Lin et al. 2016).

Hydrothermal carbonization of urban sludge is affected by temperature, mixing with or without catalysts, heavy metal content, turbidity, and COD value. It was concluded that with an increase in temperature from 180 to 300°C, yield of solid products decreases from 92.04 to 52.65%, while turbidity and COD decrease from 450 to 175 and 13 to 6.8 g/l, respectively. Hence, this will facilitate the conversion of solid waste to non-polluting energy-rich products (Xu and Jiang 2017).

Zhou et al. (2018) suggested a method using a symbiotic relationship between food waste (feedstock) and hydrothermal carbonization technology. This method, when combined with anaerobic digestion as a co-treatment method for food waste, helped in achieving maximum energy output, thereby increasing economic efficiency and reusing output.

6.4.1.6 Plasma Processing

This is a reliable, mature and well-known method for the production of plasma (Leal-Quirós 2004). This process has been used for the treatment of municipal waste, including solid organic–inorganic waste, biohazards, industrial, toxic, medical waste and nuclear waste, at atmospheric pressure and very high temperatures (5000–100 000°C). This process does not produce ash, due to disintegration of organic materials, and results in a mixture of H_2 and CO. Inorganic waste has been converted into solid-like glass materials, used in the construction industry (magmavication process), while syngas, fuel for heavy duty turbines, has been produced with organic materials such as plastics, paper, oil and biomaterials (gasification process) (Leal-Quirós 2004).

Green electric power has been generated from heavy duty turbines in an integrated plasma gasification combined cycle (IPGCC) method. In this method, heat generated from non-transferred electric plasma torch is used to gasify the municipal waste for production of syngas (Do and Letherman 2001). The syngas is then combusted in two simple cycle combustion turbines, leading to electricity production, which is exported to the electric grid.

6.4.2 Biotechnological Approaches

Biotechnological approaches have always been considered as an indispensible part of human life, and with advancements in this area, these techniques are being used by researchers and industries for the production of value-added products. Utilizing food waste as a substrate in biotechnological processes is considered as an innovative technique (Pleissner and Lin 2013). This concept of biotransformation of food waste into novel products is gaining popularity across the globe and is expected to significantly contribute in the proper utilization of food waste, while maintaining sustainable waste management techniques and valorisation processes (Jin et al. 2005). Biotechnological methods are considered most attractive for the bioconversion of food waste, as they are cost-effective, easy to control, have low energy requirements, and produce low volumes of sludge (Britz et al. 2008). Numerous products can be obtained from processing food wastes using biotechnological approaches, including single-cell proteins (SCPs), biofuels, organic acids, biopigments, and biopolymers.

The main biotechnological approaches for valorisation of food wastes include fermentation technology, enzymatic treatment, and biorefinery (Murugan et al. 2013). Table 6.2 depicts various products obtained by the valorisation of food wastes using biotechnological approaches.

6.4.2.1 Single-cell Proteins

Single-cell proteins (SCPs) are the proteins derived from dried microorganisms such as fungi, algae, bacteria, and yeast, which are cultivated on a variety of substrates and are further used as animal feed or human food (Litchfield 1983). SCPs serve as an excellent source of protein and are also rich in various other nutrients (Panesar et al. 2014). Food wastes, either solid or liquid, have proved to be a suitable substrate for the production of SCPs due to their low cost. Extensive research on the possibility of SCP production using food waste has been done by various researchers. Production of SCPs using whey (Moeini et al. 2004), lemon pulp and peels (De Gregorio et al. 2002) have been reported. Some studies have revealed the production of SCPs from wastes generated from fermentation industries such as breweries and distilleries (Hang 1979). *Saccharomyces cerevisiae* yeast has successfully been cultivated on beet molasses and SCP with approximately 70% protein content was produced (Bélafi-Bakó 2007). Apple pomace has been proven to be a suitable substrate for the cultivation of *Saccharomycopsis fibuligera* and *Candida utilis* to produce SCP (Fellows and Worgan 1987). *S. cerevisiae* was used for the production of SCP using pomegranate rind, banana skin and sweet orange peel as substrates (Adedayo et al. 2011). Whey, a byproduct from the dairy industry, can be used as a substrate for SCP production. Another research study reported using whey as a substrate for the isolation of *Kluyveromyces marxianus*, which resulted in successful production of SCP (Anvari and Khayati 2011). Kang et al. (2010) reported SCP production using *S. cerevisiae*, with papaya waste as a substrate. Rice polishings have been used as a substrate for the successful production of SCP inoculated with *Trichoderma harzianum* (Ahmed et al. 2017).

6.4.2.2 Biopolymers

Biopolymers are the polymers synthesized by various living organisms. Examples of biopolymers may include polyhydroxy alkanoates (PHAs) and polyesters of 1,3-propanediol (PD). Biopolymers are gaining in popularity because of their biodegradable nature and thermoplastic properties. Biopolymers are known for their wide range of applications, including use in paints, packaging, and adhesives, and also as a chiral substrate in the synthesis of drugs. Food wastes have been successfully used to produce

Table 6.2 *Utilization of different food wastes for various products.*

Food waste	Microorganism used	Product	Reference(s)
Single-cell protein			
Beet molasses	*Saccharomyces cerevisiae*	SCP	Bélafi-Bakó (2007)
Papaya waste	*Saccharomyces cerevisiae*	SCP	Kang et al. (2010)
Whey	*Kluyveromyces marxianus* A2	SCP	Anvari and Khayati (2011)
Wastewater from potato chip factory	*Streptomyces sp.*	SCP	Srisamai et al. (2013)
Deproteinized cheese whey	*Yarrowia lipolytica*	SCP	Taskin et al. (2015)
Olive mill wastewater	*Schwanniomyces etchelllsiii* M2 and *Candida pararugosa*	SCP	Arous et al. (2016)
Rice polishings	*Trichoderma harzianum*	SCP	Ahmed et al. (2017)
Biopolymers			
Wheat straw	*Gluconacetobacter xylinus*	Bacterial cellulose	Chen et al. (2013)
Cassava starch wastewater	*Cupriavidus sp.* KKU38	Polyhydroxyalkanoate	Poomipuk et al. (2014)
Wastewater of candied jujube	*Acetobacter xylinum*	Bacterial cellulose	Li et al. (2015)
Lignocellulosic food industry waste	*Burkholderia sacchari* DSM 17165	Polyhydroxybutyrate	Ravindran and Jaiswal (2016a,b)
Whey	*Haloferax mediterranei*	Poly-(3-hydroxybutyrate-co-3-hydroxyvalerate)	Pais et al. (2016)
Lignocellulose hydrolysates	*Pseudomonas* species	Polyhydroxyalkanoates	Diaz et al. (2018)
Organic acids			
Whey permeate	*Streptococcus lactis* and *Clostridium formicoaceticum*	Acetic acid	Hansen and Cheong (2007)
Whey	*Lactobacillus casei*	Lactic acid	Panesar et al. (2007a)
Banana peels	*Aspergillus niger*	Citric acid	Karthikeyan and Sivakumar (2010)
Waste bread	*Actinobacillus succinogenes*	Succinic acid	Leung et al. (2012)
Papaya peel	*Saccharomyces cerevisiae* and *Acetobacter acetii*	Acetic acid	Vikas and Mridul (2014)
Dried apricot waste	*Aspergillus foetidus*	Citric acid	Wadhwa et al. (2015)
Wheat bran	*Lactobacillus pentosus*	Lactic acid	Tirpanalan et al. (2015)

Table 6.2 (continued)

Food waste	Microorganism used	Product	Reference(s)
Apple pomace and peanut shell	*Aspergillus ornatus and Alternaria alternate*	Citric acid	Ali et al. (2016)
Citrus peel waste	*Actinobacillus succinogenes*	Succinic acid	Patsalou et al. (2017)
Spent coffee grounds	*Lactobacillus rhamnosus*	Lactic acid	Hudeckova et al. (2018)
Mixed fruit and vegetable waste hydrolysate	*Yarrowia lipolytica*	Succinic acid	Li et al. (2018)
Enzymes			
Apple pomace	*Aspergillus niger*	Pectinase	Joshi et al. (2006)
Grape must	*Kluyveromyces marxianus*	2-phenylethanol	Garavaglia et al. (2007)
Whey	*Kluyveromyces marxianus*	β-galactosidase	Panesar (2008)
Corn cob	*Escherichia coli*	Biovanillin	Torres et al. (2009)
Wheat bran and sugarcane bagasse	*Aspergillus oryzae*	Glucoamylase	Parbat and Singhal (2011)
Sugarcane bagasse+ wheat bran	*Trichoderma viride*	Cellulase	Lan et al. (2013)
Rice bran	*Cladosporium* sp.	Xylanase	Patel and Prajapati (2014)
Sugarcane bagasse	*Trichoderma viride*	Lactones (6-pentyl-α-pyrone)	Ladeira et al. (2010); Fadel et al. (2015)
Wheat bran	*T. reesei*	Cellulase	Idris et al. (2017)
Sugarcane bagasse and beet molasses	*Kluyveromyces marxianus*	Acetaldehyde, crotonaldehyde, ethanol, acetone, diacetyl, methyl acetate, butyl acetate	Martínez et al. (2017)
Corn bran	*Bacillus amyloliquefaciens*	α-amylase	Uygut and Tanyildizi (2018)
Rapeseed oil	*Saccharomyces cerevisiae*	2-phenylethanol	Chreptowicz and Mierzejewska (2018)
Biofuels			
Glucose and molasses	*Zymomonas mobilis*	Bioethanol	Panesar et al. (2007b)
Papaya waste	*Aspergillus niger* and *Saccharomyces cerevisiae*	Bioethanol	Akin-Osanaiye et al. (2008)
Whey	*Clostridium acetobutylicum*	Biobutanol	Foda et al. (2010)
Grape waste	*Saccharomyces cerevisiae*	Bioethanol	Raikar (2012)

(Continued)

Table 6.2 (continued)

Food waste	Microorganism used	Product	Reference(s)
Apple pomace	*Saccharomyces cerevisiae*	Bioethanol	Parmar and Rupasinghe (2013)
Rice straw	*Clostridium thermocellum* and *Clostridium saccharoperbutyl acetonicum*	Biobutanol	Kiyoshi et al. (2015)
Corn cob hydrolysate	*Candida lipolytica* and *Yarrowia lipolytica*	Biodiesel	Kahr et al. (2015)
Whey	*Aspergillus candidus*	Biodiesel	Kakkad et al. (2015)
Rice straw hydrolyzate	*Candida shehatae* TISTR 5843	Bioethanol	Yuvadetkun and Boonmee (2016)
Sugarcane molasses	*Torulaspora maleeae* and *Torulaspora globosa*	Biodiesel	Papone et al. (2016)
Apple pomace	*Clostridium beijerinckii* CECT 508	Biobutanol	Hijosa-Valsero et al. (2017)
Pineapple residues	*Saccharomyces bayanus* CECT 1926	Bioethanol	Gil and Maupoey (2018)

biopolymers. Various studies show that wastes and effluents from food industries, and molasses from the sugarcane industry, can be used in the production of biopolymers (Dahiya et al. 2017). According to Japanese research, production of PHAs from palm oil effluent is a two-stage process, where the anaerobic digestion of palm oil sludge generates organic acids in the first stage, and these produced acids are further converted into PHA by the phototrophic bacteria *Rhodobacter sphaeroides* in the second stage (Hassan et al. 1996). The *Pseudomonas* species is capable of metabolizing different sugars present in lignocellulosic hydrolysates of food industry wastes to successfully produce PHA (Diaz et al. 2018).

Meanwhile, 1,3-propanediol (PD) was developed from *Clostridium butyricum* (Papanikolaou et al. 2000) and *Klebsiella pneumonia* (Biebl et al. 1998) using glycerol. Bacterial cellulose has been produced from a variety of wastes such as konjac powder using *Acetobacteraceti xylinus* (Hong and Qiu 2008); fruit juices using *Acetobacter xylinum* (Kurosumi et al. 2009); thin stillage using *Gluconaacetobacter xylinus* (Wu et al. 2012), and wastewater from candied jujube using *A. xylinum* (Li et al. 2015).

A variety of fermented food wastes have been utilized for the production of polyhydroxybutyrate (PHB) using a pure culture of *Cupriavidus necator* (Hafuka et al. 2011). *Azohydromonas australica*, which can be easily cultivated on sucrose, has been successfully employed for the production of PHB (Gahlawat et al. 2012; Gahlawat and Srivastava 2013). Further enhancement in the concentration of PHB has been done by *A. australica*

DSM 1124 (Gahlawat and Srivastava 2018). *Burkholderia sacchari* DSM 17165 has been employed for the production of PHB because of its capability to metabolize glucose, xylose and other reducing sugars from lignocellulosic food industry wastes (Ravindran and Jaiswal 2016a,b). Development of poly-(3-hydroxybutyrateco-3-hydroxyvalerate) using whey as a substrate for the growth of *Haloferax mediterranei* has been successfully performed (Pais et al. 2016). Bacterial cellulose has been successfully produced utilizing bagasse as a substrate for the growth of *A. xylinum* (Cheng et al. 2016).

6.4.2.3 Organic Acids

Organic acids are water-soluble compounds, having acidic properties due to the presence of one or more carboxyl groups (Piero 2008). Examples of various organic acids may include citric acid, lactic acid, acetic acid, gluconic acid, and pyruvic acid. Organic acids are widely used in the food sector as preservatives, flavoring agents and acidulants, and can also be used to manufacture various products (Padmaja and Jyothi 2013). Microorganisms such as *Lactobacillus* sp. and *Aspergillus* sp., which are produced during the solid-state fermentation of different food processing wastes, are commonly used for the production of organic acids (Sauer et al. 2008). A variety of food processing wastes have been used as a substrate for the production of organic acids, including wheat bran, rice bran, residues of sweet potato, apple pomace, sugarcane bagasse, and mandarin orange waste (Marwaha et al. 2000). Citric acid is considered as one of the most versatile organic acids and has a wide range of applications in the food, chemical, pharmaceutical and beverage industries. *Aspergillus niger* has also been employed for the production of citric acid using banana peels as a substrate during Koji fermentation (Karthikeyan and Sivakumar 2010). Sugar beet molasses and chicken feather peptone (CFP) have been successfully utilized as a substrate for the production of citric acid using *A. niger* during submerged fermentation (Ozdal and Kurbanoglu 2018). Dried apricot waste has been reported to produce higher amounts of citric acid compared with other wastes such as apple pomace, rice bran or wheat bran during solid-state fermentation with *Aspergillus foetidus* ACM 3996 (Wadhwa et al. 2015).

Lactic acid is one of the most common organic acids extensively used in the food, textile, and pharmaceutical sectors. Lactic acid has been successfully produced using cane sugar molasses for the growth of *Lactobacillus delbrueckii* mutant Uc-3 in a batch fermentation process (Dumbrepatil et al. 2008). Wheat bran has been used as a substrate for the production of lactic acid while isolating *Lactobacillus pentosus* (Wadhwa et al. 2015). Waste potato was considered as an ideal substrate for the production of lactic acid using *Lactobacillus rhamnosus* (Djukić-Vuković et al. 2016). Lactic acid has been successfully produced using spent coffee grounds as a substrate using *L. rhamnosus* (Hudeckova et al. 2018).

Acetic acid, also known as vinegar, is a colorless organic acid mainly known for the production of various chemicals and esters. Vinegar has been successfully produced using orange peels and waste from the pineapple juice industry as a substrate (Ben-Gera and Kramer 1969). Acetic acid was produced using a mixture of apple pomace and molasses in a ratio of 2 : 1 supplemented with ammonium sulfate (Joshi et al. 1999). Whey has been used for the production of acetic acid, where *Saccharomyces* sp. was used for the fermentation followed by oxidation using *Acetobacter aceti* (Panesar et al. 2014). Acetic acid was successfully produced from *S. cerevisiae* and *A. aceti* using papaya peel as substrate (Vikas and Mridul 2014).

6.4.2.4 Enzymes

Production of enzymes is gaining popularity in biological sciences. Enzymes are known for their wide range of applications in industries such as the pulp and paper, food, textile, pharmaceutical, detergent, and leather industries. In order to utilize these enzymes at an industrial level, there is a need to produce enzymes in large amounts, thereby increasing the cost of industrial processes. Therefore, to reduce the cost, food waste is used as a substrate for the production of enzymes. Solid-state fermentation is preferred over submerged fermentation for enzyme production as it has been reported to be more economical (Pandey et al. 2000). Solid-state fermentation has been reported to reduce the industrial costs of cellulase production by 10-fold compared with submerged fermentation (Singhania et al. 2010).

Various fruit and vegetable waste such as apple pomace, orange pomace, orange peel, lemon peel, hazelnut shells, melon peel, and so on, have been evaluated as a substrate for the production of xylanase using *T. harzianum*, and among these wastes, melon peel was reported to produce enzymes with maximum enzymatic activity (Seyis and Aksoz 2005). Apple pomace was used for the production of pectin methylesterase (Joshi et al. 2006). Mango seed kernels were used to produce α-amylase by submerged fermentation using *Fusarium solani* (Kumar et al. 2013).

Molasses was successfully used as a substrate for the production of fumarase and aspartase using *Erwinia* species (Bagdasaryan et al. 2005). Enzyme tannase was successfully produced by *Aspergillus* and *Penicillium* species using barbados fruit and mangaba fruit as substrates (Lima et al. 2014). Whey was successfully used by various researchers for the production of β-galactosidase using *Candida pseudotropicalis, K. marxianus, K. fragilis* and *K. lactis* (Panesar 2008). A combination of agro-industrial wastes (rice bran: corncob: cassava bagasse) was used for the production of α-amylase using *Rhizopus microspores* TISTR 3531 (Trakarnpaiboon et al. 2017). Wastes from palm oil mills were successfully utilized for the production of cellulase and xylanase using *A. tubingensis* (Behera and Ray 2016).

Potato waste was utilized as a substrate for amylase production using *Bacillus amyloliquefaciens* (Abd-Elhalem et al. 2015). Wheat bran was successfully utilized for the production of cellulase using *T. reesei* during simultaneous saccharification and fermentation (SSF) (Idris et al. 2017). Food waste compost was used for the production of α-amylase using *Brevibacillus borstelensis* and *Bacillus licheniformis*, and cellulase using *Bacillus thuringiensis* and *B. licheniformis* (Awasthi et al. 2018).

6.4.2.5 Bioflavors

Flavors play an indispensible role in the food, pharmaceutical, chemical, and cosmetic industries as they determine the quality and acceptance of the product (Janssens et al. 1992). Bioflavors are compounds produced naturally that are responsible for imparting the desired aroma to the product. Bioflavors can be produced either by chemical synthesis or by biotechnological fermentation (Franco 2004). However, microbial fermentation is preferred over chemical synthesis as the bioconversion process is economical and easy, and the compounds produced have better consumer acceptance than chemically-produced compounds. Various microbial species such as *Ascoidea hylecoeti, Phellinus* sp., and *Cronartium fusiforme* have been used for the production of bioflavors (Francke and Bruemmer 1978). Production of bioflavors using food processing wastes such as citric pulp, coffee husk, and sugarcane bagasse has been reported (Bicas et al. 2010). Propionic acid, which is known for its raspberry and butter-type flavor, has been successfully produced using *Propionibacterium freudenreichii* with a combination of glucose and glycerol as a substrate (Wang and Yang 2013). Coffee husk has been considered as a potential substrate for the production of compounds with fruity smells using *Ceratocystis fimbriata* during SSF (Medeiros et al. 2006). Rapeseed oil has been demonstrated as an excellent substrate for the production of 2-phenylethanol, having a pleasant rose flavor, using *S. cerevisiae* strains (Chreptowicz and Mierzejewska 2018).

Fermented waste bread was reported as a potential substrate for the production of fruity aroma compounds using *Geotrichum candidum* ATCC 62217 (Daigle et al. 1999). A variety of volatile compounds have been produced using apple pomace as a substrate using *Rhizopus* sp. (Christen et al. 2000). Grape must be considered as an ideal substrate for the production of 2-phenylethanol using *K. marxianus* yeast (Garavaglia et al. 2007). Martínez et al. (2017), reported the production of variety of bioflavors including esters and alcohols utilizing sugarcane bagasse and beet molasses as substrates for growth of *K. marxianus.*

6.4.2.6 Biosurfactants

Biosurfactants are amphiphilic compounds responsible for reducing the surface tension and interfacial tension between individual molecules at the surface and interface, respectively. Biosurfactants are gaining popularity due to their diverse applications in various industries and their ability to eliminate issues of toxicity and environmental concerns (Clapés and Infante 2002). By utilizing microbial cells, various food industrial wastes such as plant-derived oils, oil waste, groundnut oil cake, wheat bran, rice bran, whey, and so on, can be used to produce biosurfactants (Panesar et al. 2011).

Rapeseed oil has been successfully used as substrate for the production of rhamnolipids using *Pseudomonas* sp. DSM 2874 (Trummler et al. 2003). A surface-active compound called rhamnolipid was successfully produced using a variety of food wastes such as orange peels, carrot peel waste, lime peels, and banana waste as substrates using *Pseudomonas aeruginosa* MTCC 2297 by submerged fermentation (George and Jayachandran 2009).

Whey has been successfully utilized as a substrate for the production of biosurfactants using *P. aeruginosa* BS2 (Dubey and Juwarkar 2004). Biosurfactant production has also been done utilizing potato processing effluent as a substrate using *Bacillus subtilis* ATCC 21332 (Noah et al. 2005). Glucose and soy oil were reported to produce mannosylerythritol lipid using *Candida* sp. SY16 yeast (Kim et al. 2006). Sugarcane bagasse and glycerol has been utilized to produce a biosurfactant using *Bacillus pumilus* UFPEDA448 (Slivinski et al. 2012). A variety of oil wastes including soyabean oil waste, babassu, palm, corn oil refinery waste and cottonseed waste have been utilized as a substrate for the production of rhamnolipids using *P. aeruginosa*. However, soyabean soapstock waste was reported as the best substrate, producing the highest quantity (11.7 g/l) of rhamnolipids (Nitschke et al. 2005). A potential biosurfactant was produced utilizing cassava wastewater as a substrate for the cultivation of *Candida tropicalis* UCP 1613 (Rubio-Ribeaux et al. 2017). Biosurfactant was produced by utilizing vineyard pruning waste as a substrate for the growth of *Lactobacillus paracasei*. Depending upon the substrate used and the process of production, the composition of biosurfactant changes, therefore, *L. paracasei* resulted in the production of glycopeptide biosurfactant using cellulosic sugars as a substrate, and produced glycoprotein biosurfactant using a lactose-based medium as a substrate (Vecino et al. 2017). Rhamnolipid was successfully produced using *P. aeruginosa* utilizing a variety of substrates such as clarified blackstrap molasses (Raza et al. 2014), a combination of corn steep liquor and molasses (Gudiña et al. 2015), kefir and fish meal (Kaskatepe et al. 2015), palm oil refinery waste (Radzuan et al. 2017), and olive mill waste (Ramírez et al. 2016). Vinasse and whey have been utilized as a substrate for the production of biosurfactants by *Lactococcus lactis* CECT-4434 (Vera et al. 2018). Martins and Martins (2018) reported the production of biosurfactants utilizing fish waste and sugarcane bagasse as a substrate for growth of *Corynebacterium aquaticum* and *Corynebacterium* sp. CTT 1968.

6.4.2.7 Biofuels

Stocks of fossil fuels are continuously dwindling in order to meet the demands of an increasing population, which poses a great threat to the environment. Therefore, in order to maintain the current levels of fuels and mitigate the increasing cost of these fuels, there is an urgent need for the development of sustainable and innovative strategies to meet these global targets (Lin et al. 2013). Extensive research is being done for the development of strategies with low environmental impact that rely on different food wastes as substrates for the production of biofuels via biotechnological methods.

Bioethanol is one of the most important biofuels, known for its wide range of applications in various industries. Ethanol production has been successfully reported using various food processing wastes such as banana peel, pineapple, and mangoes (Liguori et al. 2013). The fermentation of papaya waste to ethanol using *A. niger* and *S. cerevisiae* has been reported (Akin-Osanaiye et al. 2008). Successful production of ethanol using grape waste as a substrate was reported using a *S. cerevisiae* strain with an ethanol yield of 16% (Raikar 2012) and 95 g/l (Asli 2010). Ethanol can be produced utilizing rice straw hydrolyzate as a substrate using *C. tropicalis* Y-27290 (Swain and Krishnan 2015). Waste bread has been reported as a potential substrate for the production of bioethanol utilizing *S. cerevisiae* (Datta et al. 2018). María et al. (2017) reported the production of ethanol utilizing lignocellulosic residues including wood shavings, bagasse, corncob, and corn-leaves using *S. cerevisiae* in packed bed bioreactors. A byproduct of semolina milling, durum clear flours, has been reported for the successful production of bioethanol using *S. cerevisiae* (Sayaslan et al. 2018).

Biobutanol is an important fuel used for the production of various chemicals. Various microorganisms, mainly *Clostridium acetobutylicum* and *Clostridium beijerinckii*, are used for the production of biobutanol (Qureshi et al. 2006). Food wastes such as molasses, sago starch, wheat bran, palm oil waste, and various lignocellulosic materials such as wheat straw and cane molasses, have been reported as potential substrates for the production of biobutanol (Liguori et al. 2013). Whey permeate has been reported as a suitable substrate for the production of 2,3-butanediol using *K. pneumonia* during fermentation (Maddox et al. 1988). 1-butanol was successfully produced using apple pomace and waste from the bakery industry. However, production of butanol with 22 g/l was reported when supplementation of apple pomace was done with nitrogen source and minerals (Hegde et al. 2018). Apple pomace has been demonstrated as a potential substrate for the production of biobutanol by *C. beijerinckii* CECT 508 (Hijosa-Valsero et al. 2017).

6.4.2.8 Miscellaneous Products

(i) *Biopigments*: Biopigments are substances produced by living organisms which are known for imparting color resulting from selective color absorption. They play a vital role in imparting color to different food products. With the increase in health awareness among consumers, the focus has shifted from synthetic colors to biopigments as they can be produced throughout the year from natural ingredients (Carvalho et al. 2007). *Monascus*, *Penicillium*, *Rhodotorula*, and *Aspergillus* have been successfully grown on food waste for the production of a variety of biopigments (Panesar et al. 2015). Apple pomace has been used for the production of carotenoid using *Rhodotorula* sp. (Sandhu and Joshi 1997). Other research includes using apple pomace for the production of biopigments using *Micrococcus* and *Chromobacter* sp. (Attri and Joshi 2006). Utilization of brine from the fermented radish industry as a substrate has been reported for the production of β-carotene using *Rhodotorula glutinis* DM28 (Malisorn and Suntornsuk 2008). Jackfruit seed powder has been reported as a potential substrate for the production of a red-colored pigment using *Monascus ruber* during SSF (Kamalam et al. 2012). Mung bean waste flour and sweet potato extract has been utilized for the production of carotenoids using *R. glutinis* (Tinoi et al. 2005). Waste from the tomato processing industry has been utilized for the production of carotenoid using *R. glutinis* (Chandi et al. 2010). A variety of agro-industrial wastes such as banana stalks, coconut husk, rice husk and rice flour have been utilized for the production of melanin using *Streptomyces griseorubens* during SSF (Santhanalakshmi et al. 2017). Biopigments can be produced using rice, corn, sorghum grain and sorghum bran as a substrate using *Monascus purpureus* during SSF (Srianta et al. 2016). Astaxanthin, a red-colored carotenoid, was successfully produced from *Xantophylomyces dendrorhous* and *Sporidiobolus salmonicolor* using olive pomace as a substrate during SSF (Eryılmaz et al. 2016). Corn and sugarcane bagasse have been reported as a potential substrate for the production of pigments using *Monascus* sp. (Rajeswari et al. 2014).

(ii) *Amino acids*: Utilization of food wastes for the production of amino acids using biotechnological approaches has been successfully reported by researchers. Lysine has been produced by fermentation using waste generated from the fruit juice industry as a substrate (Sato 1992). Amino acids have been produced using citrus molasses as a substrate (Tsugawa et al. 1981; Joshi et al. 1999). Sugarcane bagasse has been successfully used as a substrate for the production of glutamic acid using *Brevibacterium* sp. in a solid-state fermentation technique (Nampoothiri and Pandey 1996).

6.5 Case Study: Production of Methane via Anaerobic Digestion of Food Waste

The rapid increase in urbanization and changing lifestyles, increase in population and rapid industrialization has led to the generation of huge amounts of domestic solid waste (DSW). Quantities of DSW produced depend upon food habits, standards of living, degree of commercial activities, and the season. In India, the amount of DSW is expected to increase significantly in forthcoming years (Sharma and Shah 2005; CPCB 2000; Shekdar et al. 1992). For India, per capita MSW generated has been estimated to increase at the rate of 1–1.33% annually (Pappu et al. 2007; Shekdar 1999; Bhide and Shekdar 1998).

In Malaysia, it has been reported that approximately 55% of biomass residues are derived from agricultural activities and the remaining 45% of residues are process-based or generated from mills (Griffin et al. 2014). With an increase in population and economic growth, global MSW generation rates will probably increase at the rate of 1.42 kg/capita/yr by 2025 (Hoornweg and Bhada-Tata 2012). In New Zealand, according to the Ministry for the Environment (2012), over 24 million tonnes of domestic, commercial, industrial and institutional waste has been disposed of in landfill sites. According to Al-Jarallah and Aleisa (2014), Kuwait has the highest domestic solid waste generation of 1.5 kg/person/d in the world.

Innovative mechanisms for disposal of waste in India are (i) composting, including aerobic and vermin-composting, and (ii) waste-to-energy (WTE), including incineration, palletization, and biomethanation.

6.5.1 Anaerobic Digestion

Anaerobic digestion is an effective method which utilizes anaerobic microorganisms to produce methane and carbon dioxide by burying DSW in pits. The leftover organic residue can be used as manure (Sharholy et al. 2007). This process is cost-intensive and produces low amounts of residual waste (Nasir et al. 2012; Morita and Sasaki 2012). There are three phases of anaerobic digestion: enzymatic hydrolysis, acid formation, and gas production. This method is promising and eco-friendly for waste conversion (Paritosh et al. 2017). Thermophilic digestion has been also been used commercially for biomethanation at a fast rate. It has been found that 1 tonne of DSW, by this method, produces 2–4 times as much methane within three weeks compared the amount that landfill can produce in six to seven years (Ahsan 1999; Khan 1994)

Western Paques (India) tested methane production under anaerobic conditions at a pilot scale. It was found that 14 000 m^3 of biogas with a methane content of 55–65% could be produced from 150 tonne/d DSW, resulting in generation of 1.2 MW of power. Various schemes of DSW utilization for methane production (Table 6.3) have been planned in some cities (Ambulkar and Shekdar 2004; Chakrabarty et al. 1995).

A report of the Global Methane Initiative (2016) shows that, in the past 25 years in Europe, hundreds of anaerobic digesters using domestic solid waste have been constructed to eradicate waste disposal problems and avoid high landfill fees. Several anaerobic digester plants have been developed in Canada and the US with capacities ranging from 25 000 to 130 000 tonnes per annum. It has been reported that Vancouver (Canada) became the greenest city in the year 2010 by diverting 108 600 tonnes of food waste to anaerobic digestion, while the remaining 184 000 tonnes per year will be diverted to reach a 2020 target (Davis et al. 2014). According to the United States Environmental Protection Agency (EPA) 2018 survey, there are 198 operational anaerobic digestion facilities and 34 non-operational facilities in the US (Pennington 2019). Furthermore, according to the data provided by the EPA (2020), there are 255 anaerobic digester projects operating on livestock farms in the US,

Table 6.3 List of various decentralized biomethanation plants in India using domestic waste.

State No.	State	Capacity/day	Year
1.	Andhra Pradesh		
	(a) Biogas plant based on starch industry wastes at Vensa Biotek, Samalkot, A.P.	0.70 MW eq.	1999–2000
	(b) Power generation from starch industry solid waste at M/s Vensa Biotek Ltd., E.G. District, Samalkot, A.P.	4.00 MW	2003–2004
	(c) Oil extraction industry waste-based power generation project by M/s Food Fats and Fertilizer Ltd. Tadepalligudem, Vest Godavari District, A.P.	6.00 MW	2006–2007
	(d) RDF Power Chinnaravulapally, Bibinagar (Nalgonda district), Hyderabad.	11 MW	2011
	(e) Selco, Gandamguda, Hyderabad	6.6 MW	1999–2009
	(f) Shriram Energy, Vijaywada	6 MW	2003–2008
	(g) Shalivahana (MSW) Green Energy Ltd., Karimnagar	12 MW	2010–present
2.	Delhi		
	(a) Timarpur, Delhi	3.75 MW	1986–1991
	(b) Jindal Ecopolis waste Management Company Pvt. Ltd. Okhla, Timarpur, Delhi	16 MW	2012–present
	(c) Integrated municipal waste-processing complex at Ghazipur, Delhi	10 MW	—
3.	Gujarat		
	(a) Biomethanation plant based on starch industry waste at Anil Starch, Ahmedabad, Gujarat	0.45 MW eq. (4800 m^3 biogas/d)	2001–2002
	(b) Starch industry waste-based biomethanation project by M/s Riddhi Siddhi Gluco Biols, Ltd., Riddhi Siddhi Nagar, Village – Juna Paddaar, Becharjee Road, Virmgam, Dist. Ahmedabad, Gujarat	0.458 MW eq. (5500 m^3 biogas/d)	2007–2008
	(c) Starch industry waste-based biogas to power (through 100% biogas engine) project by M/s Sayaji Industry Ltd., Ahmedabad, Gujarat	1.00 MW	2008–2009
	(d) MSW of Vadodara City, Gujarat	—	—
4.	Karnataka		
	(a) Biogas-based power project at Ugar Sugar Works, Belgaum and Karnataka	1.00 MW	1999–2000
	(b) Starch industry waste-based biomethanation project by M/s Riddhi Siddhi Gluco Boils Pvt. Ltd., Gokak, Karnataka	2.00 MWeq. (24 000 m^3 biogas/d).	2007–2008
	(c) Starch industry waste-based biomethanation project by M/s Riddhi Siddhi Gluco Boils Pvt. Ltd., Gokak, Karnataka	1.00 MWeq (12 000 m^3 biogas/d)	2008–2009
5.	Maharashtra		
	(a) Biogas-based power generation project at Brihan Sugar Syndicate Ltd., Sheerpur, Dist. Solapur, Maharashtra	1.00 MW	2000–2001
	(b) Seafood industry waste-based biomethanation project by M/s Gadre Marine Export Pvt. Ltd., Ratnagiri, Maharashtra	0.86 MW eq. (10 000 m^3 biogas/d)	2006–2007

(*Continued*)

Table 6.3 *(continued)*

State No.	State	Capacity/day	Year
	(c) Food processing industry liquid waste-based biomethanation project by M/s Saf Yeast Co. Pvt. Ltd., Ratnagiri, Maharashtra	0.36 MW eq. (4410 m^3 biogas/d)	2008–2009
	(d) Organic waste Recycling Systems Pvt. Ltd. Sholapur, Maharashtra	9.02 MW	2012–present
	(e) Rochem, Pune	10 MW	2012–present
6.	Punjab		
	(a) Starch industry waste-based biomethanation project by M/s Sukhjit Starch & Chemicals Ltd., Phagwara, Punjab	0.458 MW eq. (5500 m^3 biogas/d)	2006–2007
	(b) Biogas waste-based power project by M/s Chandigarh Distillers and Bottlers Ltd., Banur, Dist. Patiala, Punjab	8.25 MW	2007–2008
7.	Tamil Nadu		
	(a) Power generation project based on biogas from sago industry waste at Varalakshmi Starch Industry Ltd., Salem, Tamil Nadu	0.20 MW	2001–2002
	(b) Biogas generation project for tapioca processing industry wastewater at Varalaxmi Starch Industry Ltd., Salem, T.N.	0.50 MW eq. (6000 m^3 biogas/d)	2002–2003
	(c) Starch industry waste-based biomethanation project by M/s Varalakshmi Starch Industries Pvt. Ltd., Salem, T.N.	1.00 MW eq. (12 000 biogas m^3/d)	2006–2007
	(d) Starch industry liquid waste-based biomethanation project by M/s Varalakshmi Co., Mallur, Salem, District Namakkal	0.75 MW eq. (9000 m^3 biogas/d)	2008–2009
	(e) Tapioca starch industry liquid waste-based biomethanation project by M/s Spac Tapioca Products (India) Ltd., Poonachi Bhavani TK, Erode, Tamil Nadu	1.00 MW eq. (12 000 m^3 biogas/d)	2008–2009
8.	Uttar Pradesh		
	(a) Biogas-based power project at K.M. Sugar Mills at Faizabad, U.P.	1.00 MW	1997–1998
	(b) Biogas-based power generation project by M/s Pilkhani Distillery & Chemical Works, Pilkhani, District Saharanpur, U.P.	1.00 MW	2006–2007
	(c) Food industry waste-based biomethanation project by M/s Saf Yeast Co. Pvt. Ltd., 101, UPSIDC, Industrial Area, Sandila, District Hardoi, U.P.	0.73 MW eq. (8820 m^3 biogas/d)	2006–2007
	(d) Asia Bio-energy India Ltd., and Enkem Engrs Pvt. ltd., Lucknow	5 MW	2003
9.	Uttarakhand		
	(a) Starch industry liquid waste-based biomethanation project by M/s Riddhi Siddhi Gluco Biols, Udham Singh Nagar, Uttarakhand	1.52 MW eq. (18 300 m^3 biogas/d)	2007–2008
	(b) Starch industry liquid waste-based biomethanation project by M/s Gujarat Ambuja Exports Ltd., Udham Singh Nagar	1.08 MW eq. (12 960 m^3 biogas/d)	2008–2009

Source: From Krishania et al. (2013) and Dhar et al. (2017).

helping to reduce greenhouse gas emissions from methane. Of these, 107 projects have been awarded grants by the United States Department of Agriculture. In China and India, anaerobic digester plants are common in rural areas.

6.5.2 TEAM Digester for Domestic Food Waste Digestion

TERI (The Energy and Resource Institute) is an independent, multidimensional organization located in New Delhi (India). It works on the efficient utilization of Earth's resources, as well as innovative ways for the management and reuse of waste. The main goal of the organization is to enhance access to clean energy through renewable energy pathways, resulting in a transformative impact on industry as well as communities.

TERI's enhanced acidification and methanation (TEAM) digester was developed by the orgnisation. This plant had capacity for 50 kg domestic food waste per day, and was installed at TERI's Gual Pahari campus, Gurgaon, India. A TEAM digester was also installed at two other locations: NTPC India (for household waste management) and Sona Steering Ltd. Haryana, India (for canteen waste management). The TEAM digester has advantages over other reactors as it produces no slurry, has a very low water requirement due to water recycling, produces high energy and enriched manure, has a low maintenance cost, and is suitable for small and decentralized applications.

Organic solid waste is homogenized into small pieces before being fed into the digester tank. The TEAM digester has a two-phase process; first is the acidification stage, in which high organic strength liquid leachate extraction is performed in an acidification reactor. The second phase is methanation, in which biogas consisting of 50–60% methane is generated by treatment of leachate in a UASB reactor as shown in Figure 6.7.

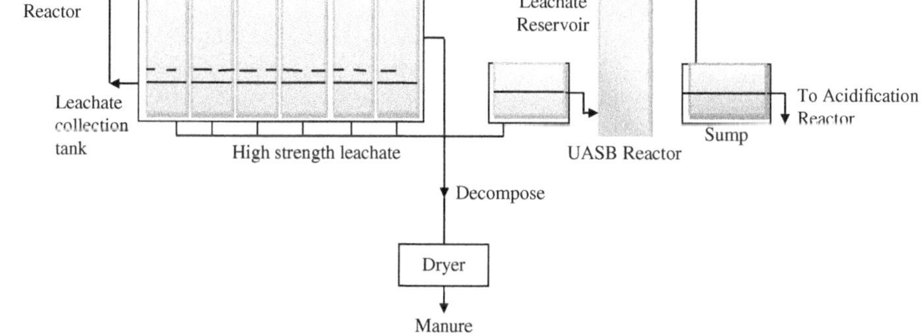

Figure 6.7 *The TEAM digester with six acidification reactors (acidification phase) and a UASB reactor (methanation phase).* Source: From Kishore and Pant (2009).

Its performance had been proven in a pilot plant since January 1999. Not only did it prevent the formation of scum, but the design was also simple and cost-effective, and required less water due to recycling. This model was suitable for small and decentralized applications. It generated large amounts of energy and a good amount of enriched manure. It was suitable for adaptation by small entrepreneurs and hence it had high market potential (Rajeshwari et al. 1998). Various domestic food wastes such as apple waste, pineapple waste, vegetable waste and mixed food waste were used for the generation of biogas, at 32.4, 13.77, 20, and 54 m^3/t respectively. TEAM digesters for treatment of various food wastes were installed in various states of India with capacities ranging from 50 to 1500 kg/d (Kishore and Pant 2009).

6.6 Conclusions and Future Trends

The rapid increase in population and industrialization has led to the development of numerous food processing industries, which result in the generation of considerable amounts of food waste. Food waste is considered to be a major environmental and societal concern. Due to the organic nature of these wastes and their poor handling/disposal problems, they pose a serious threat to the environment. Therefore, different strategies have been adopted for the proper management of food waste.

Currently, numerous strategies have been applied for the valorisation of food wastes that may include recycling, composting and related practices. However, efficient processing of these food wastes cannot be achieved by these conventional methods. There is a need for modern valorisation techniques for the proper treatment of food wastes, for recovery of nutrients, and their conversion into value-added products. These approaches include anaerobic digestion, chemical technologies that include advanced chemical separation methods, as well as biotechnological approaches that use enzymes and fermentation technologies. Since food waste is a rich source of valuable components, so the conversion of this waste into value-added products can not only reduce environmental problems but also improve the economics of the process, thereby contributing to sustainable development. Therefore, research has been carried out on the production of a variety of value-added products (biofuels, organic acids, biopolymers, enzymes, etc.) and the extraction and recovery of valuable components such as antioxidants and nutrients from these wastes.

Although various technologies have been applied for the valorisation of the food waste, a multidisciplinary approach is required in order to understand and utilize these processes efficiently. Effective and efficient systems must be developed for the proper management of food waste. Cross-industry and public–private collaborations along the food waste supply chain are required for the proper utilization of food waste as a potential raw material by industries to produce value-added products.

The success rate of recovery of valuable components and valorisation of food waste into value-added products depends upon the proper monitoring and control of the processes applied. Therefore, there is a need for the development of processes and monitoring devices specific for a particular product which can increase the yield of the desired product, and improve the efficiency of conventional valorisation methods used.

References

Abd-Elhalem, B.T., El-Sawy, M., Gamal, R.F., and Abou-Taleb, K.A. (2015). Production of amylases from *Bacillus amyloliquefaciens* under submerged fermentation using some agro-industrial by-products. *Ann. Agric. Sci.* **60** (2): 193–202.

Adedayo, M.R., Ajiboye, E.A., Akintunde, J.K., and Odaibo, A. (2011). Single cell proteins: as nutritional enhancer. *Adv. Appl. Sci. Res.* **2** (5): 396–409.

Ahmed, S., Mustafa, G., Arshad, M., and Rajoka, M.I. (2017). Fungal biomass protein production from *Trichoderma harzianum* using rice polishing. *BioMed Res. Int.*: 6232793.

Ahsan, N. (1999). Solid waste management plan for Indian megacities. *Indian J. Environ. Prot.* **19** (2): 90–95.

Akin-Osanaiye, B.C., Nzelibe, H.C., and Agbaji, A.S. (2008). Ethanol production from *Carica papaya* (pawpaw) fruit waste. *Asian J. Biochem.* **3** (3): 188–193.

Al Seadi, T., Drosg, B., Fuchs, W. et al. (2013). Biogas digestate quality and utilization. In: *The Biogas Handbook* (eds. A. Wellinger, J. Murphy and D. Baxter), 267–301. UK: Woodhead Publishing Limited.

Ali, S.R., Anwar, Z., Irshad, M. et al. (2016). Bio-synthesis of citric acid from single and co-culture-based fermentation technology using agro-wastes. *J. Radiat. Res. Appl. Sci.* **9** (1): 57–62.

Al-Jarallah, R. and Aleisa, E. (2014). A baseline study for municipal solid waste characterization for the State of Kuwait. *Waste Manage.* **34**: 952–960. http://dx.doi.org/10.1016/j.wasman.2014.02.015.

Ambulkar, A.R. and Shekdar, A.V. (2004). Prospects of bio-methanation technology in Indian context: a pragmatic approach. *J. Resour. Conserv. Recycl.* **40** (2): 111–128.

Ammar, A.S. (2013). Food processing wastes: characteristics, treatments and utilization. *J. Agri. Vet. Sci.* **7**: 71–84.

Angelidaki, I., Ellegaard, L., and Ahring, B.K. (2003). Applications of the anaerobic digestion process. In: *Biomethanation II* (eds. B.K. Ahring, I. Angelidaki, J. Dolfing, et al.), 1–33. Springer Berlin Heidelberg.

Anvari, M. and Khayati, G. (2011). Submerged yeast fermentation of cheese whey for protein production and nutritional profile analysis. *Adv. J. Food Sci. Technol.* **3** (2): 122–126.

Arena, U. (2012). Process and technological aspects of municipal solid waste gasification. A review. *Waste Manage.* **2012** (32): 625–639.

Arous, F., Azabou, S., Jaouani, A. et al. (2016). Biosynthesis of single-cell biomass from olive mill wastewater by newly isolated yeasts. *Environ. Sci. Pollut. Res.* **23** (7): 6783–6792.

Arsova, L. (2010). Anaerobic digestion of food waste: Current status, problems and an alternative product. Master's thesis. Columbia University, New York.

Arvanitoyannis, I.S. and Ladas, D. (2007). Meat waste management: treatment methods and potential uses of treated waste. In: *Waste Management for the Food Industries* (ed. I.S. Arvanitoyannis), 765–800. Burlington, MA, USA: Academic Press.

Asli, M.S. (2010). A study on some efficient parameters in batch fermentation of ethanol using *Saccharomyces cerevesiae* SC1 extracted from fermented siahe sardasht pomace. *Afr. J. Biotechnol.* **9** (20).

Attri, D. and Joshi, V.K. (2006). Optimization of apple pomace based pigment production medium and fermentation conditions for pigment production by *Chromobacter* species. *J. Food Sci. Technol.* **43**: 484–487.

Awasthi, M.K., Wong, J.W., Kumar, S. et al. (2018). Biodegradation of food waste using microbial cultures producing thermostable α-amylase and cellulase under different pH and temperature. *Bioresour. Technol.* **248**: 160–170.

Bagdasaryan, Z.N., Aleksanyan, G.A., Mirzoyan, A.M. et al. (2005). Stimulation of *Erwinia* sp. fumarase and aspartase synthesis by changing medium components. *Appl. Biochem. Biotechnol.* **125** (2): 113–126.

Bajpai, P. (2017). Basics of anaerobic digestion process. In: *Anaerobic Technology in Pulp and Paper Industry* (ed. P. Bajpai), 7–12. Singapore: Springer.

Bamelis, L., Blancke, S., Camargo-Valero, M.A., et al. (2015). *Techniques for nutrient recovery from digestate derivatives*. Project BIOREFINE: Recycling inorganic chemicals from agro- and bio-industrial waste streams. Technical Report. https://www.researchgate.net/publication/299510910_Techniques_for_nutrient_recovery_from_digestate_derivatives

Banks, C.J. and Wang, Z. (2006). Treatment of meat wastes. In: *Waste Treatment in the Food Processing Industry* (eds. L.K. Wang, Y.T. Hung, H.H. Lo and C. Yapijakis), 67–100. Boca Raton, FL: CRC Press.

Basu, P. (2013). *Biomass Gasification, Pyrolysis and Torrefaction – Practical Design and Theory*, 2e. San Diego, CA: Academic Press.

Behera, S.S. and Ray, R.C. (2016). Solid state fermentation for production of microbial cellulases: recent advances and improvement strategies. *Int. J. Biol. Macromol.* **86**: 656–669.

Bélafi-Bakó, K. (2007). Enzymatic extraction and fermentation for the recovery of food processing products. In: *Handbook of Waste Management and Co-Product Recovery in Food Processing*, vol. **1** (ed. K. Waldron), 198–216. Cambridge England: Woodhead Publishing Limited.

Ben-Gera, I. and Kramer, A. (1969). The utilization of food industries wastes. In: *Advances in Food Research*, vol. **17** (eds. E.M. Mark and G.F. Stewart), 77–152. Elsevier, Academic Press.

Berge, N.D., Ro, K.S., Mao, J. et al. (2011). Hydrothermal carbonization of municipal waste streams. *Environ. Sci. Technol.* **45**: 5696–5703.

Bhaskar, N., Modi, V.K., Govindaraju, K. et al. (2007). Utilization of meat industry by products: protein hydrolysate from sheep visceral mass. *Bioresour. Technol.* **98** (2): 388–394.

Bhaskaran, T.R. and Chakrabarty, R.N. (1966). Pilot plant for treatment of cane-sugar waste. *J. Water Pollut. Control Fed.* **38** (7): 1160–1169.

Bhide, A.D. and Shekdar, A.V. (1998). Solid waste management in Indian urban centers. *Int. Solid Waste Assoc. Times* **1**: 26–28.

Bicas, J.L., Silva, J.C., Dionísio, A.P., and Pastore, G.M. (2010). Biotechnological production of bioflavors and functional sugars. *Food Sci. Technol.* **30** (1): 07–18.

Biebl, H., Zeng, A.P., Menzel, K., and Deckwer, W.D. (1998). Fermentation of glycerol to 1, 3-propanediol and 2, 3-butanediol by *Klebsiella pneumoniae*. *Appl. Microbiol. Biotechnol.* **50** (1): 24–29.

Britz, T.J., van Schalkwyk, C., and Hung, Y.T. (2006). Treatment of dairy processing wastewaters. In: *Waste Treatment in the Food Processing Industry* (eds. L.K. Wang, Y. Hung, H.H. Lo and C. Yapijakis), 1–28. Boca Raton, FL: CRC Press, Taylor and Francis Group.

Britz, T.J., Lamprecht, C., and Sigge, G.O. (2008). Dealing with environmental issues. In: *Advanced Dairy Science and Technology* (eds. T.J. Britz and R.K. Robinson), 262–293. Oxford, UK: Blackwell Publishing Ltd.

Byun, Y., Namkung, W., Cho, M. et al. (2010). Demonstration of thermal plasma gasification/vitrification for municipal solid waste treatment. *Environ. Sci. Technol.* **44**: 6680–6684.

Campoy, M., Gomez, B.A., Ollero, P., and Nilsson, S. (2014). Gasification of wastes in a pilot fluidized bed gasifier. *Fuel Process. Technol.* **121**: 63–69.

Capodaglio, A.G., Hlavínek, P., and Raboni, M. (2015). Physico-chemical technologies for nitrogen removal from wastewaters: a review. *Rev. Ambiente Agua.* **10** (3): 481–498.

Carawan, R. E., Chambers, J. V., Zall, R. R., and Wilkowske, R. H. (1979). *Spin-off on seafood water and wastewater management*. Extension Special Report No-AM-18F. North Carolina State University, Cornell University and Purdue University, Raleigh, NC.

Carvalho, J.C., Oishi, B.O., Woiciechowski, A.L. et al. (2007). Effect of substrates on the production of Monascus biopigments by solid-state fermentation and pigment extraction using different solvents. *Indian J. Biotechnol.* **6**: 194–199.

Chakrabarty, P., Srivastava, V.K., and Chakrabarti, S.N. (1995). Solid waste disposal and the environment – a review. *Indian J. Environ. Prot.* **15** (1): 39–43.

Chandi, G.K., Singh, S.P., Gill, B.S. et al. (2010). Optimization of carotenoids by *Rhodotorula glutinis*. *Food Sci. Biotechnol.* **19** (4): 881–887.

Chandrappa, R. and Das, D.B. (2012). Waste quantities and characteristics. In: *Solid Waste Management* (eds. R. Chandrappa and D.B. Das), 47–63. Springer-Verlag: Berlin Heidelberg.

Chen, B., Zhou, D., and Zhu, L. (2008). Transitional adsorption and partition of nonpolar and polar aromatic contaminants by biochars of pine needles with different pyrolytic temperatures. *Environ. Sci. Technol.* **42** (14): 5137–5143.

Chen, L., Hong, F., Yang, X.X., and Han, S.F. (2013). Biotransformation of wheat straw to bacterial cellulose and its mechanism. *Bioresour. Technol.* **135**: 464–468.

Cheng, Z., Yang, R., and Liu, X. (2016). Production of bacterial cellulose by *Acetobacter xylinum* through utilizing acetic acid hydrolysate of bagasse as low-cost carbon source. *BioResources* **12** (1): 1190–1200.

Chreptowicz, K. and Mierzejewska, J. (2018). Enhanced bioproduction of 2-phenylethanol in a biphasic system with rapeseed oil. *New Biotechnol.* **42**: 56–61.

Christen, P., Bramorski, A., Revah, S., and Soccol, C.R. (2000). Characterization of volatile compounds produced by *Rhizopus* strains grown on agro-industrial solid wastes. *Bioresour. Technol.* **71** (3): 211–215.

Clapés, P. and Infante, M.R. (2002). Amino acid-based surfactants: enzymatic synthesis, properties and potential applications. *Biocatal. Biotransform.* **20** (4): 215–233.

Cocolo, G. (2013). Assessment of different solid–liquid separation techniques for livestock slurry. PhD thesis. Università degli Studi di Milano.

Consonni, S. and Viganò, F. (2012). Waste gasification vs. conventional waste-to-energy: a comparative evaluation of two commercial technologies. *Waste Manage.* **32**: 653–666.

CPCB. (2000). *Status of Municipal Solid Waste Generation, Collection, Treatment and Disposal in Class I Cities*. Series: ADSORBS/31/1999–2000. New Delhi: Central Pollution Control Board.

Czajczyńska, D., Anguilano, L., Ghazal, H. et al. (2017). Potential of pyrolysis processes in the waste management sector. *Therm. Sci. Eng. Prog.* **3**: 171–197.

Dahiya, S. and Joseph, J. (2015). High rate biomethanation technology for solid waste management and rapid biogas production: an emphasis on reactor design parameters. *Bioresour. Technol.* **188**: 73–78.

Dahiya, S., Kumar, A.N., Sravan, J.S. et al. (2017). Food waste biorefinery: sustainable strategy for circular bioeconomy. *Bioresour. Technol.*

Daigle, P., Gélinas, P., Leblanc, D., and Morin, A. (1999). Production of aroma compounds by *Geotrichum candidum* on waste bread crumb. *Food Microbiol.* **16** (5): 517–522.

Datta, P., Tiwari, S., and Pandey, L.M. (2018). Bioethanol production from waste breads using *Saccharomyces cerevisiae*. In: *Utilization and Management of Bioresources* (ed. S.K. Ghosh), 125–134. Springer: Singapore.

Davis, R.C, Mirza, F., Pope, D., and Breton, W. L. (2014). Anaerobic digestion: Pathways for using waste as energy in urban settings. https://sustain.ubc.ca/sites/default/files/2014-15_Using%20Waste%20as%20Energy%20in%20Urban%20Settings_Davis.pdf (accessed 25 April 2020).

De Gregorio, A., Mandalari, G., Arena, N. et al. (2002). SCP and crude pectinase production by slurry-state fermentation of lemon pulps. *Bioresour. Technol.* **83** (2): 89–94.

Demirel, B., Yenigun, O., and Onay, T.T. (2005). Anaerobic treatment of dairy wastewaters: a review. *Process Biochem.* **40** (8): 2583–2595.

Dhar, H., Kumar, S., and Kumar, R. (2017). A review on organic waste to energy systems in India. *Bioresour. Technol.* **245**: 1229–1237.

Di Iaconi, C., Pagano, M., Ramadori, R., and Lopez, A. (2010). Nitrogen recovery from a stabilized municipal landfill leachate. *Bioresour. Technol.* **101** (6): 1732–1736.

Diaz, A.B., Blandino, A., and Caro, I. (2018). Value added products from fermentation of sugars derived from agro-food residues. *Trends Food Sci. Technol.* **71**: 52–64.

Djukić-Vuković, A., Mladenović, D., Radosavljević, M. et al. (2016). Wastes from bioethanol and beer productions as substrates for l (+) lactic acid production – a comparative study. *Waste Manage.* **48**: 478–482.

Do, R.T. and Letherman, G. (2001). Renewable energy market: waste to energy utilizing plasma technology. In: *(Global Plasma Systems Corporation), Solena Presentation to the Annual Meeting of the Society of Women Engineers at PUPR*. Hato Rey, PR: Polytechnic University of Puerto Rico.

Drosg, B., Fuchs, W., Al Seadi, T. et al. (2015). Nutrient recovery by biogas digestate processing. *IEA Bioenergy*: 7–11.

Dubey, K. and Juwarkar, A. (2004). Determination of genetic basis for biosurfactant production in distillery and curd whey wastes utilizing *Pseudomonas aeruginosa* strain BS2. *Indian J. Biotechnol.* **3**: 74–81.

Dumbrepatil, A., Adsul, M., Chaudhari, S. et al. (2008). Utilization of molasses sugar for lactic acid production by *Lactobacillus delbrueckii* subsp. delbrueckii mutant Uc-3 in batch fermentation. *Appl. Environ. Microbiol.* **74** (1): 333–335.

Edo, M., Skoglund, N., Gao, Q. et al. (2017). Fate of metals and emissions of organic pollutants from torrefaction of waste wood, MSW, and RDF. *Waste Manage.* **68**: 646–652.

Environmental Protection Agency (2020). *AgSTAR data and trends March 2020*. Washington, DC: EPA https://www.epa.gov/agstar/agstar-data-and-trends (accessed 25 April 2020).

Eryılmaz, E.B., Dursun, D., and Dalgıç, A.C. (2016). Multiple optimization and statistical evaluation of astaxanthin production utilizing olive pomace. *Biocatal. Agric. Biotechnol.* **7**: 224–227.

Fadel, H.H.M., Mahmoud, M.G., Asker, M.M.S., and Lotfy, S.N. (2015). Characterization and evaluation of coconut aroma produced by *Trichoderma viride* EMCC-107 in solid state fermentation on sugarcane bagasse. *Electron. J. Biotechnol.* **18** (1): 5–9.

FAO (2011). *Global Food Losses and Food Waste: Extent Causes and Prevention*. Rome: Food and Agriculture Organization of the United Nations.

FAO (2014). *Definitional Framework of Food Loss*. Rome: Food and Agriculture Organization of the United Nations.

Fellows, P.J. and Worgan, J.T. (1987). Growth of *Saccharomycopsis fibuliger* and *Candida utilis* in mixed culture on apple processing wastes. *Enzyme Microb. Technol.* **9** (7): 434–437.

Foda, M.I., Dong, H., and Li, Y. (2010). Study the suitability of cheese whey for bio-butanol production by *Clostridia*. *J. Am. Sci.* **6** (8): 39–46.

Food Waste Reduction Alliance (FWRA) (2016). *Analysis of U.S. Food Waste among Food Manufacturers, Retailers, and Restaurants*. Washington, DC: Food Waste Reduction Alliance.

Francke, W. and Bruemmer, B. (1978). Terpenes from *Ascoidea hylecoeti*. *Planta Med.* **34** (08): 426–429.

Franco, M.R.B. (2004). *Aroma and Taste of Food: Current Issues*, 246. Varela.

Gahlawat, G. and Srivastava, A.K. (2013). Development of a mathematical model for the growth associated polyhydroxybutyrate fermentation by *Azohydromonas australica* and its use for the design of fed-batch cultivation strategies. *Bioresour. Technol.* **137**: 98–105.

Gahlawat, G. and Srivastava, A.K. (2018). Enhancing the production of polyhydroxyalkanoate biopolymer by *Azohydromonas australica* using a simple empty and fill bioreactor cultivation strategy. *Chem. Biochem. Eng. Q.* **31** (4): 479–485.

Gahlawat, G., Sengupta, B., and Srivastava, A.K. (2012). Enhanced production of poly (3-hydroxybutyrate) in a novel airlift reactor with in situ cell retention using *Azohydromonas australica*. *J. Ind. Microbiol. Biotechnol.* **39** (9): 1377–1384.

Garavaglia, J., Flôres, S.H., Pizzolato, T.M. et al. (2007). Bioconversion of L-phenylalanine into 2-phenylethanol by *Kluyveromyces marxianus* in grape must cultures. *World J. Microbiol. Biotechnol.* **23** (9): 1273–1279.

Gedam, V.V. and Regupathi, I. (2012). Pyrolysis of municipal solidwaste for syngas production by microwave irradiation. *Nat. Resour. Res.* **21**: 75–82.

George, S. and Jayachandran, K. (2009). Analysis of rhamnolipid biosurfactants produced through submerged fermentation using orange fruit peelings as sole carbon source. *Appl. Biochem. Biotechnol.* **158** (3): 694–705.

Gerardo, M.L., Zacharof, M.P., and Lovitt, R.W. (2013). Strategies for the recovery of nutrients and metals from anaerobically digested dairy farm sludge using cross-flow microfiltration. *Water Res.* **47** (14): 4833–4842.

Gil, L.S. and Maupoey, P.F. (2018). An integrated approach for pineapple waste valorisation. Bioethanol production and bromelain extraction from pineapple residues. *J. Cleaner Prod.* **172**: 1224–1231.

Global Methane Initiative (2016). *Anaerobic Digestion for Municipal Solid Waste*. https://globalmethane.files.wordpress.com/2016/08/ad-training-deck_dec2015-update1.pdf (accessed 25 April 2020).

Gong, H., Yan, Z., Liang, K.Q. et al. (2013). Concentrating process of liquid digestate by disk tube-reverse osmosis system. *Desalination* **326**: 30–36.

Gorsuch, T.T. (1986). *Food Processing Consultative Committee Report*. London: Ministry of Agriculture, Fisheries and Foods.

Griffin, W.M., Michalek, J., Matthews, H.S., and Hassan, M.N. (2014). Availability of biomass residues for co-firing in peninsular Malaysia: implications for cost and GHG emissions in the electricity sector. *Energies.* **7**: 804–823.

Gudiña, E.J., Fernandes, E.C., Rodrigues, A.I. et al. (2015). Biosurfactant production by *Bacillus subtilis* using corn steep liquor as culture medium. *Front. Microbiol.* **6**: 59.

Guo, W., Ngo, H.H., and Li, J. (2012). A mini-review on membrane fouling. *Bioresour. Technol.* **122**: 27–34.

Hack, P.J.F.M. (1985). Application of the UASB-reactor for anaerobic treatment of brewery effluent. *Water Sci. Technol.* **17** (8): 1489–1490.

Hadidi, L.A. and Omer, M.M. (2017). A financial feasibility model of gasification and anaerobic digestion waste-to-energy (WTE) plants in Saudi Arabia. *Waste Manage.* **59**: 90–101.

Hafuka, A., Sakaida, K., Satoh, H. et al. (2011). Effect of feeding regimens on polyhydroxybutyrate production from food wastes by *Cupriavidus necator*. *Bioresour. Technol.* **102** (3): 3551–3553.

Hang, Y.D. (1979). *Production of Single Cell Protein from Food Processing Waste*, 442–455. AVI, Westport, Conn.: Food Processing Management.

Hansen, C.L. and Cheong, D.Y. (2007). Fermentation, biogas and biohydrogen production from solid food processing. In: *Handbook of Waste Management and Co-Product Recovery in Food Processing* (ed. K. Waldron), 611–642. Boca Raton, FL, USA: CRC Press.

Hassan, M.A., Shirai, Y., Kusubayashi, N. et al. (1996). Effect of organic acid profiles during anaerobic treatment of palm oil mill effluent on the production of polyhydroxyalkanoates by *Rhodobacter sphaeroides*. *J. Ferment. Bioeng.* **82** (2): 151–156.

He, M., Hu, Z., Xiao, B. et al. (2009). Hydrogen rich gas from catalytic steam gasification of municipal solid waste (MSW): influence of catalyst and temperature on yield and product composition. *Int. J. Hydrogen Energy* **34**: 195–203.

He, S., Huang, Q., Zhang, Y. et al. (2015). Investigation on direct and indirect electrochemical oxidation of ammonia over Ru–Ir/TiO_2 anode. *Ind. Eng. Chem. Res.* **54** (5): 1447–1451.

Hegde, S., Lodge, J.S., and Trabold, T.A. (2018). Characteristics of food processing wastes and their use in sustainable alcohol production. *Renewable Sustainable Energy Rev.* **81**: 510–523.

Higman, C. (2014). State of the gasification industry: Worldwide Gasification Database 2014 update. In: *Gasification Technologies Conference, Washington, DC*. Sugar Land, TX: GSTC https://www.globalsyngas.org/uploads/downloads/GTC_Database_2014.pdf (accessed 13 April 2020).

Hijosa-Valsero, M., Paniagua-García, A.I., and Díez-Antolínez, R. (2017). Biobutanol production from apple pomace: the importance of pretreatment methods on the fermentability of lignocellulosic agro-food wastes. *Appl. Microbiol. Biotechnol.* **101** (21): 8041–8052.

Hjorth, M., Christensen, K.V., Christensen, M.L., and Sommer, S.G. (2011). Solid–liquid separation of animal slurry in theory and practice. In: *Sustainable Agriculture*, vol. **2** (eds. E. Lichtfouse, M. Hamelin, M. Navarrete and P. Debaeke), 953–986. Dordrecht: Springer.

Hong, F. and Qiu, K. (2008). An alternative carbon source from konjac powder for enhancing production of bacterial cellulose in static cultures by a model strain *Acetobacter aceti* subsp. xylinus ATCC 23770. *Carbohydr. Polym.* **72** (3): 545–549.

Hoornweg, D. and Bhada-Tata, P. (2012). *What a Waste: A Global Review of Solid Waste Management.* Washington, DC.: World Bank Group.

Hudeckova, H., Neureiter, M., Obruca, S. et al. (2018). Biotechnological conversion of spent coffee grounds into lactic acid. *Lett. Appl. Microbiol.* **66** (4): 306–312.

Idris, A.S.O., Pandey, A., Rao, S.S., and Sukumaran, R.K. (2017). Cellulase production through solid-state tray fermentation, and its use for bioethanol from sorghum stover. *Bioresour. Technol.* **242**: 265–271.

Ionescu, G., Rada, E.C., Ragazzi, M. et al. (2013). Integrated municipal solid waste scenario model (IMSW) using advanced pretreatment and waste to energy processes. *Energy Conversion and Management* **2013** (76): 1083–1092.

Iroba, K.L., Baik, O.D., and Tabil, L.G. (2017). Torrefaction of biomass from municipal solid waste fractions II: grindability characteristics, higher heating value, pelletability and moisture adsorption. *Biomass Bioenergy* **106**: 8–20.

Irshad, A., Sureshkumar, S., Raghunath, B.V. et al. (2016). Treatment of waste water from meat industry. In: *Integrated Waste Management in India* (eds. M. Prashanthi and R. Sundaram), 251–263. Switzerland: Springer International Publishing.

Jain, V.K. and Mishra, R.B. (1989). Methane from distillery effluent: experiments with two stage packed bed reactors. *J. Microb. Biotechnol.* **4** (1): 21–28.

Janssens, L., De Pooter, H.L., Schamp, N.M., and Vandamme, E.J. (1992). Production of flavours by microorganisms. *Process Biochem.* **27** (4): 195–215.

Jayathilakan, K., Sultana, K., Radhakrishna, K., and Bawa, A.S. (2012). Utilization of byproducts and waste materials from meat, poultry and fish processing industries: a review. *J. Food Sci. Technol.* **49** (3): 278–293.

Jin, B., Yin, P., Ma, Y., and Zhao, L. (2005). Production of lactic acid and fungal biomass by *Rhizopus* fungi from food processing waste streams. *J. Ind. Microbiol. Biotechnol.* **32** (11-12): 678–686.

Joshi, V.K. and Sharma, S.K. (2011). Food processing industrial waste: present scenario. In: *Food Processing Waste Management* (eds. V.K. Joshi and S.K. Sharma), 1–30. New Delhi: New India Publishing Agency.

Joshi, V.K., Pandey, A., and Sandhu, D.K. (1999). Fermentation technology for food industry waste utilization. In: *Biotechnology: Food Fermentation-Microbiology, Biochemistry and Technology*, vol. **2** (eds. V.K. Joshi and A. Pandey), 1291–1348. Delhi, India: Asiatech Publishers, Inc.

Joshi, V.K., Parmar, M., and Rana, N.S. (2006). Pectin esterase production from apple pomace in solid-state and submerged fermentations. *Food Technol. Biotechnol.* **44** (2): 253–256.

Jouhara, H., Ahmed, D., van den Boogaert, I. et al. (2018). Pyrolysis of domestic based feedstock at temperatures up to 300°C. *Therm. Sci. Eng. Prog.* **5**: 117–143.

Kabir, M.J., Chowdhury, A.A., and Rasul, M.G. (2015). Pyrolysis of municipal green waste: a modeling simulation and experimental analysis. *Energies.* **8**: 7522–7541.

Kahr, H., Pointner, M., Krennhuber, K. et al. (2015, 2015). Lipid production from diverse oleaginous yeasts from steam exploded corn cobs. *Agron. Res.* **13** (2): 318–327.

Kakkad, H., Khot, M., Zinjarde, S. et al. (2015). Conversion of dried *Aspergillus candidus* mycelia grown on waste whey to biodiesel by in situ acid transesterification. *Bioresour. Technol.* **197**: 502–507.

Kamalam, J.N., Anburaj, J., Kuberan, T. et al. (2012). DNA amplification and characterization of pigment producing gene from *Monascus ruber*. *Eur. J. Exp. Biol.* **2** (2): 427–435.

Kang, H.Y., Yang, P.Y., Dominy, W.G., and Lee, C.S. (2010). Bioprocessing papaya processing waste for potential aquaculture feed supplement – economic and nutrient analysis with shrimp feeding trial. *Bioresour. Technol.* **101** (20): 7973–7979.

Karthikeyan, A. and Sivakumar, N. (2010). Citric acid production by Koji fermentation using banana peel as a novel substrate. *Bioresour. Technol.* **101** (14): 5552–5556.

Karthikeyan, O.P., Mehariya, S., and Wong, J.W.C. (2017). Bio-refining of food waste for fuel and value products. *Energy Procedia* **136**: 14–21.

Kaskatepe, B., Yildiz, S., Gumustas, M., and Ozkan, S.A. (2015). Biosurfactant production by *Pseudomonas aeruginosa* in kefir and fish meal. *Braz. J. Microbiol.* **46** (3): 855–859.

Khan, R.R. (1994). Environmental management of municipal solid wastes. *Indian J. Environ. Prot.* **14** (1): 26–30.

Kim, H.S., Jeon, J.W., Kim, B.H. et al. (2006). Extracellular production of a glycolipid biosurfactant, mannosylerythritol lipid, by *Candida* sp. SY16 using fed-batch fermentation. *Appl. Microbiol. Biotechnol.* **70** (4): 391–396.

Kishore, V.V.N. and Pant, D.C. (2009). *Anaerobic Digesters in India*. TERI, India. https://www.globalmethane.org/expo-docs/india10/postexpo/ag_kishore.pdf (accessed 13 April 2020).

Kiyoshi, K., Furukawa, M., Seyama, T. et al. (2015). Butanol production from alkali-pretreated rice straw by co-culture of *Clostridium thermocellum* and *Clostridium saccharoperbutylacetonicum*. *Bioresour. Technol.* **186**: 325–328.

Klinghoffer, N.B. and Castaldi, M.J. (2013). *Gasification and Pyrolysis of Municipal Solid Waste. (MSW). Waste to Energy Conversion Technology*, 146–176. Woodhead Publishing Limited.

Kosov, V.V., Sinelshchikov, V.A., Sytchev, G.A., and Zaichenko, V.M. (2014). Effect of torrefaction on properties of solid granulated fuel of different biomass types. *High Temp.* **52**: 907–912.

Kosseva, M.R. (2013a). Sources, characterization, and composition of food industry wastes. In: *Food Industry Wastes: Assessment and Recuperation of Commodities* (eds. M.R. Kosseva and C. Webb), 37–60. San Diego, CA: Academic Press.

Kosseva, M.R. (2013b). Recent European legislation on management of wastes in the food industry. In: *Food Industry Wastes: Assessment and Recuperation of Commodities* (eds. M.R. Kosseva and C. Webb), 3–15. San Diego, CA: Academic Press.

Krishania, M., Kumar, V., Vijay, V.K., and Malik, A. (2013). Analysis of different techniques used for improvement of biomethanation process: a review. *Fuel* **106**: 1–9.

Kumar, D., Yadav, K.K., Muthukumar, M., and Garg, N. (2013). Production and characterization of [alpha]-amylase from mango kernel by *Fusarium solani* NAIMCC-F-02956 using submerged fermentation. *J. Environ. Biol.* **34** (6): 1053.

Kurosumi, A., Sasaki, C., Yamashita, Y., and Nakamura, Y. (2009). Utilization of various fruit juices as carbon source for production of bacterial cellulose by *Acetobacter xylinum* NBRC 13693. *Carbohydr. Polym.* **76** (2): 333–335.

Kuzmina, J.S., Sytchev, G.A., and Zaychenko, V.M. (2016). Torrefaction: prospects and application. *Chem. Eng. Trans.* **50**: 265–270.

La, J., Kim, T., Jang, J.K., and Chang, I.S. (2014). Ammonia nitrogen removal and recovery from swine wastewater by microwave radiation. *Environ. Eng. Res.* **19** (4): 381–385.

Ladeira, N.C., Peixoto, V.J., Penha, M.P. et al. (2010). Optimization of 6-pentyl-alpha-pyrone production by solid state fermentation using sugarcane bagasse as residue. *BioResources* **5** (4): 2297–2306.

Lan, T.Q., Wei, D., Yang, S.T., and Liu, X. (2013). Enhanced cellulase production by *Trichoderma viride* in a rotating fibrous bed bioreactor. *Bioresour. Technol.* **133**: 175–182.

Leal-Quirós, E. (2004). Plasma processing of municipal solid waste. *Braz. J. Phys.* **34** (4B): 1587–1593.

Lemmens, E., Ceulemans, J., Elslander, H. et al. (2007). *Best Available Techniques (BAT) for Animal Manure Processing*. Ghent, Belgium: Academia Press.

Leung, C.C.J., Cheung, A.S.Y., Zhang, A.Y.Z. et al. (2012). Utilisation of waste bread for fermentative succinic acid production. *Biochem. Eng. J.* **65**: 10–15.

Li, X., Li, L., Zheng, M. et al. (2009). Anaerobic co-digestion of cattle manure with corn stover pretreated by sodium hydroxide for efficient biogas production. *Energy Fuels* **23** (9): 4635–4639.

Li, R., Chen, S., and Li, X. (2010, 2010). Biogas production from anaerobic co-digestion of food waste with dairy manure in a two-phase digestion system. *Appl. Biochem. Biotechnol.* **160** (2): 643–654.

Li, J.F., Liao, S.Y., Dan, W.Y. et al. (2012). Experimental study on catalytic steam gasification of municipal solid waste for bioenergy production in a combined fixed bed reactor. *Biomass Bioenergy* **46**: 174–180.

Li, Z., Wang, L., Hua, J. et al. (2015). Production of nano bacterial cellulose from waste water of candied jujube-processing industry using *Acetobacter xylinum*. *Carbohydr. Polym.* **120**: 115–119.

Li, C., Yang, X., Gao, S. et al. (2018). Hydrolysis of fruit and vegetable waste for efficient succinic acid production with engineered *Yarrowia lipolytica*. *J. Cleaner Prod.* **179**: 151–159.

Liang, Y. and Hultin, H.O. (2003). Functional protein isolates from mechanically deboned turkey by alkaline solubilization with isoelectric precipitation. *J. Muscle Foods* **14** (3): 195–205.

Libra, J.A., Ro, K.S., Kammann, C. et al. (2011). Hydrothermal carbonization of biomass residuals: A comparative review of the chemistry, processes and applications of wet and dry pyrolysis. *Biofuels* **2**: 71–106.

Liguori, R., Amore, A., and Faraco, V. (2013). Waste valorisation by biotechnological conversion into added value products. *Appl. Microbiol. Biotechnol.* **97** (14): 6129–6147.

Lima, J.S.D., Cruz, R., Fonseca, J.C. et al. (2014). Production, characterization of tannase from *Penicillium montanense* URM 6286 under SSF using agroindustrial wastes, and application in the clarification of grape juice (*Vitis vinifera L.). Sci. World J.* **2014**: 1–9.

Lin, C.S.K., Pfaltzgraff, L.A., Herrero-Davila, L. et al. (2013). Food waste as a valuable resource for the production of chemicals, materials and fuels. Current situation and global perspective. *Energy Environ. Sci.* **6** (2): 426–464.

Lin, Y., Ma, X., Peng, X. et al. (2016). Combustion, pyrolysis and char CO_2-gasification characteristics of hydrothermal carbonization solid fuel from municipal solid wastes. *Fuel* **181**: 905–915.

Lipinski, B., Hanson, C., Lomax, J. et al. (2013). Reducing food loss and waste. *World Resour. Inst.* **22**.

Litchfield, J.H. (1983). Single-cell proteins. *Science* **219** (4585): 740–746.

Lombardi, L., Carnevale, E., and Corti, A. (2015). A review of technologies and performances of thermal treatment systems for energy recovery from waste. *Waste Manage.* **37**: 26–44.

Lopes, E.J., Queiroz, N., Yamamoto, C.I., and da Costa Neto, P.R. (2018). Evaluating the emissions from the gasification processing of municipal solid waste followed by combustion. *Waste Manage.* **73**: 504–510.

Luo, S., Xiao, B., Hu, Z. et al. (2010). Influence of particle size on pyrolysis and gasification performance of municipal solid waste in a fixed bed reactor. *Bioresour. Technol.* **101**: 6517–6520.

Maddox, I.S., Qureshi, N., and McQueen, J.M. (1988). Continuous production of 2, 3-butanediol from whey permeate using cells of *Klebsiella pneumoniae* immobilized on to bonechar. *N. Z. J. Dairy Sci. Technol.* **23** (2): 127–132.

Mahony, T., O'Flaherty, V., Colleran, E., et al. (2002). *Feasibility Study for Centralised Anaerobic Digestion for Treatment of Various Wastes and Wastewaters in Sensitive Catchment Areas*. Final Project Report. Ireland: Environmental Protection Agency.

Malisorn, C. and Suntornsuk, W. (2008). Optimization of β-carotene production by *Rhodotorula glutinis* DM28 in fermented radish brine. *Bioresour. Technol.* **99**: 2281–2287.

Marangon, F., Tempesta, T., Troiano, S., and Vecchiato, D. (2014). Food waste, consumer attitudes and behaviour. A study in the North-Eastern part of Italy. *Rivista di Economia Agraria/Italian Review of Agricultural Economics* **69** (2-3): 201–209.

María, A.E.L., Mussatto, S.I., Peñuela, M., and Teixeira, J.A. (2017). Physicochemical characterization of the yeast cells and lignocellulosic waste used in cell immobilization for ethanol production. In: *Yeast-Industrial Applications* (eds. A. Morata and I. Loira), 229–248. Croatia: InTech.

Martínez, O., Sánchez, A., Font, X., and Barrena, R. (2017). Valorisation of sugarcane bagasse and sugar beet molasses using *Kluyveromyces marxianus* for producing value-added aroma compounds via solid-state fermentation. *J. Cleaner Prod.* **158**: 8–17.

Martins, P.C. and Martins, V.G. (2018). Biosurfactant production from industrial wastes with potential remove of insoluble paint. *Int. Biodeterior. Biodegrad.* **127**: 10–16.

Marwaha, S.S., Arora, J.K., and Grover, R. (2000). Biomanagement of food industry waste. In: *Food Processing: Biotechnological Applications* (eds. S.S. Marwaha and J.K. Arora), 295–347. New Delhi, India: Asiatech Publishers.

Maschio, G., Koufopanos, C., and Lucchesi, A. (1992). Pyrolysis, a promising route for biomass utilization. *Bioresour. Technol.* **42**: 219–231.

Matsakas, L. and Christakopoulos, P. (2015). Ethanol production from enzymatically treated dried food waste using enzymes produced on-site. *Sustainability* **7**: 1446–1458.

Matsakas, L., Rova, U., and Christakopoulos, P. (2016). Strategies for enhanced biogas generation through anaerobic digestion of forest material – an overview. *BioResources* **11**: 5482–5499.

Matsakas, L., Gao, Q., Jansson, S. et al. (2017). Green conversion of municipal solid wastes into fuels and chemicals. *Electron. J. Biotechnol.* **26**: 69–83.

McDonough, W. and Braungart, M. (2010). *Cradle to Cradle: Remaking the Way we Make Things*. Eboo: MacMillan.

Medeiros, A.B., Pandey, A., Vandenberghe, L.P. et al. (2006). Production and recovery of aroma compounds produced by solid-state fermentation using different adsorbents. *Food Technol. Biotechnol.* **44** (1): 47–51.

Ministry for the Environment. (2012). Quantity of solid waste sent to landfill indicator update (INFO 654). New Zealand: Ministry for the Environment. https://www.mfe.govt.nz/publications/waste-environmental-reporting/quantity-solid-waste-sent-landfill-indicator-update (accessed 25 April 2020).

Mishra, N., Bakr, A.A.E.A., and Niranjan, K. (2006). Environmental aspects of food processing. In: *Food Processing Handbook* (ed. J.J. Brennan), 385–398. Weinheim, Germany: Wiley-VCH Verlag GmbH & Co. KGaA.

Miskolczi, N., Ateş, F., and Borsodi, N. (2013). Comparison of real waste (MSW and MPW) pyrolysis in batch reactor over different catalysts. Part II: contaminants, char and pyrolysis oil properties. *Bioresour. Technol.* **144**: 370–379.

Moeini, H., Nahvi, I., and Tavassoli, M. (2004). Improvement of SCP production and BOD removal of whey with mixed yeast culture. *Electron. J. Biotechnol.* **7** (3): 06–07.

Mohan, D., Pittman, C.U., and Steele, P.H. (2006). Pyrolysis of wood/biomass for bio-oil: a critical review. *Energy Fuels* **20**: 848–889.

Møller, H.B. (2001). Anaerobic digestion and separation of livestock slurries. Danish Experiences. In: *Manuscript for Matresa*, 2e. Horsens, Denmark: Danish Institute of Agricultural Sciences, Research Centre Bygholm.

Møller, H.B., Lund, I., and Sommer, S.G. (2000). Solid–liquid separation of livestock slurry: efficiency and cost. *Bioresour. Technol.* **74** (3): 223–229.

Møller, H.B., Sommer, S.G., and Ahring, B.K. (2002). Separation efficiency and particle size distribution in relation to manure type and storage conditions. *Bioresour. Technol.* **85** (2): 189–196.

Möller, K., Stinner, W., Deuker, A., and Leithold, G. (2008). Effects of different manuring systems with and without biogas digestion on nitrogen cycle and crop yield in mixed organic dairy farming systems. *Nutr. Cycling Agroecosyst.* **82** (3): 209–232.

Monfet, E., Aubry, G., and Ramirez, A.A. (2018). Nutrient removal and recovery from digestate: a review of the technology. *Biofuels* **9** (2): 247–262.

Morita, M. and Sasaki, K. (2012). Factors influencing the degradation of garbage in methanogenic bioreactors and impacts on biogas formation. *Appl. Microbiol. Biotechnol.* **94** (3): 575–582.

Murugan, K., Chandrasekaran, V.S., Karthikeyan, P., and Al-Sohaibani, S. (2013). Current state-of-the-art of food processing by-products. In: *Valorisation of Food Processing By-Products* (ed. M. Chandrasekaran), 35–62. Boca Ratón, FL, USA: CRC Press.

Nampoothiri, K.M. and Pandey, A. (1996). Solid state fermentation for L-glutamic acid production using *Brevibacterium* sp. *Biotechnol. Lett.* **18** (2): 199–204.

Nandy, T., Shastry, S., and Kaul, S.N. (2002). Wastewater management in a cane molasses distillery involving bioresource recovery. *J. Environ. Manage.* **65**: 25–38.

Nasir, M., Ghazi, T.I.M., and Omar, R. (2012). Production of biogas from solid organic wastes through anaerobic digestion: a review. *Appl. Microbiol. Biotechnol.* **95** (2): 321–329.

Nayono, S.E. (2009). Anaerobic digestion of organic solid waste for energy production. PhD Thesis. Universität Karlsruhe, Germany.

Nges, I.A., Wang, B., Cui, Z., and Liu, J. (2015). Digestate liquor recycle in minimal nutrients-supplemented anaerobic digestion of wheat straw. *Biochem. Eng. J.* **94**: 106–114.

Nishio, N. and Nakashimada, Y. (2007). Recent development of anaerobic digestion processes for energy recovery from wastes. *J. Biosci. Bioeng.* **103** (2): 105–112.

Nitschke, M., Costa, S.G., Haddad, R. et al. (2005). Oil wastes as unconventional substrates for rhamnolipid biosurfactant production by *Pseudomonas aeruginosa* LBI. *Biotechnol. Progr.* **21** (5): 1562–1566.

Noah, K.S., Bruhn, D.F., and Bala, G.A. (2005). Surfactin production from potato process effluent by *Bacillus subtilis* in a chemostat. In: *Twenty-Sixth Symposium on Biotechnology for Fuels and Chemicals* (ed. B.H. Davison), 465–473. Humana Press.

Otles, S., Despoudi, S., Bucatariu, C., and Kartal, C. (2015). Food waste management, valorisation, and sustainability in the food industry. In: *Food Waste Recovery* (ed. C.M. Galanakis), 3–23. San Diego, CA: Academic Press.

Ozdal, M. and Kurbanoglu, E.B. (2018). Citric acid production by *Aspergillus niger* from agro-industrial by-products: molasses and chicken feather peptone. *Waste Biomass Valorisation* **10**: 631–640.

Padmaja, G. and Jyothi, A.N. (2013). Roots and tubers. In: *Valorisation of Food Processing By-Products* (ed. M. Chandrasekaran), 377–454. Boca Raton: CRC Press.

Pais, J., Serafim, L.S., Freitas, F., and Reis, M.A. (2016). Conversion of cheese whey into poly (3-hydroxybutyrate-co-3-hydroxyvalerate) by *Haloferax mediterranei*. *New Biotechnol.* **33** (1): 224–230.

Pandey, A., Soccol, C.R., and Mitchell, D. (2000). New developments in solid state fermentation: I-bioprocesses and products. *Process Biochem.* **35** (10): 1153–1169.

Panepinto, D., Tedesco, V., Brizio, E., and Genon, G. (2015). Environmental performances and energy efficiency for MSW gasification treatment. *Waste Biomass Valorisation* **6**: 123–135.

Panesar, P.S. (2008). Production of β-D-Galactosidase from whey using *Kluyveromyces marxianus*. *Res. J. Microbiol.* **3** (1): 24–29.

Panesar, P.S., Rai, R., and Marwaha, S.S. (1999). Biological treatment of dairy industry wastewater. *Asian J. Microbiol., Biotechnol. Environ. Sci.* **1**: 67–72.

Panesar, P.S., Kennedy, J.F., Gandhi, D.N., and Bunko, K. (2007a). Bioutilisation of whey for lactic acid production. *Food Chem.* **105** (1): 1–14.

Panesar, P.S., Marwaha, S.S., and Kennedy, J.F. (2007b). Comparison of ethanol and temperature tolerance of *Zymomonas mobilis* strain in glucose and molasses medium. *Indian J. Biotechnol.* **2007** (6): 74–77.

Panesar, R., Panesar, P.S., and Bera, M.B. (2011). Development of low cost medium for the production of biosurfactants. *Asian J. Biotechnol.* **3** (4): 388–396.

Panesar, P.S., Marwaha, S.S., and Panesar, R. (2014). Value addition of agro-industrial wastes and residues. In: *Biotechnology in Agriculture and Food Processing: Opportunities and Challenges* (eds. P.S. Panesar and S.S. Marwaha), 557–605. Boca Raton, USA: CRC Press.

Panesar, R., Kaur, S., and Panesar, P.S. (2015). Production of microbial pigments utilizing agro industrial waste: a review. *Curr. Opin. Food Sci.* **1**: 70–76.

Papanikolaou, S., Ruiz-Sanchez, P., Pariset, B. et al. (2000). High production of 1, 3-propanediol from industrial glycerol by a newly isolated *Clostridium butyricum* strain. *J. Biotechnol.* **77** (2-3): 191–208.

Papone, T., Kookhunthod, S., Paungbut, M., and Leesing, R. (2016). Producing of microbial oil by mixed culture of microalgae and oleaginous yeast using sugarcane molasses as carbon substrate. *J. Clean Energy Technol.* **4** (4): 253–256.

Pappu, A., Saxena, M., and Asokar, S.R. (2007). Solid waste generation in India and their recycling potential in building materials. *J. Build. Environ.* **42** (6): 2311–2324.

Parbat, R. and Singhal, B. (2011). Production of glucoamylase by *Aspergillus oryzae* under solid state fermentation using agro industrial products. *Int. J. Microbiol. Res.* **2** (3): 204–207.

Paritosh, K., Kushwaha, S.K., Yadav, M. et al. (2017). Food waste to energy: an overview of sustainable approaches for food waste management and nutrient recycling. *BioMed Res. Int.* **2017**: 1–19.

Parmar, I. and Rupasinghe, H.V. (2013). Bio-conversion of apple pomace into ethanol and acetic acid: enzymatic hydrolysis and fermentation. *Bioresour. Technol.* **130**: 613–620.

Patel, K. and Prajapati, K. (2014). Xylanase production by *Cladosporium* sp. from agricultural waste. *Int. J. Curr. Res. Acad. Rev.* **2** (12): 84–90.

Patsalou, M., Menikea, K.K., Makri, E. et al. (2017). Development of a citrus peel-based biorefinery strategy for the production of succinic acid. *J. Cleaner Prod.* **166**: 706–716.

Pavlostathis, S.G. and Giraldo-Gomez, E. (1991). Kinetics of anaerobic treatment: a critical review. *Crit. Rev. Env. Sci. Technol.* **21** (5-6): 411–490.

Pennington, M. (2019). *Anaerobic Digestion Facilities Processing Food Waste in the United States.* EPA/903/S-19/001. https://www.epa.gov/sites/production/files/2019-09/documents/ad_data_report_v10_-_508_comp_v1.pdf (accessed 25 April 2020).

Pfaltzgraff, L.A., Cooper, E.C., Budarin, V., and Clark, J.H. (2013). Food waste biomass: a resource for high-value chemicals. *Green Chem.* **15** (2): 307–314.

Piero, R. (2008). Organic acids. In: *Laboratory Guide to the Methods in Biochemical Genetics* (eds. N. Blau, M. Duran and K. Michael Gibson), 137–170. Springer, Berlin.

Pleissner, D. and Lin, C.S.K. (2013). Valorisation of food waste in biotechnological processes. *Sustainable Chem. Processes* **1** (1): 21.

Poletti, A., Poletti, L., Santini, S., and Poletti, R. (2012). Renewable energy from ammonium rich anaerobic wastewaters and liquid sludges. *Virt&L-Comm.* **2**: 2–2012.

Poomipuk, N., Reungsang, A., and Plangklang, P. (2014). Poly-β-hydroxyalkanoates production from cassava starch hydrolysate by *Cupriavidus* sp. KKU38. *Int. J. Biol. Macromol.* **65**: 51–64.

Pratape, V.M., Singh, V., and Sashikala, V.B. (2011). Grain processing waste management. In: *Food Processing Waste Management* (eds. V.K. Joshi and S.K. Sharma), 53–72. New Delhi: New India Publishing Agency.

Prawisudha, P., Namioka, T., and Yoshikawa, K. (2012). Coal alternative fuel production from municipal solid wastes employing hydrothermal treatment. *Appl. Energy* **90**: 298–304.

Qureshi, N., Li, X.L., Hughes, S. et al. (2006). Butanol production from corn fiber xylan using *Clostridium acetobutylicum*. *Biotechnol. Progr.* **22** (3): 673–680.

Radzuan, M.N., Banat, I.M., and Winterburn, J. (2017). Production and characterization of rhamnolipid using palm oil agricultural refinery waste. *Bioresour. Technol.* **225**: 99–105.

Raikar, R.V. (2012). Enhanced production of ethanol from grape waste. *Int. J. Environ. Sci.* **3** (2): 776.

Rajeshwari, K.V., Pant, D.C., Lata, K., and Kishore, V.V.N. (1998). Studies on biomethanation of vegetable market waste. *Biogas Forum* **3**: 4–11.

Rajeswari, T.R., Ponnusami, V., and Sugumaran, K.R. (2014). Production of monascus pigment in low cost fermentation. *Int. J. Chem. Technol. Res.* **6**: 2929–2932.

Ramírez, I.M., Vaz, D.A., Banat, I.M. et al. (2016). Hydrolysis of olive mill waste to enhance rhamnolipids and surfactin production. *Bioresour. Technol.* **205**: 1–6.

Rathinaraj, K. and Sachindra, N.M. (2013). Meats, poultry and eggs. In: *Valorisation of Food Processing By-Products* (ed. M. Chandrasekaran), 617–648. Boca Ratón, FL, USA: CRC Press.

Ravindran, R. and Jaiswal, A.K. (2016a). A comprehensive review on pre-treatment strategy for lignocellulosic food industry waste: challenges and opportunities. *Bioresour. Technol.* **199**: 92–102.

Ravindran, R. and Jaiswal, A.K. (2016b). Exploitation of food industry waste for high-value products. *Trends Biotechnol.* **34** (1): 58–69.

Raza, Z.A., Ahmad, N., and Kamal, S. (2014). Multi-response optimization of rhamnolipid production using grey rational analysis in Taguchi method. *Biotechnol. Rep.* **3**: 86–94.

Romero-Güiza, M.S., Astals, S., Chimenos, J.M. et al. (2014). Improving anaerobic digestion of pig manure by adding in the same reactor a stabilizing agent formulated with low-grade magnesium oxide. *Biomass Bioenergy* **67**: 243–251.

Rubio-Ribeaux, D., da Silva Andrade, R.F., da Silva, G.S. et al. (2017). Promising biosurfactant produced by a new *Candida tropicalis* UCP 1613 strain using substrates from renewable-resources. *Afr. J. Microbiol. Res.* **11** (23): 981–991.

Russ, W. and Schnappinger, M. (2007). Waste related to the food industry: a challenge in material loops. In: *Utilization of By-Products and Treatment of Waste in the Food Industry* (eds. V. Oreopoulou and W. Russ), 1–13. New York: Springer.

Sandhu, D.K. and Joshi, V.K. (1997). Development of apple pomace medium, optimization of conditions for pigment production by *Rhodotorula*. *Adv. Food Res.* **19**: 31–34.

Santhanalakshmi, K., Natarajan, E., Muthukumar, B. et al. (2017). Fermentative production of melanin pigment from *Streptomyces griseorubens* DKR4 from agro waste products. *Int. J. Appl. Res.* **3** (3): 284–288.

Sato, A. (1992). Lysine manufacture from fruit waste. Japanese Patent 92-214004.

Sauer, M., Porro, D., Mattanovich, D., and Branduardi, P. (2008). Microbial production of organic acids: expanding the markets. *Trends Biotechnol.* **26** (2): 100–108.

Sayaslan, A., Koyuncu, M., Türker, S. et al. (2018). Use of durum wheat clear flour in vital gluten and bioethanol production. *J. Cereal Sci.* **80**: 50–56.

Schieber, A., Stintzing, F.C., and Carle, R. (2001). By-products of plant food processing as a source of functional compounds – recent developments. *Trends Food Sci. Technol.* **12** (11): 401–413.

Schoumans, O.F., Rulkens, W.H., Oenema, O., and Ehlert, P.A.I. (2010). *Phosphorus recovery from animal manure: Technical opportunities and agro-economical perspectives*. Alterra Report 2158. Wageningen, Netherlands: Wageningen UR.

Seyis, I. and Aksoz, N. (2005). Xylanase production from *Trichoderma harzianum* 1073 D 3 with alternative carbon and nitrogen sources. *Food Technol. Biotechnol.* **43** (1): 37–40.

Shah, S., Ramnan, V.V., and Sharma, S.K. (2011). Impact of food industry waste on environment. In: *Food Processing Waste Management: Treatment and Utilization Technology* (eds. V.K. Joshi and S.K. Sharma), 31–52. New Delhi, India: New India Publishing Agency.

Sharholy, M., Ahmad, K., Vaishya, R.C., and Gupta, R.D. (2007). Municipal solid waste characteristics and management in Allahabad, India. *J. Waste Manage.* **27** (4): 490–496.

Sharma, S. and Shah, K.W. (2005). Generation and disposal of solid waste in Hoshangabad. In: *Book of Proceedings of the Second International Congress of Chemistry and Environment*, 749–751. Indore, India.

Sheets, J.P., Yang, L., Ge, X. et al. (2015). Beyond land application: emerging technologies for the treatment and reuse of anaerobically digested agricultural and food waste. *Waste Manage.* **44**: 94–115.

Shekdar, A.V. (1999). Municipal solid waste management – the Indian perspective. *J. Indian Assoc. Environ. Manage.* **26** (2): 100–108.

Shekdar, A.V., Krshnawamy, K.N., Tikekar, V.G., and Bhide, A.D. (1992). Indian urban solid waste management systems – jaded systems in need of resource augmentation. *J. Waste Manage.* **12** (4): 379–387.

Singhania, R.R., Sukumaran, R.K., Patel, A.K. et al. (2010). Advancement and comparative profiles in the production technologies using solid-state and submerged fermentation for microbial cellulases. *Enzyme Microb. Technol.* **46** (7): 541–549.

Slivinski, C.T., Mallmann, E., de Araújo, J.M. et al. (2012). Production of surfactin by *Bacillus pumilus* UFPEDA 448 in solid-state fermentation using a medium based on okara with sugarcane bagasse as a bulking agent. *Process Biochem.* **47** (12): 1848–1855.

Speece, R.E. (1983). Anaerobic biotechnology for industrial wastewater treatment. *Environ. Sci. Technol.* **17** (9): 416A–427A.

Srianta, I., Zubaidah, E., Estiasih, T., and Yamada, M. (2016). Comparison of *Monascus purpureus* growth, pigment production and composition on different cereal substrates with solid state fermentation. *Biocatal. Agric. Biotechnol.* **7**: 181–186.

Srisamai, S., Srikhampa, P., and Pathom-Aree, W. (2013). Production of probiotic streptomyces biomass from starchy wastewater. *Chiang Mai J. Sci.* **40**: 294–298.

Swain, M.R. and Krishnan, C. (2015). Improved conversion of rice straw to ethanol and xylitol by combination of moderate temperature ammonia pretreatment and sequential fermentation using *Candida tropicalis*. *Ind. Crops Prod.* **77**: 1039–1046.

Taghizadeh-Alisaraei, A., Hosseini, S.H., Ghobadian, B., and Motevali, A. (2017). Biofuel production from citrus wastes: a feasibility study in Iran. *Renew. Sust. Energ. Rev.* **69**: 1100–1112.

Tambone, F., Scaglia, B., D'Imporzano, G. et al. (2010). Assessing amendment and fertilizing properties of digestates from anaerobic digestion through a comparative study with digested sludge and compost. *Chemosphere* **81** (5): 577–583.

Tambone, F., Terruzzi, L., Scaglia, B., and Adani, F. (2015). Composting of the solid fraction of digestate derived from pig slurry: biological processes and compost properties. *Waste Manage.* **35**: 55–61.

Taskin, M., Saghafian, A., Aydogan, M.N., and Arslan, N.P. (2015). Microbial lipid production by cold-adapted oleaginous yeast *Yarrowia lipolytica* B9 in non-sterile whey medium. *Biofuel. Bioprod. Biorefin.* **9** (5): 595–605.

Teglia, C., Tremier, A., and Martel, J.L. (2011). Characterization of solid digestates: part 2, assessment of the quality and suitability for composting of six digested products. *Waste Biomass Valorisation* **2** (2): 113–126.

Thakur, N.S., Joshi, V.K., and Slathia, S. (2011). Fruit and vegetable processing waste management. In: *Food Processing Waste Management* (eds. V.K. Joshi and S.K. Sharma), 73–96. New Delhi: New India Publishing Agency.

Tikariha, A. and Sahu, O. (2014). Study of characteristics and treatments of dairy industry waste water. *J. Appl. Environ. Microbiol.* **2**: 16–22.

Tinoi, J., Rakariyatham, N., and Deming, R.L. (2005). Simplex optimization of carotenoid production by *Rhodotorula glutinis* using hydrolyzed mung bean waste flour as substrate. *Process Biochem.* **40** (7): 2551–2557.

Tirpanalan, Ö., Reisinger, M., Smerilli, M. et al. (2015). Wheat bran biorefinery – an insight into the process chain for the production of lactic acid. *Bioresour. Technol.* **180**: 242–249.

Tobiassen, A. (2016). Socio-demographic determinants and health related behaviors associated with sustainable food related choices. Master's thesis. Universitetet i Agder, Norway.

Torres, B.R., Aliakbarian, B., Torre, P. et al. (2009). Vanillin bioproduction from alkaline hydrolyzate of corn cob by *Escherichia coli* JM109/pBB1. *Enzyme Microb. Technol.* **44** (3): 154–158.

Trakarnpaiboon, S., Srisuk, N., Piyachomkwan, K. et al. (2017). Enhanced production of raw starch degrading enzyme using agro-industrial waste mixtures by thermotolerant *Rhizopus microsporus* for raw cassava chip saccharification in ethanol production. *Prep. Biochem. Biotechnol.* **47** (8): 813–823.

Trummler, K., Effenberger, F., and Syldatk, C. (2003). An integrated microbial/enzymatic process for production of rhamnolipids and L-(+)-rhamnose from rapeseed oil with *Pseudomonas* sp. DSM 2874. *Eur. J. Lipid Sci. Technol.* **105** (10): 563–571.

Tsugawa, R., Nakamura, A., Takeyama, T., et al. (1981). *Production of purified citrus molasses for the raw materials of amino acid fermentation.* Abstract at International Citrus Congress, Tokyo, Japan, 2, 893-896.

Tumuluru, J.S., Hess, J.R., Boardman, R.D. et al. (2012). Formulation, pretreatment, and densification options to improve biomass specifications for co-firing high percentages with coal. *Ind. Biotechnol.* **8** (3): 113–132.

Uçkun, K.E., Trzcinski, A.P., Ng, W.J., and Liu, Y. (2014). Bioconversion of food wastes to energy: a review. *Fuel* **134**: 389–399.

Uslu, A., Faaij, A.P.C., and Bergman, P.C.A. (2008). Pre-treatment technologies, and their effect on international bioenergy supply chain logistics. Techno-economic evaluation of torrefaction, fast pyrolysis and pelletisation. *Energy* **33**: 1206–1223.

Uygut, M.A. and Tanyildizi, M.Ş. (2018). Determination of effective parameters for alpha-amylase production in a modified rotating drum bioreactor. *Arabian J. Sci. Eng.* **43**: 3381–3391.

Uysal, A., Yilmazel, Y.D., and Demirer, G.N. (2010). The determination of fertilizer quality of the formed struvite from effluent of a sewage sludge anaerobic digester. *J. Hazard. Mater.* **181** (1-3): 248–254.

Vaneeckhaute, C., Lebuf, V., Michels, E. et al. (2017). Nutrient recovery from digestate: systematic technology review and product classification. *Waste Biomass Valorisation* **8** (1): 21–40.

Vecino, X., Rodríguez-López, L., Gudiña, E.J. et al. (2017). Vineyard pruning waste as an alternative carbon source to produce novel biosurfactants by *Lactobacillus paracasei*. *J. Ind. Eng. Chem.* **55**: 40–49.

Vera, E.C.S., de Azevedo, P.O.D.S., Domínguez, J.M., and de Souza Oliveira, R.P. (2018). Optimization of biosurfactant and bacteriocin-like inhibitory substance (BLIS) production by *Lactococcus lactis* CECT-4434 from agroindustrial waste. *Biochem. Eng. J.* **133**: 168–178.

Vikas, O.V. and Mridul, U. (2014). Bioconversion of papaya peel waste in to vinegar using *Acetobacter aceti*. *Int. J. Sci. Res.* **3** (11): 409–411.

Vivanpatarakij, S. and Assabumrungrat, S. (2013). Thermodynamic analysis of combined unit of biomass gasifier and tar steam reformer for hydrogen production and tar removal. *Int. J. Hydrogen Energ.* **38**: 3930–3936.

Wadhwa, M., Bakshi, M.P.S., and Makkar, H.P.S. (2015). Wastes to worth: value added products from fruit and vegetable wastes. *CAB Int.*: 1–25.

Wang, Z. and Yang, S.T. (2013). Propionic acid production in glycerol/glucose co-fermentation by *Propionibacterium freudenreichii* subsp. *shermanii*. *Bioresour. Technol.* **137**: 116–123.

Weiland, P. (2006). State of the art of solid-state digestion – recent developments. In: *Solid-State Digestion – State of the Art and Further R&D Requirements*, vol. **24** (ed. F.N. Rohstoffe), 22–38. Gulzower Fachgespräche.

Wiedner, K., Rumpel, C., Steiner, C. et al. (2013). Chemical evaluation of chars produced by thermochemical conversion (gasification, pyrolysis and hydrothermal carbonization) of agro-industrial biomass on a commercial scale. *Biomass Bioenergy* **59**: 264–278.

Williams, P.T. (2005). *Waste Treatment and Disposal*, 9. England: Wiley.

Williams, P.T. and Barton, J. (2011). Demonstration scale flash pyrolysis of municipal solid waste. *Proc. Inst. Civ. Eng.* **164**: 205–210.

World Economic Forum and Ellen MacArthur Foundation (2014). *Towards the Circular Economy*, Accelerating the Scale-up Across Global Supply Chains. Geneva: World Economic Forum.

Wu, F., Zhang, Y., Liu, L., and Yao, J. (2012). Synthesis and characterization of a novel cellulose-g-poly (acrylic acid-co-acrylamide) superabsorbent composite based on flax yarn waste. *Carbohydr. Polym.* **87** (4): 2519–2525.

Wu, B., Zhang, X., Xu, Y. et al. (2015). Assessment of the energy consumption of the biogas upgrading process with pressure swing adsorption using novel adsorbents. *J. Cleaner Prod.* **101**: 251–261.

Xu, X. and Jiang, E. (2017). Treatment of urban sludge by hydrothermal carbonization. *Bioresour. Technol.* **238**: 182–187.

Xu, F., Li, Y., Ge, X. et al. (2018). Anaerobic digestion of food waste – challenges and opportunities. *Bioresour. Technol.* **247**: 1047–1058.

Yang, L., Xu, F., Ge, X., and Li, Y. (2015). Challenges and strategies for solid-state anaerobic digestion of lignocellulosic biomass. *Renew. Sust. Energ. Rev.* **44**: 824–834.

Yin, C. (2012). Microwave-assisted pyrolysis of biomass for liquid biofuels production. *Bioresour. Technol.* **120**: 273–284.

Young, G.C. (2010). *Municipal Solid Waste to Energy Conversion Processes: Economic, Technical, and Renewable Comparisons*. Wiley.

Yuvadetkun, P.R. and Boonmee, M. (2016). Ethanol production capability of *Candida shehatae* in mixed sugars and rice straw hydrolysate. *Sains Malaysiana* **45** (4): 581–587.

Zaichenko, V.M. and Sytchev, G.A. (2015). Energy recycling of wood waste. *Energetik* **10**: 48–50.

Zhou, Y., Engler, N., and Nelles, M. (2018). Symbiotic relationship between hydrothermal carbonization technology and anaerobic digestion for food waste in China. *Bioresour. Technol.* **260**: 404–412.

7

Bioconversion of Processing Waste from Agro-Food Industries to Bioethanol: Creating a Sustainable and Circular Economy

Deepak Kumar[1] **and Vijay Singh**[2]

[1] *Department of Chemical Engineering, State University of New York College of Environmental Science and Forestry, Syracuse, NY, USA*
[2] *Agricultural and Biological Engineering Department, University of Illinois at Urbana-Champaign, Urbana, IL, USA*

7.1 Introduction

Increasing energy demand, limited fossil-fuel reserves, and environmental concerns associated with production and use of fossil fuels necessitate the identification of alternative renewable and green energy resources. In 2018, the United States consumed a record high of 101.3 quadrillion Btu of energy. About a third of the total energy used in the US is consumed only in the transportation sector, and 92% of this is derived from petroleum (70% of total petroleum use) (EIA 2019). Bioethanol, a liquid fuel produced from the fermentation of sugars, is considered as the most promising renewable transportation fuel. Ethanol can be blended with gasoline at low concentrations, such as E10 (10% ethanol v/v), and used in the existing gasoline vehicles, or at high concentrations, such as E85 (85% ethanol

Waste Valorisation: Waste Streams in a Circular Economy, First Edition.
Edited by Carol Sze Ki Lin, Guneet Kaur, Chong Li and Xiaofeng Yang.

v/v), to use in modified vehicles, also known as "flexible fuel vehicles." Ethanol has a high blending octane rating (a measure of its ability to resist "knocking" in the engine) of 114, which increases the octane rating of E10 mix to 87, compared with the octane rating of 84 for regular gasoline (RFA 2019; Zabed et al. 2017). With 16.1 billion gallons of ethanol produced in 2018, the US is the biggest producer of bioethanol in the world (56% of total world production). The numbers of ethanol plants have increased from 56 (total production capacity 1.78 billion gal) to 210 (total production capacity 16.5 billion gal) in the last two decades (RFA 2019). Brazil is the second major ethanol producer (28%) in the world, and produced 7.95 billion gal in 2018. The European Union, China, and Canada produced 5%, 4%, and 2% of total ethanol production in world, respectively. Bioethanol plays a critical role in the US nation's dependence on oil and petroleum imports and its energy security. In the absence of 16.1 billion gallons of ethanol production, the US dependence on oil imports could increase from 14% to 20% (an additional 594 million barrels of imported crude oil) (RFA 2019). As the ethanol production from corn starch using the dry grind process produces 40–45% less greenhouse gas (GHG) emissions compared with gasoline, use of ethanol-blend gasoline in the US resulted in a decrease of approximately 55 million metric tonnes of CO_2-equivalent GHG emissions (equivalent to taking 11.7 million cars from the road for an entire year) (RFA 2019; Wang et al. 2012a).

Even with extensive research efforts, only a limited number of feedstocks have been found to be successful on a commercial scale. Among the different feedstocks (corn starch, sorghum starch, cellulosic biomass, food waste), corn starch is used to produce more than 94% of total ethanol in the US, mostly using the dry grind ethanol process explained later in this chapter (Kumar and Singh 2019a,b). Feedstock purchase cost is observed to be the major contributor to the total production cost. Moreover, use of food crops like corn or sugarcane create "food vs. fuel" issues, require intensive agriculture inputs and arable land, and have limited capacity to replace a large amount of gasoline use. These challenges with first-generation fuels have led to a continuous surge in interest in biofuel production from non-food sources, such as agricultural residues, energy crops, forestry wastes, agro-food processing industry wastes, and municipal solid wastes. Agricultural wastes from the field (e.g., corn stover, wheat straw, etc.) or agro-food processing industry (e.g., citrus peels, spent grains, etc.) are abundant, non-food, and low-value sources for bioethanol production. This chapter will focus on processing waste from agro-food industries, and technologies to produce bioethanol using these wastes. Agro-food industries generate several million tonnes of various wastes and low-value co-product streams every year during their operations. Traditionally, most of these streams are either sold as low-value product (mainly as animal feed) or are disposed of at landfills along with municipal solid wastes. Disposal of this waste in a landfill is high operating cost to food industries, and at the same time causes potential environmental problems, with direct and indirect emissions of GHGs, and contamination of groundwater (Huang et al. 2015; Jin et al. 2018; Zhang et al. 2016). Several wastes from agro-processing industries are rich in carbohydrates, protein, and other value compounds (Table 7.1). Use of agro-food processing waste for biofuel production provides a low-value raw material as well as solving the environmental issues. As these wastes are a mixture of starch, cellulose, hemicellulose, pectin, essential oils, proteins and other constituents, efficient bioconversion and utilization of these wastes requires a mix of the first-generation and second-generation ethanol production technology platforms.

Table 7.1 *Chemical composition of agro-industrial wastes (dry basis).*

Feedstock	Cellulose	Hemicellulose	Lignin	Starch	Free sugars	Protein	Others	References
Citrus peel waste (mix)	22	11	2.19	—	23	6	25% pectin	Pourbafrani et al. (2010)
Lemon peel	22.8	22.4	8.3	—	6.5	7	9% pectin; 1.5% fat	Boluda-Aguilar and López-Gómez (2013)
Mandarin peel	20.8	17.2	8.9	—	10.1	7.5	14.2% pectin; 1.6% fat	Boluda-Aguilar et al. (2010)
Orange peel	22.0	19.9	8.4		9.6	9.1	12.3% pectin; 4% fat	Boluda-Aguilar et al. (2010)
Sweet lime peel	25.4	9.4	23.6	—	—	—	17.1% pectin	John et al. (2017)
Potato peel	33.5	5.5	4.7	42.0–52.14	1.0	8	2.6% fat	Ben Taher et al. (2017), Arapoglou et al. (2010)
Cassava peel	6.37	6.27	6.7	44.48	—	—	—	Aruwajoye et al. (2019)
Soybean hull	38.4	10.2	2.8	—	—	10.7	—	Mielenz et al. (2009)
Olive pulp	9.6	12.5	22.8	—	8.3	—	—	Georgieva and Ahring (2007)
Coffee husk	16	11	9	—	—	8.24	3.5% oil	Gouvea et al. (2009)
Barley spent Grains	14.4–18.2	25.1–34.21	3.93–27.8	—	—	15.2–21.2	8.6% fat	Liguori et al. (2015), Mussatto and Roberto (2005), Wilkinson et al. (2014), Rojas-Chamorro et al. (2018)
Grape pomace	8.04	8.2	44.46	—	1.42	12.68	—	Jin et al. (2018)
Mill feed (pasta waste)	—	—	—	19	—	19	7% oil; 32% NDF	Kumar and Singh (2019b)
Germ meal	—	—	—	14%	—	26	5% oil; 42% NDF	Kumar and Singh (2019b)
DDGS	13.5–16	13.5–16.4	—	4.7–5.2	—	24.9–31.1	—	Kim et al. (2010)

NDF, neutral detergent fiber.

7.2 Bioconversion Technologies for Bioethanol Production

Bioconversion technologies to produce ethanol vary significantly depending on the type of feedstock. Sugar-based feedstocks, such as sugarcane, are easy to process and require mainly an extraction step to obtain a sugar solution, which can then be fermented to ethanol. Since the yeast, the most commonly used fermentation microorganism in the ethanol industry, cannot directly ferment carbohydrate polymers presented in starchy and lignocellulosic feedstocks, these polymers are first hydrolyzed to sugar monomers, which are subsequently fermented into ethanol. Due to the structural and compositional differences among various feedstocks, the sugar production process differs for starchy and lignocellulosic feedstocks. In the case of starch-based feedstocks, starch can be directly hydrolyzed to glucose in one or multiple steps, whereas a pretreatment step is necessary in the case of lignocellulosic biomass to gain access to cellulose polymer before hydrolysis (Figures 7.1 and 7.2). The processes for ethanol production from both starch- and cellulose-based feedstocks are discussed in the following sections.

7.2.1 Ethanol Production from Starchy Feedstock (First-Generation Bioethanol)

Corn is the major starch-based feedstock used for ethanol production, and bioconversion of corn to ethanol is a mature technology. Dry grind and wet milling are the two conventional

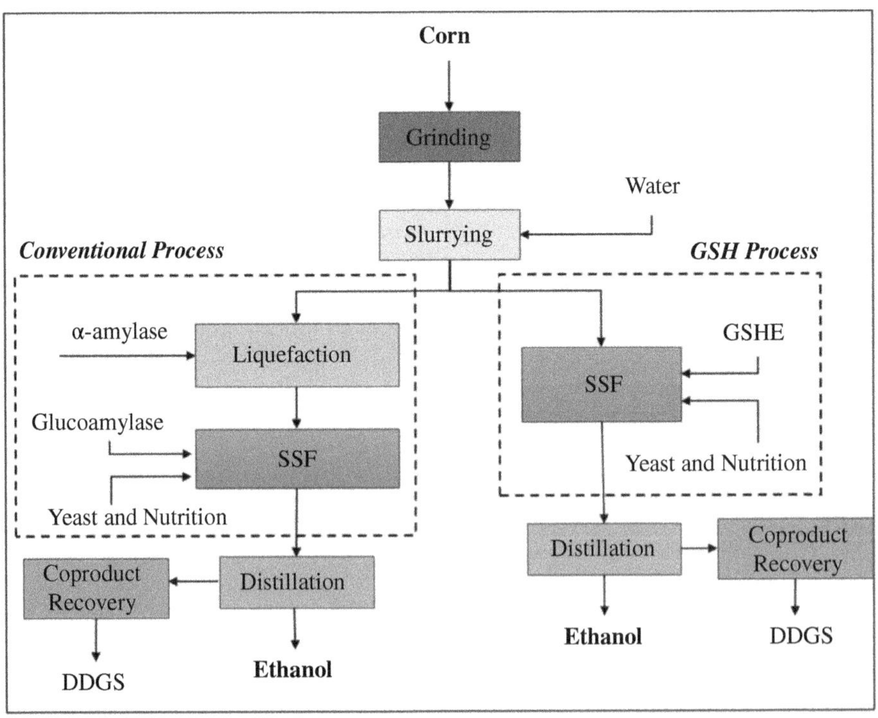

Figure 7.1 Schematic of the dry grind ethanol production process from corn. Source: Adapted from Juneja et al. (2017), reproduced with permission of Springer Nature.

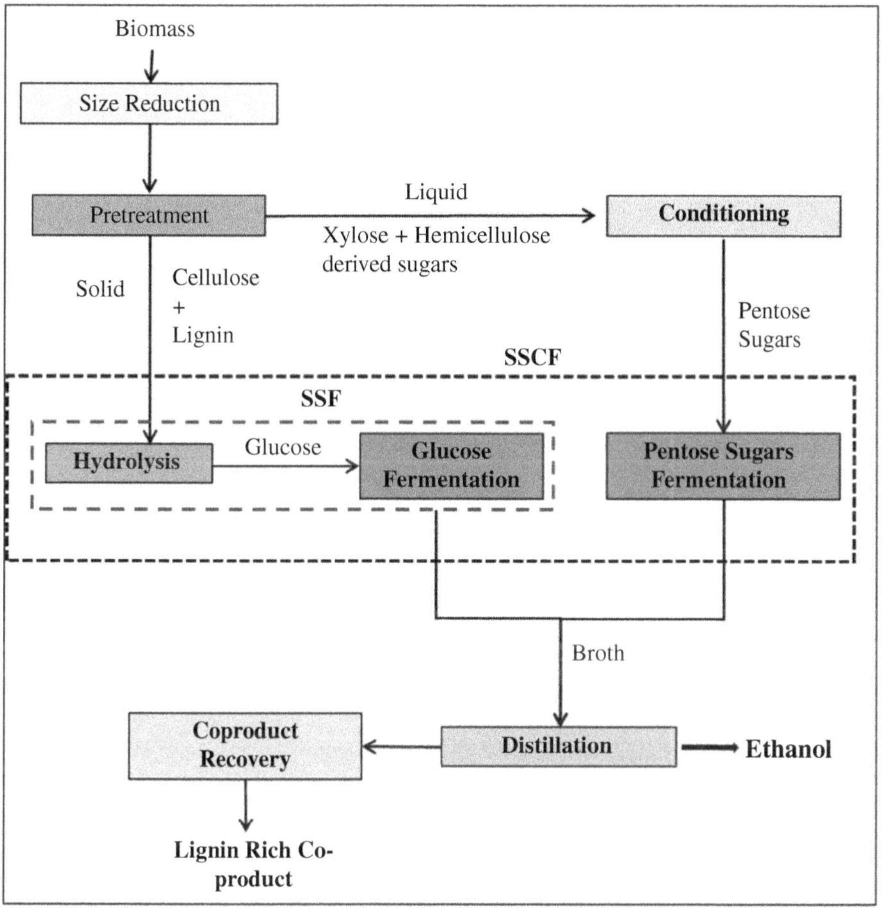

Figure 7.2 Schematic of the ethanol production process from lignocellulosic feedstock using a dilute acid pretreatment process.

processes used for ethanol production from corn on a commercial scale. Due to relatively low capital investment (about four times lower than wet milling) and incentive policies, most of the corn ethanol is produced using the dry grind process (91% in the year 2018) (Bothast and Schlicher 2005; Kumar and Singh 2019a; RFA 2019). In the dry grind process, whole corn kernels are processed to produce ethanol, whereas in the case of the wet milling process, corn is fractionated into basic components after soaking, such as starch, germ, fiber, and solubles, in which only starch is processed for ethanol production. Only the dry grind process is discussed in this chapter. Details of the wet milling process can be found elsewhere (Rausch et al. 2019).

A simple schematic of steps involved in the conventional dry grind process is presented in Figure 7.1. The process would be similar for other starch-based feedstocks. Cleaned corn is ground into "meal" to reduce particle size, and ground corn flour is mixed with water to form a corn slurry (27–37% solid content). In the case of corn ethanol, recycled thin stillage is also used along with water to make a slurry. The pH of corn slurry is adjusted as per the recommended pH for the enzymes used in the subsequent liquefaction process. Starch liquefaction involves the conversion of starch into short-chain saccharides (dextrins) using α-amylase enzymes (1,4- α-D-glucan glucanohydrolase enzymes (EC 3.2.1.1)) and heat. This process also reduces viscosity of the slurry, which helps in proper agitation and pumping. In the commercial process, starch is first gelatinized by heating the slurry to high temperatures (80–85°C), followed by cooking at high pressure in jet cookers at 104–107°C (Singh et al. 2010). Mechanical shear at high temperature and pressure during the jet cooking process ruptures starch granules and solubilizes starch. During the liquefaction process, the α-amylase, an endoenzyme, acts randomly to hydrolyze α-1,4-glucosidic bonds in starch and produce dextrins. In the next step, known as saccharification, soluble dextrins are converted to glucose using glucoamylase (GA) (amyloglucosidase or AMG) enzymes. GA belongs to a class of exoenzymes and acts on the non-reducing end of dextrins to release glucose. Although the GA enzyme hydrolyzes both α-1,4-glucosidic bonds and α-1,6-glucosidic bonds, the rate of hydrolysis of α-1,4-glucosidic bonds is significantly (up to 50 times) faster. Over recent years, an alternative process, known as cold starch hydrolysis or granular starch hydrolysis, has gained popularity. This process does not require separate liquefaction and saccharification of starch and directly converts raw starch into glucose using the synergistic action of several enzymes (α-amylases, glucoamylases, α-glucosidases, isoamylases, β-amylases, and maltogenic β-amylases) included in the granular starch hydrolyzing enzymes (GSHEs) complex. As the raw starch is insoluble in water, the GSHEs act on the solid phase by getting adsorbed onto the surface. The starch to sugar conversion process is completed in several steps including diffusion, adsorption, and catalysis (Cinelli et al. 2015). Due to the simultaneous conversion of starch to glucose and fermentation of glucose to ethanol, the glucose concentrations remain low throughout the process and eliminate the risk of osmotic stress to yeast and the formation of byproducts. Ethanol yields from the GSH process have been reported to be higher due to the absence of unwanted reactions, such as the Maillard reaction at high temperatures required in the conventional process. Pradyawong et al. (2018) reported 2.8% higher final ethanol concentrations for cassava starch using GSH process compared to conventional dry grind process.

The next step in this bioconversion process is fermentation of glucose to ethanol by yeast (usually of the *Saccharomyces* genus). The fermentation can be performed either separately (separate hydrolysis and fermentation or SHF) or combined with the saccharification step (simultaneous saccharification and fermentation, SSF). In the case of using GSH enzymes, the process is sometimes referred to as simultaneous liquefaction, saccharification, and fermentation (SLSF). The SSF process provides several advantages over the SHF process, including reduced equipment cost, slow release of glucose (avoids the osmotic shock to the yeast due to glucose buildup at the start of the fermentation), and lower risk of microbial contamination due to the presence of ethanol. Nutrients in the form of urea or ammonium sulfate are commonly added externally during fermentation as nitrogen sources for yeast. Pure ethanol is recovered from the fermented slurry, known as "beer," through a combination of a distillation process and molecular sieves. About 95% (v/v) pure ethanol is recovered after a series of distillation steps, which cannot be further purified using distillation due to the formation of an azeotrope. In most of the plants, dehydrated (100%) ethanol is recovered from this ethanol–water azeotrope using size-based separation technologies, such as molecular sieves. Some other techniques, such as a chemical dehydration process, extractive distillation processes, membrane processes, adsorption processes, and a diffusion distillation process, are also being used to recover pure ethanol (Cutzu and Bardi 2017). In the case of corn ethanol, the slurry (stillage) recovered from the bottom of the distillation columns containing non-fermentable components and unconverted starch is processed to produce distiller's dried grains with solubles (DDGS), which is primarily used as ruminant ingredient that is used in animal diets. Approximately a third of corn grains processed are recovered as animal feed. Distiller's corn oil is another major co-product from the dry grind industry. More details of downstream processing of the dry grind process can be found in our earlier publication (Kumar and Singh 2019a).

7.2.2 Ethanol from Lignocellulosic Biomass (Second-Generation Bioethanol)

The overall process steps for ethanol production from lignocellulosic feedstocks are similar to the starch-based ethanol production process: biomass grinding, conversion of carbohydrate polymers to sugars, fermentation of sugars to ethanol, and ethanol and co-product recovery. However, unlike starch-based feedstocks, extraction of sugars from structural carbohydrate in lignocellulosic biomass is challenging because of its highly crystalline and chemically rigid structure, which makes it highly resistant to chemical and biochemical degradation. Hydrolysis of cellulose and hemicellulose to sugar monomers can be performed through acid or enzymatic catalysts. However, in general, the enzymatic hydrolysis process is preferred over the use of acid due to lower process energy requirements, mild operating conditions (temperature and pH), high sugar yields (better conversion and no sugar degradation), and lower capital and maintenance cost of equipment (no need for expensive corrosive-resistant equipment). This chapter will discuss only the enzymatic hydrolysis process, and details of the acid hydrolysis process can be found elsewhere (Taherzadeh and Karimi 2007). Cellulose, hemicellulose, and lignin form a complex matrix, where lignin acts as the glue that provides a physical barrier, holding the polysaccharide fibers together to make the structure stable and fairly impervious to enzyme degradation. Therefore, a pretreatment step is required to reduce this recalcitrance and modify the structure to increase access of enzymes to polysaccharides. A generic schematic diagram of ethanol production from lignocellulosic biomass using a dilute acid pretreatment process is shown in Figure 7.2.

Pretreatment is the first major operation in the process after initial biomass preparation, which includes biomass cleaning and size reduction to facilitate biomass handling. Although this step is critical to the overall conversion process, it is highly energy and cost intensive. This poses a major disadvantage in the commercial success of the process. The aim of the pretreatment step is to open up the complex matrix of carbohydrates and lignin, depolymerize lignin and hemicellulose (depending on the type of pretreatment), increase the porosity of the matrix (increase the surface area), and consequently improve the subsequent hydrolysis yields (3–10 fold) (Figure 7.3). A large number of pretreatment technologies have been developed in the last few decades and investigated on a wide range of feedstocks to improve the sugar yields during hydrolysis. Pretreatment techniques can be broadly classified as physical (e.g., disc refining, irradiation), chemical (e.g., dilute acid, alkali, liquid hot water, ionic liquid, solvents, etc.), physicochemical (combination of physical and chemical, e.g., steam explosion (SE) and ammonia fiber explosion (AFEX)), and biological pretreatments (e.g., using white rot fungi) (Kumar and Murthy 2011; Tomás-Pejó

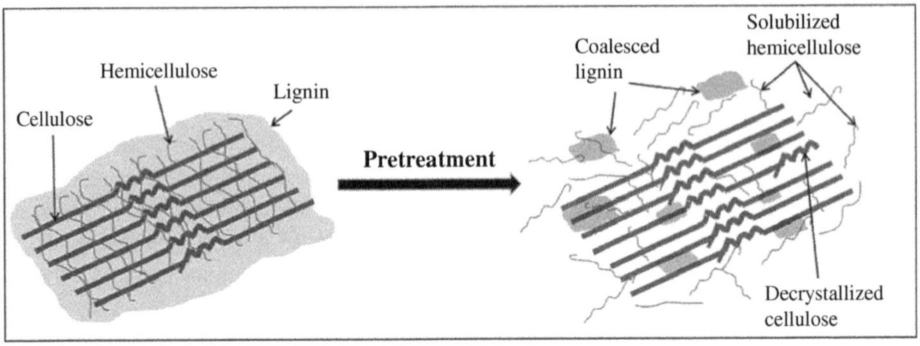

Figure 7.3 *Illustration of pretreatment effect on biomass structure.* Source: Adapted from Kumar (2014).

et al. 2011). Use of two or more pretreatments consecutively to leverage the synergistic effects of different methods is another approach that has gained attention (Kim et al. 2016; Wang et al. 2018).

Physical pretreatments mainly open up the biomass structure, increase porosity and the surface area for better accessibility to enzymes in the subsequent hydrolysis process, without altering the chemical composition. Chemical pretreatments involve use of chemicals, usually at high temperatures and pressure (depends on the type of chemical and its concentration) for a period of time (few seconds to several hours), to achieve depolymerization of hemicellulose and redistribution of lignin (Kumar and Murthy 2011). Among the various chemical pretreatments, dilute acid (mostly using sulfuric acid) pretreatment is the most studied process and has been found effective on a wide variety of lignocellulose substrates (Mosier et al. 2005). Most of the hemicellulose is solubilized during this process, which improves the accessibility of enzymes to cellulose chains. The major disadvantages of acid pretreatment are sugar degradation, generation of inhibitory compounds, and the requirement for corrosion-resistant reactors. Alkali or high pH pretreatments use alkaline reagents such as ammonia and sodium hydroxide, and mainly entail delignification. These reagents are also believed to act by swelling cellulose fibrils, which increase the internal surface area, and alter the lignin–carbohydrate linkages (ester bonds) to improve the cellulose accessibility to enzymes during hydrolysis (Kumar and Murthy 2011; Wilkinson et al. 2014). Hot water pretreatment (neutral pH) is another choice which does not use any chemical but is an "auto-catalyzed" process, where self-ionization of water at elevated temperatures and acetic acid released from hemicellulose act as catalysis. However, the sugar yields from hot water pretreatment are lower compared with acid pretreatment (Kim et al. 2016; Kumar and Murthy 2011). The yields can be improved significantly by performing mechanical pretreatment such as disc refining of hot water pretreated biomass (Kim et al. 2016; Wang et al. 2018). Physicochemical pretreatments combine the effects of chemical and mechanical forces in a single step. During processes such as steam explosion and AFEX, biomass is heated under pressure in the presence or absence of chemicals or catalysts (e.g., NH_3, H_2SO_4, SO_2) similar to chemical pretreatment; however, the pressure is released very rapidly to cause an explosive breakdown of biomass structure. An organosolvant pretreatment process uses organic or aqueous solvents (e.g., ethanol, methanol, acetone, ethylene glycol, etc.) to delignify the biomass to improve accessibility of enzymes and result in high hydrolysis efficiency (Tomás-Pejó et al. 2011). The solvents are removed from the biomass after pretreatment using adequate extraction and separation techniques (e.g., evaporation and condensation) and recycled in the process. Use of ionic liquids for biomass pretreatment has received much attention over the last two decades, especially because these can address the challenge of environmental pollution caused by the use of traditional organic solvents (Tian et al. 2018; Tomás-Pejó et al. 2011). These ionic liquids, containing salts, composed of large organic cations and small inorganic anions, are liquids at relatively low temperatures (melting point <100°C), break down the extensive hydrogen-bonding network in the biomass matrix, and cause its solubilization. The solvent properties (e.g., chemical and thermal stability, non-flammability, and the range of temperatures where these salts remain in liquid form) can be changed by altering the anion and alkyl constituents of the cation. The ionic liquids are also known as "green solvents" because no explosive or toxic gases are produced during the process (Mood et al. 2013; Tomás-Pejó et al. 2011). Efficient recovery and recycling of these ionic liquids is critical for commercial-scale application of this process. In addition to reducing costs, recovery of these salts after the pretreatment process is also important because of their toxicity to enzymes and fermentative microorganisms. Biological pretreatments use microorganisms such as white, brown and soft rot fungi, which mainly degrade lignin and hemicellulose and a small amount of cellulose. Biological pretreatments are environmentally friendly compared with chemical pretreatments; however, the process requires a very long time (several days to weeks) and subsequent hydrolysis yields are low (Tomás-Pejó et al. 2011).

Choice of pretreatment process is critical because in addition to direct energy use and cost (up to 33% of total process cost), it is the source of chemicals or inhibitory compound formation that negatively affects the enzymes and fermenting microorganisms, and also determines enzyme loading and sugar yields (Kim et al. 2016; Tomás-Pejó et al. 2011). In general, an effective pretreatment method should be of low cost (low energy use and avoiding specialized equipment), effective on a wide range of lignocellulosic materials, ensure high sugar yields during subsequent enzyme hydrolysis, require minimum preparation, and minimize the sugar degradation and formation of compounds that inhibit hydrolysis and fermentation processes (Ben Taher et al. 2017; Kim et al. 2016; Kumar and Murthy 2011).

After pretreatment of biomass, polysaccharides are converted to sugar monomers by the synergistic action of multiple enzymes. Enzymatic hydrolysis is a rate-limiting and cost-intensive step. Due to its highly crystalline structure, conversion of cellulose to glucose requires a much higher enzyme dosage compared with starch hydrolysis (40–100 times that for starch hydrolysis) (Kumar and Singh 2019a). The three main classes of enzymes required for cellulose hydrolysis are endoglucanases (EG) (EC3.2.1.4), exoglucanases (also known as cellobiohydrolase [CBH]) (EC3.2.1.91), and β-glucosidase (BG) (EC3.2.1.21). EG enzymes bind randomly along a glucose chain on the surface of cellulose and hydrolyze one/few accessible bonds to produce shortened chains. Exoglucanases act from the chain end (CBH I act on reducing end and CBH II act on non-reducing end) and work in a sequence to produce a cellobiose molecule (a dimer of glucose). Exoglucanases, in general, are processive in nature, which means that the enzymes remain bound to the glucose chain even after releasing a cellobiose molecule and continue to cleave cellobiose units until a minimum chain length is reached (Barr et al. 1996; Kumar and Murthy 2013). BG enzymes act in the solution on cellobiose and short-chain oligomers (mainly DP2-6) and produce glucose. Although these enzymes have a different mode of action and preferences, however, they exhibit a synergism for efficient hydrolysis of cellulose. For example, EG enzymes act effectively in the amorphous region and produces chain ends, whereas exocellulase enzymes can bind and produce cellobiose (eno-exo synergism). Cellobiose is a strong inhibitor to exocellulase enzymes. BG enzymes convert cellobiose to glucose and reduce this inhibition (exo-BG synergism). An in-depth discussion of the enzymatic hydrolysis process, including the synergism, inhibition, other challenges, and approaches to reducing the cost of the hydrolysis process, can be found in numerous recent review papers and book chapters (Kumar and Murthy 2016; Lynd et al. 2002; Wang et al. 2012b; Zhang and Lynd 2004).

The sugars obtained from the hydrolysis and pretreatment are fermented into ethanol by microorganisms such as *Saccharomyces cerevisiae* (most commonly used), *Escherichia coli*, *Zymomonas mobilis*, or *Klebsiella oxytoca* (Mielenz et al. 2009). Similar to the corn ethanol process, pure ethanol is recovered from the fermented broth using combinations of distillation columns and molecular sieves post-fermentation. As discussed in the previous section, the SSF process is preferred over the SHF. In the case of SSF, it provides the advantage of eliminating glucose inhibition to cellulase enzymes and improving overall process efficiency. If the hydrolysis, fermentation of glucose (from cellulose) and fermentation of pentose sugars (from hemicellulose) are performed together in a single step, the process is called simultaneous saccharification and co-fermentation (SSCF). The unfermented components containing lignin, protein, ash, and residual carbohydrates are recovered in the downstream process, and can be used as an energy source for in-house production of steam and electricity.

7.3 Use of Processing Waste to Produce Ethanol

Waste generated from the industrial and agro-food processes are renewable feedstock, mostly having zero economic value and containing high amounts of carbohydrates, that can be reused and processed to produce bioethanol and other value-added products. This section of the chapter discusses the research efforts and major findings during processing of these wastes to ethanol, using the technologies discussed in Section 7.2.

7.3.1 Citrus Peel Waste (CPW)

The citrus processing industry is a major part of the agro-industrial sector. More than 124 million metric tonnes of citrus fruits were produced worldwide in 2016, and 40–50% of these were used for industrial processing (Zema et al. 2018). Citrus peel waste (CPW), solid/semisolid residues containing 50–70% weight of processed fruits, consists of the peel, segment membranes, and seeds left over after extraction of juice from citrus fruits such as oranges and grapefruit (Pourbafrani et al. 2010; Zema et al. 2018). According to Wilkins et al. (2007), more than 4.5 million metric tonnes of CPW is produced annually in the state of Florida alone. CPW from various fruits contains a high proportion of structural carbohydrates (about 35–45%), free sugars (6–23%) and a low percentage of lignin (2–10%), which make it an attractive feedstock for bioethanol production (Table 7.1). Due to a relatively low lignin content compared with typical lignocellulosic feedstocks, such as corn stover (16–23% lignin) and wheat straw (5–19% lignin), processing of CPW requires only mild pretreatment conditions (Boluda-Aguilar et al. 2010; Zema et al. 2018). The major challenge of using CPW for biofuel production is the presence of D-limonene, a monoterpene that inhibits yeast and other microorganisms even at very low concentrations (0.01% w/v) (Pourbafrani et al. 2010; Wilkins et al. 2007). The actual inhibitory concentration varies depending on the strain used. Some strains, such as *Mucor indicus* and *Rhizopus* sp., have shown tolerance of 2% of D-limonene (Zema et al. 2018). However, the performance of *S. cerevisiae* was inhibited by 0.14% (v/v) of D-limonene during fermentation of CPW (Wilkins et al. 2007; Zema et al. 2018). In another study on bioethanol production from lemon peel wastes (*Citrus limon* L.), the performance of *S. cerevisiae* was negatively affected by presence of essential oil (of which D-limonene is the primary constituent) concentrations above 0.025% (Boluda-Aguilar and López-Gómez 2013). Various approaches have been investigated to remove limonene from hydrolysate before fermentation, to achieve high fermentation efficiency. Wilkins et al. (2007) used steam explosion pretreatment to partially (more than 90%) remove D-limonene, and converted the softened peel waste into ethanol using SSF by *S. cerevisiae*. Boluda-Aguilar and López-Gómez (2013) also used a steam explosion process to reduce limonene concentrations, and were able to achieve more than 60 L of ethanol per tonne of wet lemon peel waste. Similar yields (50–60 L/tonne wet biomass) were observed with steam explosion pretreatment and fermentation of mandarin (*Citrus reticulate* L.) citrus peel wastes (Boluda-Aguilar et al. 2010). In another alternative, peel waste was hydrolyzed using a cocktail of enzymes (cellulase, pectinase, and β-glucosidase) and D-limonene was removed by filtering the hydrolysate, as the solids were found to contain most of the D-limonene (Grohmann et al. 1994). Pourbafrani et al. (2010) used a dilute acid explosion process (0.5% v/v acid; 130–170°C temperature; 3–9 min duration) for the hydrolysis of CPW, and separated the limonene by separating solids (containing limonene and non-fermentable components) and liquid (containing sugars) by centrifugation. Hydrolysis at 150°C for 6 min resulted in maximum sugar yields of 42.05% and 99% removal of limonene with final concentrations of only 0.0035% w/v, lower than minimum inhibitory levels. The liquid stream was fermented using *S. cerevisiae*. John et al. (2017) also investigated acid catalyzed steam pretreatment, and optimized the process conditions (solid loadings, time, and acid concentrations) for maximizing hydrolysis yield from sweet lime peel for bioethanol production. It was observed that the pretreatment process enhanced the hydrolysis efficiency, and under optimized conditions, a maximum sugar yield of 7.09 mg reducing sugar/mL of hydrolysate can be achieved (John et al. 2017). Çalışkan Eleren et al. (2018) collected and analyzed the data on waste production from the fruit juice production industry in Turkey, and estimated that 127 tons (161 000 L) of ethanol can be produced from this waste every year.

7.3.2 Peel Residue Waste from Other Food Industries

Potatoes are one of the most common vegetables used around the world. In addition to direct consumption as a vegetable, potatoes are processed into a variety of products such as chips, fries, mashed potatoes, and dehydrated products, leaving about 20–50% of the raw product as wastes (Khawla et al. 2014). A large portion of this waste is potato peel, which is considered of zero value and cause disposal challenges to the processing industry (Khawla et al. 2014). Over 100 million tonnes of potato peels are produced as a result of the 300 million tonnes of potatoes that are produced worldwide (Dos Santos et al. 2016).

Starch (52%, dry basis) is the major components in potato peels for bioethanol production (Arapoglou et al. 2010; Khawla et al. 2014). Ben Taher et al. (2017) reported 42% starch, 33.5% cellulose, 5.5% hemicellulose, and 4.7% lignin in the potato peel residue collected from the local restaurants in Tunisia. Using hydrolysates (6% w/v) obtained from potato peel, Khawla et al. (2014) were able to achieve ethanol concentrations in the range of 19.6–21 g/L, depending upon the enzyme combinations used. The ethanol yield was higher (21 g/l) for the process using a combination of onsite-produced liquefaction enzyme (UEB-S amylase, produced from *Bacillus* sp. UEB-S) and amyloglucosidase, compared with the yield (19.6 g/l) from a process using a combination of commercial enzymes (Termamyl 120L/amyloglucosidase). In addition to a high efficiency of UEB-S amylase enzymes compared with commercial enzymes, the use of these enzymes also provided the advantage of hydrolyzing starch at a lower temperature (70°C vs. 90°C) which can reduce the process cost. Another study on bioethanol production from potato peels from the potato chip industry used acid hydrolysis to produce sugars from starch and achieved up to 2.17% ethanol concentrations using commercial *S. cerevisiae* yeast (Sheikh et al. 2016). Arapoglou et al. (2010) used a three-stage process using Termamyl, Celluclast, and Viscosyme enzymes to extract sugars, and achieved 18.5 g/l reducing sugars and produced 7.6 g/L of ethanol. Ben Taher et al. (2017) investigated various pretreatment options (hydrothermal, alkali, and acid pretreatments; 10% solids, 1% acid or alkali, at 121°C for 30 min) to enhance the enzymatic hydrolysis efficiency during bioconversion of potato peel residues, and found that hydrothermal pretreatment was most effective in increasing sugar yields. Ethanol yield of up to 30 g/L was observed under optimum process conditions, and supplementing culture medium with ammonium sulfate, about 126 L of ethanol could be retrieved per tonne of potato peel waste processing (Ben Taher et al. 2017).

Cassava peel is another processing waste rich in carbohydrates (more than 50%, mostly starch), with an annual production of more than 14 million tonnes worldwide (Aruwajoye et al. 2019). Aruwajoye et al. (2019) investigated the ethanol potential from thermally pretreated cassava peel in South Africa. Peels were soaked in 3.68% (v/v) HCl at 69.68°C for 2.57 h before thermal pretreatment (121°C and 5 min). The process was optimized to achieve a sugar yield of 0.58 g/g biomass (78.66% saccharification efficiency) by varying substrate concentration and enzyme loadings (α-amylase, glucoamylase, and cellulase). Moshi et al. (2015) investigated various approaches for bioconversion of cassava peel to bioethanol, biogas, or both. Based on the process yields, it was observed that the integrated process producing both bioethanol and biogas resulted in a higher fuel energy value compared with separate bioethanol or biogas production processes.

7.3.3 Waste from the Brewing Industry

Beer is the oldest and most popular alcoholic beverage consumed worldwide. Although the exact beer-making process varies from brewery to brewery, almost all breweries generate large quantities of organic waste materials such as spent grain, spent hops, and residual yeast. Among all the streams, spent grains, the solid fraction consisting of exhausted grain husks, obtained after mashing and lautering is the most promising feedstock for biofuel and value-added products. More than 30% of barley grains end up as spent grains, with a high carbohydrate (up to 50%) and protein (up to 30%) content. In other words, about 20 kg of spent grains (wet) are estimated to be produced for every 100 L of beer (Rojas-Chamorro et al. 2018). The actual composition of spent grains depends on several variables, such as barley cultivar, harvest year, percentage of adjuncts, malting, and processing conditions. Currently, most of the spent grain residues are used as animal feed (Romaní et al. 2018).

Similar to other lignocellulosic feedstocks, pretreatment is the key process in efficient utilization of spent grains for ethanol production. Liguori et al. (2015) investigated the fermentation performance of five ethanologenic microorganisms for ethanol production from spent grains using acid–alkali pretreatment and enzymatic hydrolysis with commercial enzymes. Spent grains were pretreated by dilute acid (1.25% (v/v) H_2SO_4; 120°C for 17 min), followed by neutralizing by washing and treatment with alkali (2% (v/v) NaOH; 120°C for 90 min). Pretreatment solubilized most of the hemicellulose and other components, and increased cellulose content in the perpetrated biomass by more than five-fold (87% w/w). This pretreatment process was found to be highly effective and resulted in 97% cellulose conversion during hydrolysis, producing a hydrolysate with a final glucose concentration equal to 75 g/L. Final ethanol yields after fermentation ranged from 0.26 to 0.28 g ethanol/g substrate in brewers' spent grain (BSG) hydrolysate, depending on the nitrogen source supplementation. Wilkinson et al. (2014) used alkaline pretreatment for ethanol production from spent grains, and investigated the effects of alkali reagent and process conditions on the hydrolysis efficiency, using an excess of cellulase enzymes. Both the tested alkali reagents, alkaline peroxide and sodium hydroxide, were found to achieve efficient hydrolysis of pretreated spent grains. The pretreatment using 5% alkaline peroxide at 100°C for five hours and 25% solids resulted in achieving 87% of the theoretical glucose yield. The glucose yield (92%) was higher for the spent grains pretreated with 5% NaOH at 25% (w/v) solids loading for 12 h at 50°C. Although saccharification yields for both reagents were high, fermentation of hydrolysates using *S. cerevisiae* was superior for the NaOH pretreated samples (81% yield) compared with alkaline peroxide (43% yield). The reason for low fermentation efficiency could be the presence of residual hydroperoxyl radicals (HOO*) or other peroxide-mediated reaction products that disrupt yeast metabolism and negatively affect its performance (Wilkinson et al. 2014). White et al. (2008) used the dilute acid pretreatment (HNO_3) and determined the optimum process conditions for maximum hydrolysis efficiency of spent grains. Spent grains pretreated with 0.16 N HNO_3 (20% solids) at 121°C for 15 min resulted in maximum sugar recovery (48 g sugars/100 g spent grains). Fermentation of the hydrolysate obtained under these optimum pretreatment conditions, by *Pichia stipitis*, resulted in 8.3 g/L ethanol, corresponding to an ethanol conversion yield of 0.32 g ethanol/g substrate (White et al. 2008). Rojas-Chamorro et al. (2018) investigated the effectiveness of dilute phosphoric acid pretreatment to enhance sugar recovery from spent grains and achieve high ethanol yields. Using this pretreatment under optimum conditions (2% H_3PO_4; 155°C), 92% of total sugars were recovered during pretreatment and enzymatic hydrolysis process. Fermentation of this hydrolysate using *E. coli* SL100 resulted in high fermentation efficiency, with overall yields of 17.9 g ethanol per 100 g raw spent grains (0.23 L/kg spent grains) (Rojas-Chamorro et al. 2018).

7.3.4 Other Processing Wastes

Soybean hull is the hard shell or hull of the soybean which is removed mechanically during soybean processing and accounts for about 5–8% (more than 100 million tonnes) of the total soybean production in the US. Soybean hulls contain a high amount of structural carbohydrates (29–51% cellulose and 10–20% hemicellulose) and protein (9–14%), but low levels of lignin (1–4%), making it an attractive feedstock for bioethanol production (Mielenz et al. 2009). It has been reported that soybean hulls are relatively easy to process compared with conventional lignocellulosic feedstocks, and do not require pretreatment to achieve high sugar yields (81–85% cellulose conversion) and high titers of ethanol in the broth during the SSF process. At these conversions, up to 50 gal of ethanol can be produced from the processing of one tonne of soybean hulls (Mielenz et al. 2009). In addition to ethanol production, due to the conversion of carbohydrates to ethanol, the protein content increases significantly (from 10% to 25%) in the solid residues, which can be used as a high-value feed in the livestock industry. Gouvea et al. (2009) investigated ethanol production potential from coffee husks, a major waste generated during coffee bean handling and processing. One kilogram of husk is generated for every kilogram of coffee beans produced (Gouvea et al. 2009). It was observed that coffee husk can be efficiently fermented into bioethanol, and the fermentation yields (7.67 g/100 g dry basis) are similar from whole and ground coffee husk.

Olive pulp is another potential processing waste material that is produced during a two-stage centrifugation oil extraction process from olives. The amount of pulp produced during the process is as high as 800 kg/t of olive processed, generating more than 8 million tonnes of olive pulp every year in Southern Europe (Georgieva and Ahring 2007). Georgieva and Ahring (2007) observed that olive pulp can be hydrolyzed at very high solids content (30% solids) and using a low dosage of cellulase enzymes (7.5 FPU/g glucans), however, the sugar yields are low and do not increase with the change in enzyme loadings. Based on these yields, it was concluded that pure bioethanol production might not be economically feasible, but it can be efficiently used in a biorefinery approach with co-production of high-value products from non-fermented fractions.

7.4 Use of Processing Waste to Enhance Ethanol Yields

Although several processing waste streams could have high potential to produce bioethanol, their production levels might not be high enough to set up an economically feasible ethanol industry using solely these feedstocks. In such cases, these streams can be used to enhance ethanol yields in existing processes, either by providing nutrition to fermenting microorganisms, or from the conversion of additional polysaccharides. This section discusses a few such approaches investigated by our research group and other researchers.

7.4.1 Improving Fermentation of Dry Fractionated Corn

With developments of new technologies and low-profit margins from only-ethanol production, ethanol refineries are trying to operate like petroleum refineries with the production of multiple products. Currently, DDGS is the major co-product produced in the dry grind ethanol industry, which is of limited use for ruminant animals because of the high fiber content. One potential approach to generating co-products in the dry grind ethanol process is front-end dry fractionation of corn to separate germ, pericarp, and corn endosperm. Dry fractionation technologies are significantly less capital-intensive compared with wet fractionation technologies (Juneja et al. 2017). Corn germ and pericarp can be used to produce high-value products (e.g., corn fiber oil from pericarp, and corn oil from germ), and the endosperm fraction (corn grits) can be converted to ethanol. Due to the removal of pericarp, the DDGS obtained in this process would have high nutritional value with a wider market because it can be used as food for non-ruminants such as poultry and swine. The corn grits contain a relatively high amount of starch (due to the removal of other components) and can potentially increase fermenter capacity. Corn fractionation at the front end also provides handling advantages with lower slurry volumes, reduced drying energy in downstream processes (due to the low amount of stillage), and higher ethanol concentrations in beer (Kumar and Singh 2019a). However, the germ is a vital source of oil and nutrients required for yeast growth and maintenance, especially in stress conditions (e.g., high ethanol concentration). Separation of germ at the front end results in the removal of these nutrients and leads to nutrient-deficient slurry during fermentation. This deficiency negatively impacts the yeast performance and results in sluggish and incomplete fermentation (Juneja et al. 2017; Murthy et al. 2006a,b). Corn grits fermentation, even at low solid loadings of 25%, leaves about 4.5% (w/v) glucose unfermented, which results in a low overall starch-to-ethanol conversion efficiency (about 65%) (Juneja et al. 2017; Kumar and Singh 2019b). This problem can be addressed by adding external nutrients, such as yeast extract, B-vitamins or lipid supplementation (Murthy et al. 2006a,b). Addition of B-vitamins provides coenzymes involved in carbohydrate metabolism and metabolic precursors required for yeast growth, which improves its performance and fermentation efficiency. Yeast performance can also be improved by the addition of protease enzymes, which helps in breaking the endosperm-associated proteins in the protein matrix and releasing free amino nitrogen (FAN) into solution. Juneja et al. (2017) addressed this issue of lack of nutrition by using germ soak water to make corn slurry. Use of water obtained from germ soaked under optimum conditions (12 h soaking at 30°C) resulted in very high fermentation rates and complete fermentation. However, use of expensive vitamins, lipids, or protease enzymes would increase the operating cost of the process and counter the benefits of fractionation. Use of germ soak water requires additional equipment and time for germ soaking, filtration, and germ drying. As an alternative, Kumar and Singh (2019b) demonstrated the use of two low-value byproducts from agro-food industries, germ meal and pasta mill feed, as the nutrient source for yeast, to improve the fermentation of dry-grind fractionated corn endosperm and achieve high ethanol concentrations. Germ meal, a low-value byproduct, is the solid residue after oil extraction from corn germ during the corn wet milling process. It contains about 25% protein and 5% oil on dry basis. About 858 000 tonnes of corn germ meal was produced from wet milling industries in the US in the year 2017. Mill feed, also known as mill bin waste, is a byproduct produced at the front end of the pasta manufacturing process and contains about 19% protein and 7% oil on dry basis. About 240 kg of mill feed is produced from the milling of one metric tonne of durum wheat during the pasta manufacturing process (Bevilacqua et al. 2007; Kumar and Singh 2019b).

Various mixtures (2.5–10% by weight) of either germ meal or mill feed mixed with corn grits were investigated for their fermentation performance during the dry grind process. The addition of germ meal or mill feed at all levels resulted in higher final ethanol concentrations compared with grits only. Addition of 2.5% mill feed or 5% germ meal in corn grits resulted in significantly higher initial ethanol production rates (0.435 vs. 0.28 v/v-h during 4–24 hours) and 24% higher final ethanol concentrations compared with a control (Figures 7.4 and 7.5). In both cases, there was no residual glucose at the end of fermentation,

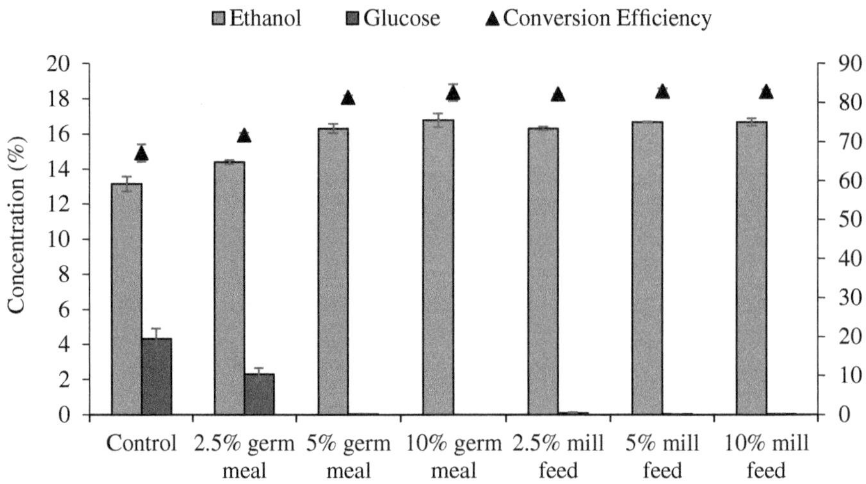

Figure 7.4 *Final ethanol (%v/v) and glucose (%w/v) concentrations, and starch to ethanol conversion efficiencies during fermentation of dry-fractionated corn (control) and corn grits mixed with different ratios (2.5–10%) of germ meal and mill feed (Kumar and Singh, 2019a).*

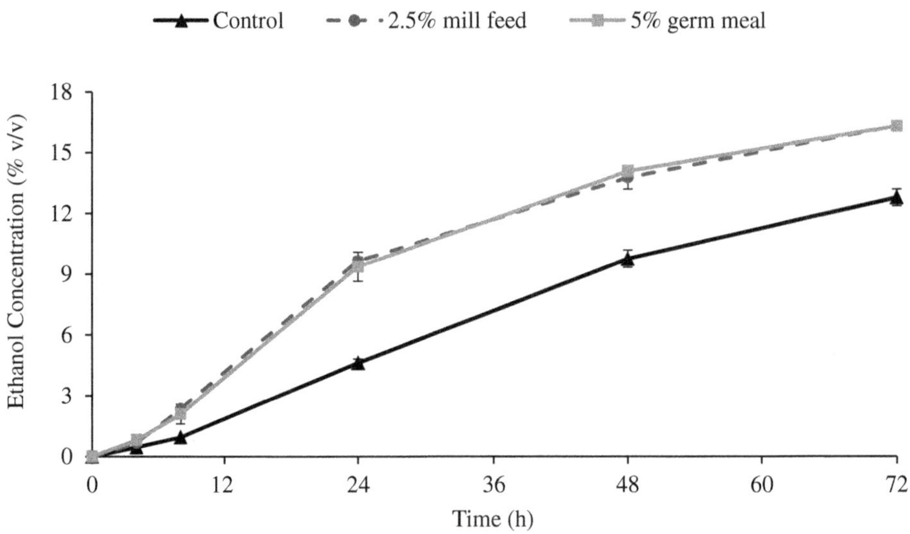

Figure 7.5 *Effect of mill feed and germ meal addition (2.5% of corn grits) on ethanol concentrations during fermentation of corn grits (Kumar and Singh, 2019b).*

indicating complete fermentation and a high starch-to-ethanol conversion efficiency (about 83%). This increase in ethanol concentration was similar to that observed with the use of protease enzymes or excess B-vitamins by Juneja et al. (2017), indicating these low-value products could replace the use of expensive nutrients. The initial ethanol production rates were even higher with use of a high percentage of mill feed (5% and 10%) and germ meal

(10%); however, the final ethanol concentrations were the same because all of the glucose was already converted to ethanol at lower mixtures.

7.4.2 Processing of DDGS to Enhance Ethanol Yields

Currently, DDGS, the major co-product from the dry grind ethanol industry, is used as animal feed. However, it is also important to note that it contains about 30–35% of the total energy in corn kernels (as carbon). Corn fiber, which accounts for about 4% of total carbohydrates in corn, remains unutilized in the conventional dry grind process and ends up in DDGS. Conversion of this stream to biofuel or biochemicals could be a more attractive form of conversion and would result in improving process economics and GHG emission savings. Extensive efforts have been made over the last few years to develop various approaches and technologies for efficient utilization and bioconversion of DDGS, allowing biorefineries to increase bioethanol yields and produce fuel from both starch and cellulosic ethanol from the corn kernel. The composition of DDGS is very similar to lignocellulosic feedstock, including cellulose (13–17%), xylan (10–13%), arabinan (6–8%) and lignin (7–10%), which has a potential to be further converted to ethanol using second generation conversion technologies. DDGS can be used as a sole feedstock to produce ethanol (Bryant et al. 2013; Kim et al. 2010; Noureddini and Byun 2010) or can be mixed with another high carbohydrate source for enhanced ethanol production (Li et al. 2019). Similar to lignocellulosic feedstock and other processing waste conversions, the first critical step for biological conversion of DDGS to ethanol is the pretreatment process. Various pretreatments, such as liquid hot water (Kim et al. 2010; Li et al. 2019), dilute acid (Noureddini and Byun 2010), dilute alkali (Li et al. 2019), AFEX (Kim et al. 2010), and steam explosion (SE) (Bryant et al. 2013) processes have been investigated to improve sugar release and ethanol production from DDGS. Process conditions and the corresponding maximum hydrolysis yields investigated in some of these studies are summarized in Table 7.2.

Li et al. (2019) used hot water pretreatment (160°C) and was able to achieve 100% conversion for cellulose from DDGS; however, the xylose yields were lower (about 15%). The xylose yields did not improve even with an increase in pretreatment time (20–30 minutes). The xylose yields can be improved using chemically-assisted pretreatment, such as dilute acid pretreatment, with which high xylose yields are reported in addition to 100% glucose yields. The solid loading, acid concentration, temperature, and residence time are important parameters that can affect the sugar yields. The highest yields of monomeric sugar were observed at lower substrate loadings and higher sulfuric acid concentration, temperature, and residence time (Noureddini and Byun 2010). Li et al. (2019) reported similar results, where dilute acid pretreatment of DDGS with 1% sulfuric acid and 10% solid loading at 140°C with a residence time of 20–30 minutes provided 100% glucose yield and more than 67% xylose yield. With the increase in solid loadings, lower yields are observed due to various factors including inefficient mixing of the slurry due to its high viscosity, inhibition of cellulose activity due to the buildup of glucose, and reduced water activity affecting the ability of the enzyme to diffuse and bind to the cellulose efficiently. Higher yields with an increase in acid concentration, temperature, and residence time are attributed to the increase of severity factor, and hence enhanced breakdown of lignocellulosic structure.

Table 7.2 Summary of pretreatment technologies, process conditions, and maximum sugar yields during bioconversion of DDGS.

Pretreatment	Feedstock	Pretreatment conditions	Maximum glucan yields (%)	Maximum xylan yields (%)	References
Liquid hot water	DDGS (wet cake)	• 15% solids loading (w/w) • 160°C for 20 min	80%	—	Kim et al. (2010)
	DDGS	• 10% solids loading (w/w) • 140–160°C, 10–30 min	101.9%	15.9%	Li et al. (2019)
Dilute acid	DDG (without solubles)	• 5–20% solid loading (w/w) • 0.5–1.5% acid conc. • 120–140°C, 0–60 min	57% (combined yield of monomeric sugars)		Noureddini and Byun (2010)
	DDGS	• 10% solids loading (w/w) • 1% H_2SO_4 conc. • 140–160°C, 10–30 min	100.2%	67.9%	Li et al. (2019)
Dilute alkali	DDGS	• 10% solid loading (w/w) • 2.5% NaOH concentration • 121°C for 10–30 min	100%	17.3%	Li et al. (2019)
AFEX	DDGS	• 40 g dry weight at 60% moisture content • Ammonia loading 0.80 : 1 g/g dry biomass • 90°C, 350–430 psi	77%	—	Kim et al. (2010)

7.5 Conclusions and Future Trends

The agro-food processing industries produce a large amount of waste or low-value byproducts each year, and the disposal of these streams is costly, both financially and environmentally. Bioconversion of these residues to bioethanol can be a sustainable approach to address these challenges, result in a financial gain to the industry, contribute to the ethanol industry development and enhance the nation's energy security. However, more research efforts and suitable techniques are required to identify the most promising streams for efficient utilisation. Pretreatment is the key step to achieving high yields during bioconversion of processing wastes to ethanol. In comparison with traditional lignocellulosic feedstocks, milder but efficient approaches are needed for the pretreatment of agro-food industrial processing wastes to avoid the degradation of starch, free sugars, and other high-value compounds. The current contribution of food and beverage waste to produce ethanol in the US is only 0.2%, but with development of efficient technologies and novel approaches to bioconversion of these wastes, this contribution can increase significantly in the future.

References

Arapoglou, D., Varzakas, T., Vlyssides, A., and Israilides, C. (2010). Ethanol production from potato peel waste (PPW). *Waste Manage.* **30** (10): 1898–1902.

Aruwajoye, G.S., Faloye, F.D., and Kana, E.G. (2019). Process optimisation of enzymatic saccharification of soaking assisted and thermal pretreated cassava peels waste for bioethanol production. *Waste Biomass Valorisation*: 1–12.

Barr, B.K., Hsieh, Y.-L., Ganem, B., and Wilson, D.B. (1996). Identification of two functionally different classes of exocellulases. *Biochemistry* **35** (2): 586–592.

Ben Taher, I., Fickers, P., Chniti, S., and Hassouna, M. (2017). Optimization of enzymatic hydrolysis and fermentation conditions for improved bioethanol production from potato peel residues. *Biotechnol. Progr.* **33** (2): 397–406.

Bevilacqua, M., Braglia, M., Carmignani, G., and Zammori, F.A. (2007). Life cycle assessment of pasta production in Italy. *J. Food Qual.* **30** (6): 932–952.

Boluda-Aguilar, M. and López-Gómez, A. (2013). Production of bioethanol by fermentation of lemon (*Citrus limon* L.) peel wastes pretreated with steam explosion. *Ind. Crops Prod.* **41**: 188–197.

Boluda-Aguilar, M., García-Vidal, L., del Pilar González-Castañeda, F., and López-Gómez, A. (2010). Mandarin peel wastes pretreatment with steam explosion for bioethanol production. *Bioresour. Technol.* **101** (10): 3506–3513.

Bothast, R. and Schlicher, M. (2005). Biotechnological processes for conversion of corn into ethanol. *Appl. Microbiol. Biotechnol.* **67** (1): 19–25.

Bryant, D.N., Firth, E., Kaderbhai, N. et al. (2013). Monitoring real-time enzymatic hydrolysis of distillers dried grains with solubles (DDGS) by dielectric spectroscopy following hydrothermal pre-treatment by steam explosion. *Bioresour. Technol.* **128**: 765–768.

Çalışkan Eleren, S., Öziş Altınçekiç, Ş., and Altınçekiç, E. (2018). Biofuel potential of fruit juice industry waste. *J. Hazard., Toxic, and Radioactive Waste* **22** (4): 05018002.

Cinelli, B.A., Castilho, L.R., Freire, D.M., and Castro, A.M. (2015). A brief review on the emerging technology of ethanol production by cold hydrolysis of raw starch. *Fuel* **150**: 721–729.

Cutzu, R. and Bardi, L. (2017). Production of bioethanol from agricultural wastes using residual thermal energy of a cogeneration plant in the distillation phase. *Fermentation* **3** (2): 24.

Dos Santos, R.G., Ventura, P., Bordado, J.C., and Mateus, M.M. (2016). Valorizing potato peel waste: an overview of the latest publications. *Rev. Environ. Sci. Biotechnol.* **15** (4): 585–592.

EIA (2019). *US Energy Facts Explained*. Washington, DC: US Energy Information Administration.

Georgieva, T.I. and Ahring, B.K. (2007). Potential of agroindustrial waste from olive oil industry for fuel ethanol production. *Biotechnol. J.* **2** (12): 1547–1555.

Gouvea, B., Torres, C., Franca, A. et al. (2009). Feasibility of ethanol production from coffee husks. *Biotechnol. Lett* **31** (9): 1315–1319.

Grohmann, K., Baldwin, E.A., and Buslig, B.S. (1994). Production of ethanol from enzymatically hydrolyzed orange peel by the yeast *Saccharomyces cerevisiae*. *Appl. Biochem. Biotechnol.* **45** (1): 315.

Huang, H., Qureshi, N., Chen, M.-H. et al. (2015). Ethanol production from food waste at high solids content with vacuum recovery technology. *J. Agric. Food. Chem.* **63** (10): 2760–2766.

Jin, Q., Neilson, A.P., Stewart, A.C. et al. (2018). Integrated approach for the valorisation of red grape pomace: production of oil, polyphenols, and acetone–butanol–ethanol. *ACS Sustainable Chem. Eng.* **6** (12): 16279–16286.

John, I., Yaragarla, P., Muthaiah, P. et al. (2017). Statistical optimization of acid catalyzed steam pretreatment of citrus peel waste for bioethanol production. *Resour.-Effic. Technol.* **3** (4): 429–433.

Juneja, A., Kumar, D., and Singh, V. (2017). Germ soak water as nutrient source to improve fermentation of corn grits from modified corn dry grind process. *Bioresour. Bioprocess.* **4** (1): 38.

Khawla, B.J., Sameh, M., Imen, G. et al. (2014). Potato peel as feedstock for bioethanol production: a comparison of acidic and enzymatic hydrolysis. *Ind. Crops Prod.* **52**: 144–149.

Kim, Y., Hendrickson, R., Mosier, N.S. et al. (2010). Effect of compositional variability of distillers' grains on cellulosic ethanol production. *Bioresour. Technol.* **101** (14): 5385–5393.

Kim, S.M., Dien, B.S., and Singh, V. (2016). Promise of combined hydrothermal/chemical and mechanical refining for pretreatment of woody and herbaceous biomass. *Biotechnol. Biofuels* **9** (1): 97.

Kumar, D. (2014). Biochemical conversion of lignocellulosic biomass to ethanol: experimental, enzymatic hydrolysis modeling, techno-economic and life cycle assessment studies. PhD Dissertation. Oregon State University.

Kumar, D. and Murthy, G.S. (2011). Impact of pretreatment and downstream processing technologies on economics and energy in cellulosic ethanol production. *Biotechnol. Biofuels* **4** (1): 27.

Kumar, D. and Murthy, G.S. (2013). Stochastic molecular model of enzymatic hydrolysis of cellulose for ethanol production. *Biotechnol. Biofuels* **6** (1): 63.

Kumar, D. and Murthy, G.S. (2016). Enzymatic hydrolysis of cellulose for ethanol production: fundamentals, optimal enzyme ratio, and hydrolysis modeling. In: *New and Future Developments in Microbial Biotechnology and Bioengineering*, 65–78. Elsevier.

Kumar, D. and Singh, V. (2019a). Bioethanol production from corn. In: *Corn*, 3e (ed. S.O. Serna-Saldivar), 615–631. Elsevier.

Kumar, D. and Singh, V. (2019b). Improving dry-fractionated corn fermentation by supplementation of corn germ meal and pasta mill feed from agro-food industries. *Cereal Chem.* **96** (2): 243–251.

Li, X., Chen, S., Yu, Y. et al. (2019). Ethanol production from mixtures of distiller's dried grains with solubles (DDGS) and corn. *Ind. Crops Prod.* **129**: 59–66.

Liguori, R., Soccol, C., De Souza Vandenberghe, L.P. et al. (2015). Second generation ethanol production from brewers' spent grain. *Energies* **8**: 2575–2586.

Lynd, L.R., Weimer, P.J., Van Zyl, W.H., and Pretorius, I.S. (2002). Microbial cellulose utilization: fundamentals and biotechnology. *Microbiol. Mol. Biol. Rev.* **66** (3): 506–577.

Mielenz, J.R., Bardsley, J.S., and Wyman, C.E. (2009). Fermentation of soybean hulls to ethanol while preserving protein value. *Bioresour. Technol.* **100** (14): 3532–3539.

Mood, S.H., Golfeshan, A.H., Tabatabaei, M. et al. (2013). Lignocellulosic biomass to bioethanol, a comprehensive review with a focus on pretreatment. *Renewable Sustainable Energy Rev.* **27**: 77–93.

Moshi, A.P., Temu, S.G., Nges, I.A. et al. (2015). Combined production of bioethanol and biogas from peels of wild cassava *Manihot glaziovii*. *Chem. Eng. J.* **279**: 297–306.

Mosier, N., Wyman, C., Dale, B. et al. (2005). Features of promising technologies for pretreatment of lignocellulosic biomass. *Bioresour. Technol.* **96** (6): 673–686.

Murthy, G.S., Singh, V., Johnston, D.B. et al. (2006a). Evaluation and strategies to improve fermentation characteristics of modified dry-grind corn processes. *Cereal Chem.* **83** (5): 455–459.

Murthy, G.S., Singh, V., Johnston, D.B. et al. (2006b). Improvement in fermentation characteristics of degermed ground corn by lipid supplementation. *J. Ind. Microbiol. Biotechnol.* **33** (8): 655–660.

Mussatto, S.I. and Roberto, I.C. (2005). Acid hydrolysis and fermentation of brewer's spent grain to produce xylitol. *J. Sci. Food Agric.* **85**: 2453–2460.

Noureddini, H. and Byun, J. (2010). Dilute-acid pretreatment of distillers' grains and corn fiber. *Bioresour. Technol.* **101** (3): 1060–1067.

Pourbafrani, M., Forgács, G., Horváth, I.S. et al. (2010). Production of biofuels, limonene and pectin from citrus wastes. *Bioresour. Technol.* **101** (11): 4246–4250.

Pradyawong, S., Juneja, A., Sadiq, M. et al. (2018). Comparison of cassava starch with corn as a feedstock for bioethanol production. *Energies* **11**: 3476.

Rausch, K.D., Hummel, D., Johnson, L.A., and May, J.B. (2019). Wet milling: the basis for corn biorefineries. In: *Corn* (ed. S.O. Serna-Saldivar), 501–535. Elsevier.

Renewable Fuels Association. (2019). *Powered with Renewed Energy: 2019 Ethanol Industry Outlook*. Washington, DC: RFA. https://ethanolrfa.org/publications/ (accessed 13 April 2020).

Rojas-Chamorro, J.A., Cara, C.B., Romero, I. et al. (2018). Ethanol production from brewers' spent grain pretreated by dilute phosphoric acid. *Energy Fuels* **32** (4): 5226–5233.

Romaní, A., Michelin, M., Domingues, L., and Teixeira, J.A. (2018). Valorisation of wastes from agrofood and pulp and paper industries within the biorefinery concept: southwestern Europe scenario. In: *Waste Biorefinery* (eds. T. Bhaskar et al.), 487–504. Elsevier.

Sheikh, R.A., Al-Bar, O.A., and Soliman, Y.M.A. (2016). Biochemical studies on the production of biofuel (bioethanol) from potato peels wastes by *Saccharomyces cerevisiae*: effects of fermentation periods and nitrogen source concentration. *Biotechnol. Biotechnol. Equip.* **30** (3): 497–505.

Singh, V., Johnston, D.B., Rausch, K.D., and Tumbleson, M.E. (2010). Improvements in corn to ethanol production technology using *Saccharomyces cerevisiae*. In: *Biomass to Biofuels: Strategies for Global Industries*, 185–198. Wiley Online Library.

Taherzadeh, M.J. and Karimi, K. (2007). Acid-based hydrolysis processes for ethanol from lignocellulosic materials: a review. *BioResources* **2** (3): 472–499.

Tian, S.-Q., Zhao, R.-Y., and Chen, Z.-C. (2018). Review of the pretreatment and bioconversion of lignocellulosic biomass from wheat straw materials. *Renewable Sustainable Energy Rev.* **91**: 483–489.

Tomás-Pejó, E., Alvira, P., Ballesteros, M., and Negro, M. (2011). Pretreatment technologies for lignocellulose-to-bioethanol conversion. In: *Biofuels*, 149–176. Elsevier.

Wang, M., Han, J., Dunn, J.B. et al. (2012a). Well-to-wheels energy use and greenhouse gas emissions of ethanol from corn, sugarcane and cellulosic biomass for US use. *Environ. Res. Lett.* **7** (4): 045905.

Wang, M., Li, Z., Fang, X. et al. (2012b). Cellulolytic enzyme production and enzymatic hydrolysis for second-generation bioethanol production. In: *Biotechnology in China III: Biofuels and Bioenergy*, 1–24. Springer.

Wang, Z., Dien, B.S., Rausch, K.D. et al. (2018). Fermentation of undetoxified sugarcane bagasse hydrolyzates using a two stage hydrothermal and mechanical refining pretreatment. *Bioresour. Technol.* **261**: 313–321.

White, J.S., Yohannan, B.K., and Walker, G.M. (2008). Bioconversion of brewer's spent grains to bioethanol. *FEMS Yeast Res.* **8** (7): 1175–1184.

Wilkins, M.R., Widmer, W.W., and Grohmann, K. (2007). Simultaneous saccharification and fermentation of citrus peel waste by *Saccharomyces cerevisiae* to produce ethanol. *Process Biochem.* **42** (12): 1614–1619.

Wilkinson, S., Smart, K.A., and Cook, D.J. (2014). Optimisation of alkaline reagent based chemical pre-treatment of brewers spent grains for bioethanol production. *Ind. Crops Prod.* **62**: 219–227.

Zabed, H., Sahu, J., Suely, A. et al. (2017). Bioethanol production from renewable sources: current perspectives and technological progress. *Renewable Sustainable Energy Rev.* **71**: 475–501.

Zema, D., Calabrò, P., Folino, A. et al. (2018). Valorisation of citrus processing waste: a review. *Waste Manage. (Oxford)* **80**: 252–273.

Zhang, Y.H.P. and Lynd, L.R. (2004). Toward an aggregated understanding of enzymatic hydrolysis of cellulose: noncomplexed cellulase systems. *Biotechnol. Bioeng.* **88** (7): 797–824.

Zhang, Z., O'Hara, I.M., Mundree, S. et al. (2016). Biofuels from food processing wastes. *Curr. Opin. Biotechnol.* **38**: 97–105.

8

Challenges with Biomass Waste Valorisation

Guihua Yan, Yunchao Feng, Sishi Long, Xianhai Zeng, Yong Sun, Xing Tang and Lu Lin

College of Energy, Fujian Engineering and Research Center of Clean and High-valued Technologies for Biomass, and Xiamen Key Laboratory of Clean and High-valued Applications of Biomass, Xiamen University, Xiamen 361102, PR China

8.1 Introduction

Modern civilization depends on energy and materials from fossil fuel resources. But in a world of increasing consumption and decreasing resources, how long will fossil resources last? One solution to this problem is the development and utilization of renewable resources. Globally, the total renewable energy and materials production and consumption have been gradually increasing in recent years, with biomass production being a major contributor. Thus, to further explore potential biomass energy is an important challenge.

Biomass waste is defined as any organic material apart from renewable biogenic sources (Tuck et al. 2012). Depending on the source, it can be divided into lignocellulosic biomass and food wastes (Figure 8.1). Lignocellulosic biomass is the most abundant and the only currently sustainable source of organic carbon (Tuck et al. 2012). It has been reported that global dry biomass production has reached 200 billion metric tons/yr

Waste Valorisation: Waste Streams in a Circular Economy, First Edition.
Edited by Carol Sze Ki Lin, Guneet Kaur, Chong Li and Xiaofeng Yang.

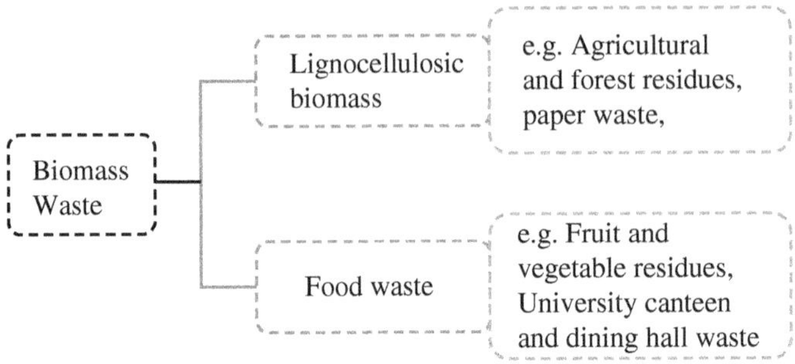

Figure 8.1 *The main components of biomass waste.*

(Sawatdeenarunat et al. 2016). Transforming lignocellulosic biomass and its residues to biofuels and chemicals is a potentially sustainable route for biomass valorisation. Food waste containing lipids, carbohydrates, and proteins is currently generated in daily human activities, with almost a third of the worldwide food production is discarded as waste (Karmee 2016). It has been reported that the global food waste generation is approximately 1.3 billion metric tons/yr (Carrero et al. 2011).

In past centuries, biomass waste has normally been treated by landfilling, discarding, incineration, and composting, which would increase the environmental concerns such as greenhouse gas emission, air and water pollution. In fact, waste can be a lucrative business, if viewed as misplaced resources; in other words, it can be used as cost-free, abundant, and renewable raw material to produce various value-added chemicals, fuels, and materials. This is the process of waste valorisation. For instance, lignocellulose is used as a feedstock to produce bioethanol, while food waste is applied for biodiesel production.

8.2 The Pre-Preparation Technologies of Biomass Waste

Biomass is the most abundant resource on Earth and is mainly divided into three parts: cellulose, hemicellulose, and lignin. They are natural macromolecules and are insoluble in water or common organic solvents. In order to dissolve a certain component of biomass, it is necessary to change the functional groups to improve compatibility and reduce the molecular weight to improve the solubility of the component. These often require more intense chemical processes. Therefore, based on the separation strategy of conventional solvent systems, we need to consider the following questions: Can the required conditions be realized? Can the chemical reagents be recovered? Are the existing forms of separated components convenient for downstream conversion and utilization? Are the processes involved environmentally friendly? Due to the complexity of raw materials and diversity of separation methods, it is extremely important to determine the separation priority of one or two components. If there is no selectivity for each component when dissolved in solvents, it is necessary to use a certain anti-solvent for selective precipitation of each component in the regeneration process, thereby realizing the effective separation of the three components. Hence, four separation strategies are summarized below, including cellulose-first processing, lignin-first processing, lignin and hemicellulose-first processing, and cellulose and hemicellulose-first processing.

8.2.1 "Cellulose-First" Biorefinery Technologies

At present, almost all separation processes of the biomass components consume chemicals, but only a few focus on separation using a single physical method. In combination with the prior art of drying, Chuetor et al. (2015) attempted to separate the biomass components by purely physical means. The biomass raw materials were subjected to superfine crushing, and then separated by a turbine classifier and an electrostatic separation cavity; the enzymolysis efficiency of the separated components was improved. The main solid components were crystalline cellulose, lignin-xylan mixtures, and ash. This technique was based on particle size and surface charge to achieve separation of components, but the whole process requires a very high energy consumption.

Furthermore, many reports on biomass pretreatment and its transformation and utilization use ionic liquids. After being dissolved by an ionic liquid and selectively precipitated, the mixed components of cellulose, hemicellulose, and lignin were obtained. The dissolution of lignin and the separation of hemicellulose in lye can be realized by dissolution and regeneration treatment. Finally, the separation and purification of biomass components can be achieved as shown in Figure 8.2. This technique is based on expensive ionic liquids that are difficult to recover.

8.2.2 "Lignin-First" Biorefinery Technologies

The hydrocarbon content in biomass is low and is generally considered not to be suitable as a biological fuel, but the low oxygen content and high calorific value of lignin make

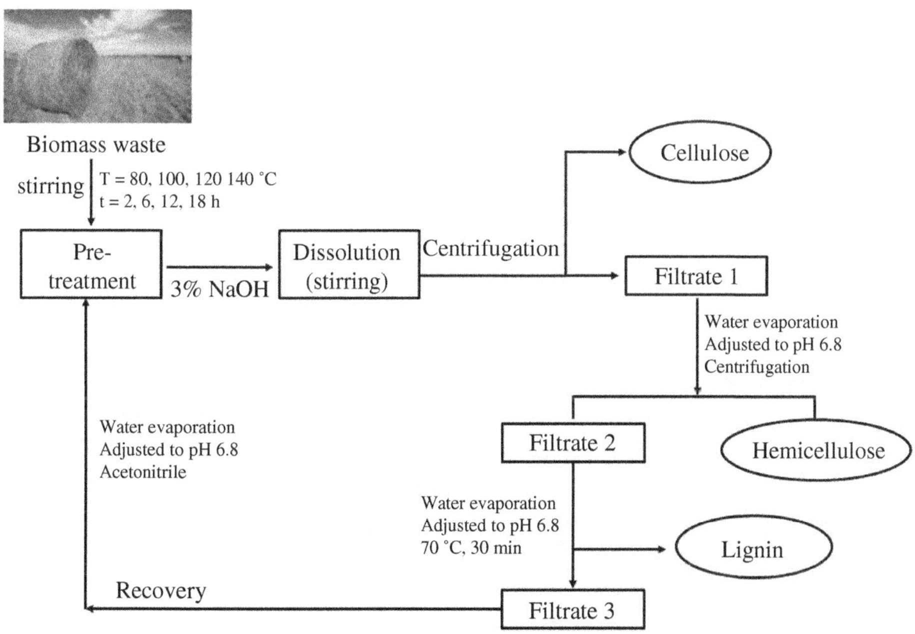

Figure 8.2 *Fractionation method for biomass using ionic liquids.*

it more potential as a fuel substitute. Based on this, two technical problems need to be overcome: (i) breaking the natural barrier of the cell wall to obtain the sugar and the phenolic intermediate; (ii) reducing the oxygen content of the intermediate. The chemical catalytic reduction method breaks through the two levels. By means of reductive catalytic fractionation (RCF), the simultaneous separation and degradation of the lignin can be conducted to obtain a lignin oil product. Furthermore, the addition of a small amount of phosphoric acid in the catalytic reduction process can improve the separation efficiency of the components and achieve the separation, conversion and utilization of the three components, as shown in Figure 8.3 (Renders et al. 2016).

Due to the strong polarity and hydrogen bonding of biomass components, ionic liquids have strong solubility to each component, which has great potential in a lignin-priority separation strategy. In particular, Sun et al. (2016) synthesized three kinds of protonic ionic liquids ([Py]Ac, [Mim]Ac, and [Pyrr]Ac), which have excellent solubility of lignin, but the solubility of cellulose is less than 1%, which can give a higher lignin extraction efficiency. In addition, the pyruvate phosphate method can also completely dissolve the three components of biomass. Although the product is still a mixture, it effectively destroys the bond and structure between the components, and is convenient for further separation (Zhang et al. 2010).

8.2.3 "Lignin and Hemicellulose-First" Biorefinery Technologies

An acid–organic solvent method is a common strategy in the preferential separation of hemicellulose and lignin. The acid–water method is mainly used to extract hemicellulose, while organic solvent is used to extract lignin selectively. Acid–organic solvent can extract hemicellulose and lignin simultaneously. Compared with inorganic acid (sulfuric acid), solid acid (phosphate) and organic acid (formic acid) have higher yields. At present, the sulfite-cooking method based on aqueous phase is commonly used. After SO_2–ethanol–water (SEW) treatment (Su et al. 2015), sulfur dioxide (95–97%) was thrown into the liquid phase and reused by simple distillation. It is especially suitable for dealing with cork biomass with a high lignin content, but the production process can cause sulfur pollution.

The traditional biomass cooking method using alkali (NaOH) can cleanly and efficiently separate hemicellulose and lignin. However, the degradation of hemicellulose is serious, the acid products are many, and the amount of alkali is large, which limits the application of this method. Hydrogen peroxide (H_2O_2) can effectively separate the three components under suitable alkaline conditions, and the reaction conditions are mild and the alkali consumption is low, which expands the application of biomass (Su et al. 2015).

8.2.4 "Cellulose and Hemicellulose-First" Biorefinery Technologies

The strategy for simultaneous separation of cellulose and hemicellulose is to hydrolyze the carbohydrates to water-soluble sugars (e.g., oligosaccharides, monosaccharides) under acidic conditions, or further conversion into chemicals (e.g., furfural, levulinic acid), to achieve separation from the lignin components. Two-step hydrolysis of the dilute acid method has become a standard method for the determination of the three components of the biomass at the National Renewable Energy Laboratory of the United States (Sluiter et al. 2008). Mats et al. (2014) used physical grinding to process the water-soluble sugar

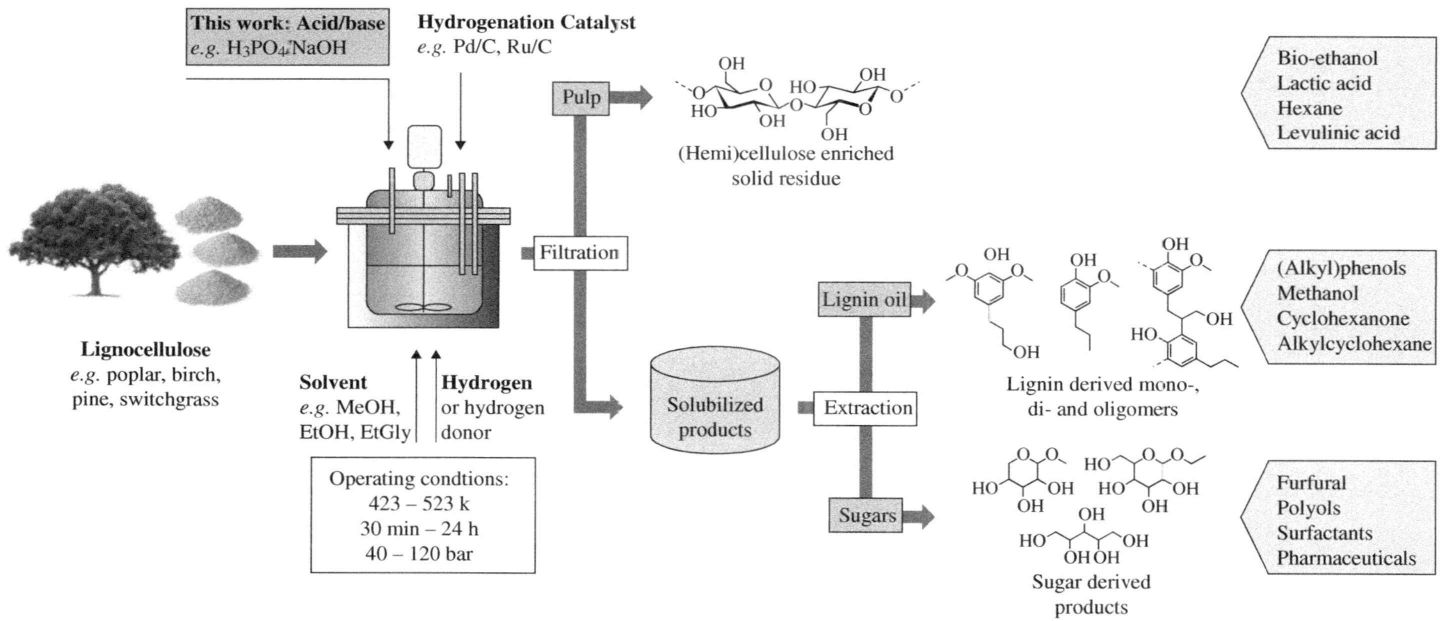

Figure 8.3 *Schematic representation of the catalytic reductive fractionation of lignocellulose.* Source: Reproduced with permission of the American Chemical Society, from Renders et al. (2015).

and lignin. The main advantages of this method are the high sugar recovery rates and its effectiveness for separation.

The conversion and utilization methods of biomass are usually combined with chemical and biological methods. In cellulose ethanol production, enzyme-catalyzed saccharification is widely used because of its mild reaction conditions, but the cellulase hydrolysis efficiency is still not satisfactory. Hayes et al. (2005) have put forward the whole chemical refining process, which can be divided into two stages to produce biopolymer chemicals (acetylpropionic acid, furfural, formic acid) and biochar, which improves production efficiency and product yield.

8.3 Handling of Emerging Biomass Wastes by Newly Developed Techniques

In recent years, high-value utilization of biomass waste has attracted extensive attention. As shown in Figure 8.4, various techniques have been developed to produce chemicals, materials, and energy. Those techniques can be grouped into three major categories: (i) catalytic chemistry (e.g., hydrolysis), (ii) thermochemistry (e.g., gasification), and (iii) biochemistry (e.g., fermentation).

8.3.1 Catalytic Chemistry Technologies

Catalytic chemistry involves the use of catalysts to transform biomass waste into valuable materials. Hydrolysis is a widely used conversion technique for catalytic degradation of biomass waste to 5-hydroxymethylfurfural (HMF), furfural, levulinic acid (LA) and aromatic monomers that serve as important platform compounds (Corma et al. 2007). The hydrolysis of biomass waste to HMF is a typical route for waste valorisation (Figure 8.5). Various solvents, biofuels, polymers, and medicines can be synthesized from HMF. Due to their recycling ability, heterogeneous catalysts, such as molecular sieves, ion-exchange

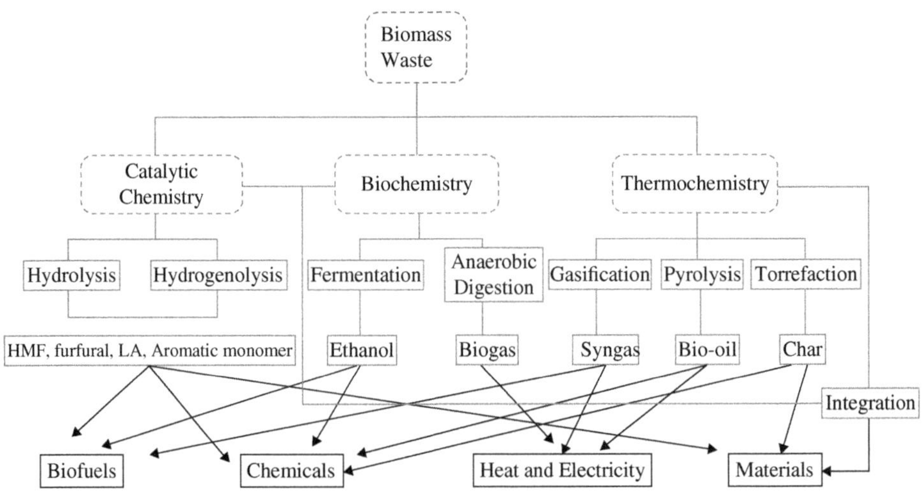

Figure 8.4 Biomass waste conversion technology.

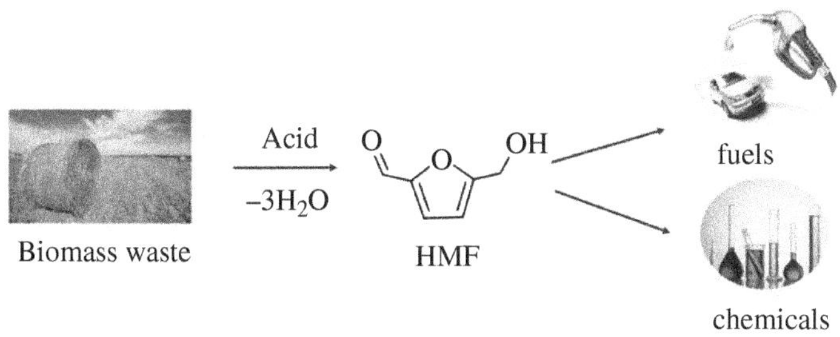

Figure 8.5 The conversion of biomass waste to products via HMF.

resins, metal oxides, and functionalized metal–organic frameworks, have been examined for the production of HMF (Feng et al. 2018, 2019a,2019b,2019c). However, the heterogeneously catalyzed conversion of biomass waste generally has disadvantages of (i) low yields of HMF, (ii) low concentrations of the substrate, (iii) excessive catalyst requirements, and (iv) reduced activity of the recycled catalyst. Homogeneous catalysis could also be a promising route for waste valorisation with the emergence of biphasic solvent systems (ionic liquid/organic solvent or water/organic) because catalysts can be recovered and reused. $AlCl_3$/betaine-based catalysts can be regarded as green and effective applied to biomass waste valorisation, achieving satisfying HMF and furfural yields (Feng et al. 2019a).

Lignin is typically the waste stream in most current biorefinery processes that is combusted to produce heat and power for the biorefinery. Some chemicals such as thiobarbituric acid, phloroglucinol, and barbituric acid can be obtained by the hydrolysis of lignin using different types of acid (Li et al. 2015).

Oxidative and reductive biomass waste depolymerization can also afford a much higher yield of useful chemicals and monomers. For example, terephthalic acid could be obtained in an outstanding yield from corn stover after depolymerization (Song et al. 2019).

8.3.2 Thermochemical Conversion Technologies

Thermochemical conversion technologies for biomass waste valorisation mainly include gasification, pyrolysis, and torrefaction (Figure 8.6). These technologies are temperature-dependent and are all performed in the absence of oxygen or with partial oxidation. Based on the various reaction conditions (temperature, oxidation, reaction rate), the final products of these processes are different. For example, gas products are generally formed when biomass is partially oxidized via a high-temperature gasification reaction; chars are easily formed via a low-temperature torrefaction process. Generally, the temperatures required need to be separated in specially designed reactors. However, all these conversion technologies are not completely separated processes; for example, pyrolysis can be regarded as a partial gasification; torrefaction can be considered a primary low-temperature stage of gasification and pyrolysis. Without the temperature and oxygen control systems, the thermochemical conversion favors the process of combustion or incineration, leaving ash, CO_2, water vapor and heat as products. In detail, gasification is a partial oxidation process with temperatures ranging from 800–1800°C to produce

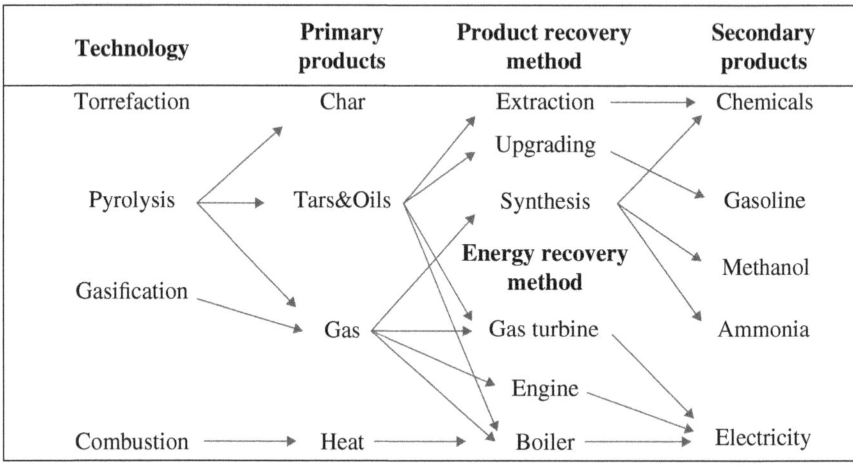

Figure 8.6 *Thermo-conversion processes and products.* Source: Adapted from Bridgwater (1994) and Matsakas et al. (2017).

syngas; pyrolysis is a thermochemical degradation process in the absence of oxygen with temperatures ranging from 300–900°C to produce syngas oil (bio-oil) and char; torrefaction is a mild thermochemical treatment process in the absence of oxygen with temperatures ranging from 200–350°C to produce char.

8.3.3 Biochemical Technologies

Biochemical conversion technologies provide an environmentally friendly route for waste valorisation, which involve the use of enzymes derived from microorganisms to convert organic fractions into energy and products. Fermentation for bioethanol production and anaerobic digestion for biogas production are major directions for biomass waste valorisation.

Ethanol is an important biofuel with the potential to replace gasoline. Global ethanol production reached 25 billion gallons in 2017, with USA and Brazil as the top producers. Lignocellulosic biomass such as corn, sugar beet, rice straw, and algae are suitable raw materials for ethanol production. Generally, the production of ethanol can be simplified into three distinct steps: (i) prepare a fermentable sugar solution; (ii) produce ethanol by fermentation; and (iii) recover and separate ethanol from the fermented solution.

Anaerobic digestion (AD) is another promising biochemical conversion technology for converting various biomass waste including lignocellulosic biomass, food waste, and animal manure to biogas, with concomitant generation of digestate (i.e., solid residue and liquid effluent). The digestate can also be further converted into valuable chemicals and products. Generally, the process of AD can be simplified into four distinct steps: (i) hydrolysis; (ii) acidogenesis, (iii) acetogenesis, and (iv) methanogenesis (Gumisiriza et al. 2017).

8.3.4 Integration with Existing Technologies and Economic Viability

Integrated technologies for biomass waste valorisation involve (i) the use of various feedstocks to produce chemicals, energy, and biofuels; or (ii) the use of a biochemical, or

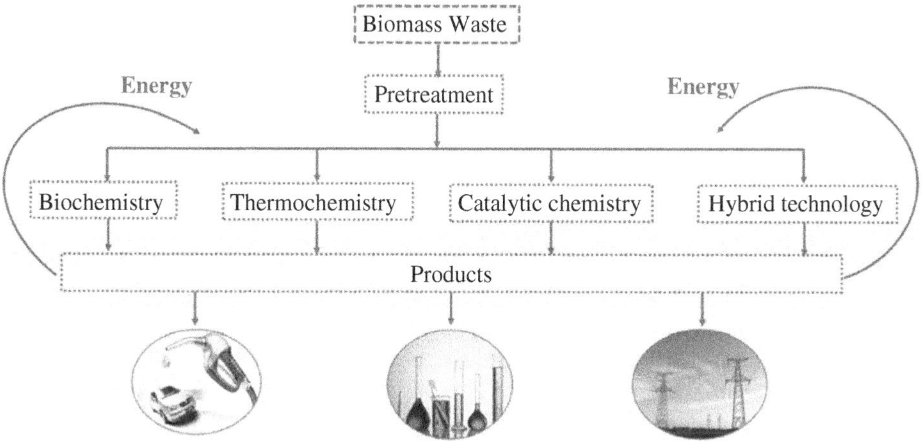

Figure 8.7 A simplified concept of integrated technologies. Various biomass wastes can be used to generate products using a biochemical, or thermochemical, or catalytic chemistry, or a combination of two or three technologies.

thermochemical, or catalytic chemical method, or a combination of two or three technologies, to degrade the raw materials to products (Figure 8.7). Typically, the generated energy is supplied for its operation. The purpose of the integrated technologies is to maximize the value of biomass waste.

Techno-economic analysis of various feedstocks and emerging technologies is crucial for waste valorisation. The cost of raw materials is determined by their source and type, but normally, biomass wastes are low-cost and much cheaper than energy crops. In addition, the cost of biomass waste collection, pretreatment, storage, and transportation are important, and are closely related to the distance traveled and the moisture content (Jeswani et al. 2011).

It has been reported that the capital costs of biochemical technology are lower than those of thermochemical technology, and the integration of multiple technologies can lower investment costs by around 25% (Boakye-Boaten et al. 2017). Cardona Alzate et al. (2018) demonstrated that the implementation of catalytic processes into biorefineries (biochemical and thermochemical technology) is a cost-effective option because more products can be produced. Bastidas-Oyanedel and Schmidt (2018) indicated that the profitability of biomass waste could be increased several times when efforts were focused on various products rather than just biofuels. Another economic evaluation showed that integrated technologies can reduce the generation of waste, minimize energy demand, and increase the output of products (Demichelis et al. 2018).

8.4 Transforming Biomass Waste to Cellulose by New Techniques

In addition to the production of chemicals, there is a strong interest in the green, sustainable production of polymer materials from biomass. In particular, cellulose, as a biodegradable natural polymer material, widely exists and as an inexhaustible resource in nature, and can thus play a major role in replacing petrochemical-based polymers and reducing dependence on fossil-fuel sources. Compared with synthetic polymers, cellulose has the advantages of complete biodegradation, good biocompatibility, non-toxicity, and easy modification, and is expected to replace fossil energy as a sustainable biomass resource. Different cellulose extraction techniques from different biomass resources directly affect the types of cellulose, including the aspect ratio and crystallinity, and thereby the later application of cellulose.

8.4.1 Cellulose Extraction or Purification Techniques from Biomass Waste

For most cellulose source materials, the pretreatment is similar and consists of techniques for pulping and papermaking. In practice, the presence of lignin may hinder the separation of biomass components, so delignification is a necessary step. Siqueira et al. (2010) gave an excellent description of the pulping and bleaching process. Usually, hemicelluloses and related impurities were removed by alkali treatment. In addition, except for chemical treatment, some physical methods (e.g., steam explosion techniques, high-pressure homogenization, sonication, and ball-milling) were carried out as auxiliary methods to further reduce the particle size. Table 8.1 lists the current preparation methods for cellulose as well as other waste sources. In general, on the basis of conventional processes, researchers not only carry out physical treatments (low-/high-pressure, ball-milling, and steam explosion), but also add chemical treatments (oxidation, bleaching by H_2O_2 or NaClO, delignification by NaOH, and enzymatic pretreatment) before hydrolysis, as shown in Figure 8.8.

8.4.2 Cellulose Micro/Nanomerization Technologies

Various commercial applications are being studied according to the different types of cellulose particles. Cellulose has created widespread interest in recent years as a preferred material due to the reuse of biomass from agricultural waste. Numerous articles and reviews have been reported to describe the applications of cellulose-based materials in various fields. At present, cellulose mainly exists in four different structural forms, namely microcrystalline cellulose (MCC), nanocrystalline cellulose (NCC), nanofibrillated cellulose (NFC)

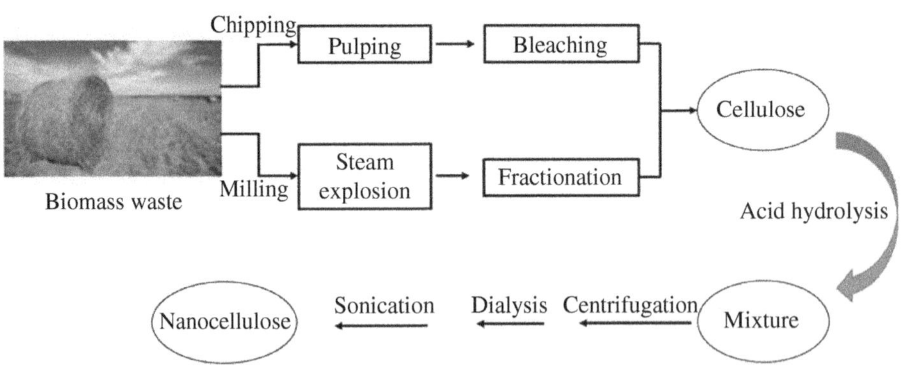

Figure 8.8 Scheme of main steps needed to extract cellulose from biomass.

Table 8.1 *Preparation methods for cellulose as well as other waste sources.*

Waste source	Method	References
Kenaf fibers and wheat straw	Ball milling combined with chemical treatment	Siqueira et al. (2010)
Wheat microcrystalline cellulose	Ultrasonic-assisted enzymatic hydrolysis	Cui et al. (2016)
Oil palm lignocellulosic biomass	Sono-assisted TEMPO oxidation	Rohaizu and Wanrosli (2017)
Tetra Pak waste	Conventional with hydrapulping	Diop and Lavoie (2017)
Zizyphus spinachristi	(i) Extract the cellulose from chipped wood via benzene-EtOH; (ii) digest the material, (iii) bleach by H_2O_2 and HOAc, (iv) extract nanocrystalline cellulose by 64 wt% H_2SO_4 at 70°C	Ogundare et al. (2017)
Rice husk	Low-pressure alkaline delignification process	Islam et al. (2017)
Cellulose-containing material	Oxidation-hydrolysis strategy (solution plasma treatment)	Surov et al. (2018)
Cotton linters	Enzymatic pretreatment, acid hydrolysis using 64 wt% H_2SO_4	Beltramino et al. (2018)
Enteromorpha prolifera	Alkali and bleaching treatments	Kazharska et al. (2019)
Wood biomass	Pretreatment (mainly hydrothermal), extract lignin, milling, and enzymolysis by cellulolytic enzymes	Nitsos et al. (2019)
Lignocellulosic biomass	CO_2-assisted hydrolysis pretreatment, catalytic hydrogenolysis of the pretreated feedstock, and alkaline hydrogen peroxide	Zhai et al. (2019)

and bacterial nanocellulose (BNC). Microcrystalline cellulose in the range of 10–50 μm is commercially available in the pharmaceutical and food industry, as well as the starting material for laboratory preparation of NCC.

In fact, although it has great application prospect, it is challenged by high cost and limited availability for practical application. Moreover, the size, morphology, and crystallinity of cellulose depend on the type of biomass and the conditions under which it is prepared, as well as the experimental techniques used, as shown in Table 8.2. In addition, the size of NCC is uneven because of the random cutting of the cellulose chains in the process of acid hydrolysis. Table 8.2 lists the four types of cellulose, including their sources, preparation methods, particle sizes, and applications.

8.4.2.1 Mechanical–Chemical Technologies

Typically, low-/high-pressure, ball-milling, and steam explosion are the most commonly used methods. Cellulose is also extracted by hypochlorite bleaching with H_2O_2 and HOAc (1 : 1 mixture) at 60°C, alkaline deacetylation, and sulfuric acid hydrolysis (Ogundare et al. 2017). Cellulose suspensions after complete mixing were oxidized by 64 wt% H_2SO_4 at 70°C, and gave excellent quality cellulose with an average crystal size of 8.4 nm. Similarly, Rohaizu and Wanrosli (2017) reported a sono-assisted TEMPO (2, 2,

Table 8.2 Sources, preparation methods and applications of cellulose – MCC, NCC, NFC, and BNC.

Type	Sources	Preparation methods	Particle size	Application
MCC	Cellulose microcrystal (CMC)	Mechanical, untreated	Length: several μm	Adsorption dyes (Wei et al. 2018), antibacterial material (Jeon et al. 2018)
NCC	Cellulose whisker, nanowhisker, nanowire, cellulose nanocrystal (CNC)	Acid hydrolysis	Length: 100–250 nm; diameter: 5–70 nm	Heavy metal adsorbent (Khoo et al. 2018), antibacterial material (Tavakolian et al. 2018)
NFC	Nanofibril, nanofiber, cellulose microfibril	Mechanical (high pressure homogenization, ball-milling), chemical, enzymatic	Length: several μm; diameter: 5–60 nm	Food products (Pares et al. 2018), papermaking (Pares et al. 2018), gels (Yan et al. 2019)
BNC	Biocellulose, microbial cellulose, bacterial cellulose	Bacterial synthesis	Diameter: 20–100 nm	Tissue engineering (Ullah et al. 2016), microbial control, ultrafiltration membrane (Xu et al. 2018), controlled drug delivery (Alkhatib et al. 2017)

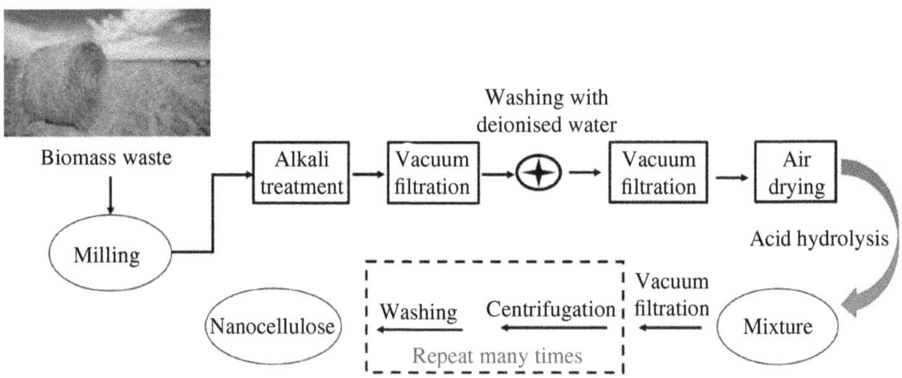

Figure 8.9 Schematic of nanocrystalline cellulose production from biomass on a conventional industrial scale.

6, 6-tetramethylpiperidine-1-oxyl radical) oxidation method to extract cellulose from oil palm lignocellulose biomass, which acted as a flame retardant, as shown in Figure 8.9. Although this method was time-consuming, the chemicals used in the production process could cause serious environmental problems, so it did not meet the needs of a green economy.

A strong acid (~64 wt% H_2SO_4) hydrolyzes the glycoside bond in the polysaccharide chain, resulting in surface functionality, such as negative Zeta potential. The

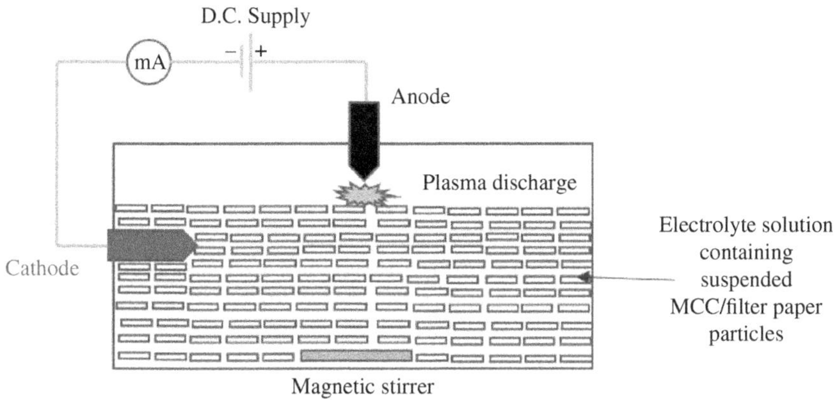

Figure 8.10 *Schematic diagram of a plasma-chemical process to obtain NCC.* Source: Reproduced with permission of Springer, from Surov et al. (2018).

process parameters, such as reaction temperature and time, determine the final product characteristics so that they affect the utility in various fields. At the end of the reaction, a large amount of water is introduced to stop the reaction, followed by dialysis to recover/remove acid. In fact, other acids such as HCl, HBr, and H_3PO_4 are also used to hydrolyze biomass; 64 wt% H_2SO_4 is the preferred choice because of the uniform suspensions, but the whole reaction time is relatively long, which results in a large loss of cellulose. Furthermore, the amount of wastewater produced in the process is another issue. In addition, to reduce the reaction time, dissolution of cellulose (by ionic liquids, NaOH/water at low temperature, tetra-butylammoniumhydroxide [TBAH]/urea, etc.) is reported to optimize hydrolysis. However, transformation from the laboratory scale to an industrial scale is a technical problem that is difficult to tackle given the current technology. All the above-mentioned methods seems to show that cellulose hydrolysis to obtain NCC is not yet a profitable investment.

While seeking a green way to extract NCC, researchers have encountered bottlenecks, leading some to think that the blind pursuit of "green" is counterproductive. Therefore, some researchers have boldly tried some extreme methods, such as chemical oxidation or chemical–mechanical combinations, in order to achieve the desired results. Many studies found that inefficiencies in the delignification process prevented higher product yield, and put forward a low-pressure alkaline delignification stage, which might be a good approach for the industrial production of NCC (Rohaizu and Wanrosli 2017). In recent work on developing NCC, an oxidation–hydrolysis strategy was used to product NCC, using the particularly active hydroxyl radical (·OH), as shown in Figure 8.10. Three different cellulose treatment modes were shown, such as plasma-chemical processing after H_2SO_4, plasma-chemical processing in H_2SO_4, and initial plasma-chemical processing followed by H_2SO_4. This technique of producing NCC at an industrial scale remains to be tested, but it does seem to overcome some limitations of the traditional hydrolysis method.

8.4.2.2 Biochemical Technologies

Recently, in line with the general use of biomass to produce NCC, enzymatic pretreatment before the hydrolytic processes has been reported as a novel strategy. Enzymatic pretreatment was used to remove monosaccharide as an initial step, followed by bleaching and delignification. A series of studies successfully proved that this strategy was a "green step" to extract NCC from biomass and could enhance the overall industrial process (Beltramino et al. 2018). Cui et al. (2016) reported that nanocellulose could be prepared by ultrasonic-assisted enzymatic hydrolysis from wheat MCC. Enzymatic hydrolysis took place at 50°C for 72–120 hours, together with ultrasonic treatment for 60 minutes every 12 hours. Cellulose fabricated by this time-consuming process exhibited a rod-like structure with controlled size and increased crystallinity. Furthermore, steam explosion was put forward to optimize conditions during NCC extraction from biomass as an initial step, followed by bleaching to remove hemicellulose and lignin (Sukyai et al. 2018). The purified cellulose was obtained by acid hydrolysis, centrifugation, and dialysis. The composite films containing NCC were proven to have excellent mechanical properties.

8.4.2.3 Other Technologies

Although there are several technical and economic issues, commercial cellulose is mainly produced by acid hydrolysis (mainly H_2SO_4). Researchers have explored various possible methods to extract cellulosic materials from biomass and agricultural waste. The sugar-containing acid solution produced during hydrolysis of the biomass presents challenges for the treatment or recovery of the acid. Dialysis membrane technologies cannot completely remove the sugar, which complicates the recovery of the acid. The recommended dose of activated carbon is 50 g/l, which can reduce the cost and avoid loss of acid. According to the experimental evaluation, activated carbon could be recycled at least five times, which further reduces the production costs of NCC and has a positive impact on commercial benefits.

Furthermore, the combination of mechanical, enzymatic, and acidic methods to synthesize NCC has more unique advantages. In this compositing process, cellulose is subjected to a series of treatments, including mechanical treatments (microwave, ball-milling, ultrasound-assisted pretreatment), alkali and incubation (in DMSO or TEMPO solution), followed by filtration, washing with deionized water, and H_2SO_4 acid hydrolysis. The obtained NCC is spherical with an average diameter of 30–40 nm and has excellent thermal stability as an intermediate product for functional materials. This approach is a step forward in the direction of "green" transformation of biomass and discarded materials/waste.

8.4.2.4 Commercial Cellulose: Towards Sustainability

Despite cellulose having excellent properties for a number of industries, such as energy storage, food, biomedicine, environmental restoration, and so on, its production cost is a major obstacle to its industrial application. The problems that cause high costs are manifold. For NCC, acid hydrolysis (~64 wt% H_2SO_4) with mechanical processing (high-pressure homogenizers or ball-milling) is the common method, and this works well on a laboratory scale. However, the high temperature together with the low production rate affects the final yield of the product. All these hinder its industrial-scale production and commercial availability, despite the great market demand. The current cellulose market mainly focuses on the development and utilization of NFC for plastics, resins, functional materials, and so on. The main attractions of cellulose are low-cost, sustainable raw material sources, as well as non-toxic, biocompatible products with potential application value. Effective management of biomass to produce high value-added products can further reduce our carbon footprint, which is closer to the circular economy than the current oil-based economy.

At present, researchers have made great efforts and progress in order to overcome the problems of cellulose extraction and micro/nanomerization, such as long processing times, environmental pollution, and high cost. However, technology transfer from the laboratory to industrial-scale production of NCC remains the most important issue. Therefore, cooperation between engineers and scientific researchers is very important.

8.5 Transforming Biomass Waste to Lignin by New Technologies

Lignin is an essential plant component, and the second most abundant natural polymer after cellulose, accounting for 30% of the total plant biomass (Beckham et al. 2016). Lignin carries a large number of functional groups, such as carbonyl, methoxyl, phenolic hydroxyl, and benzyl alcohol groups, which gives it structural complexity and diversity (Liu et al. 2019). In brief, lignin contains three main structures: p-coumaryl alcohol, coniferyl alcohol, and sinapyl alcohol. These structures contained in lignin are the main source of its heterogeneity and instability, and are a great challenge to the depolymerization and further application of lignin.

Among the recent technologies for lignin degradation, depolymerization by chemical means and conversion by microorganisms are the main methods that hold great potential (Ragauskas et al. 2014). Despite significant progress, output of lignin is still low which seriously restricts its application in various fields. In order to find feasible solutions with the problems existing in lignin extraction and utilization, researchers have carried out a series of investigations, mainly from the chemical and biological point of view.

For biotransformation of lignin, using bioinformatics to search for ligniniolytic enzymes and aromatic degradation pathways in various bacteria showed promising application prospects (Hassan et al. 2018; Tian et al. 2017). In brief, the changes in lignin content during enzymatic hydrolysis were determined through colorimetric assays in the ground wood. A large number of bacteria were determined to degrade the lignin during the experiment; it is likely that there are still many pathways and enzymes that remain unknown. With continued screening of enzymes, efficient candidates can be found to improve the biological value of lignin and produce value-added products, and also expand our knowledge of lignin degradation. Several fungi proved to be effective in processing lignin to produce biofuels and chemicals, such as *Rhodococcus genus*, *Pseudomonas putida*, yeasts, and so on. Table 8.3 lists the microbial degradation of lignin in various biomass sources.

Considering the ability of a single bacterium to degrade lignin, the synergistic effect of multiple microorganisms was particularly important, providing a considerable potential for the depolymerization and degradation of lignin. Microbial colonies have the advantages of low cost and minimal environmental effects, which is expected to reduce the operating cost involved in lignin degradation. In addition, lignin biodegradation is often accompanied by the simultaneous utilization of cellulose and hemicellulose in the biomass. The existence of cellulose in the degradation process is beneficial to the catalase oxidation of lignin produced by microorganisms. This is the main result of the removal of carbohydrates during biomass separation, resulting in lignin residue. In recent technologies, several strains were chosen to depolymerize lignin while keeping the cellulose structure intact.

Table 8.3 Genetically engineered microorganisms for lignin degradation in various biomass resources.

Biomass	Fungi/bacteria	Degradation time (days)	Lignin degradation	References
Wheat straw	*L. betulinus, D. squalens*	14	28.3%, 34%	Hassan et al. (2018), Tian et al. (2017)
Bamboo	Unidentified TUFC20057, *Dictyopanus sp. TUFC20090*	84	53%, 27.3%	Suhara et al. (2012)
Rye straw	*Galactomyces geotrichum, Myrothecium verrucaria*	60	48%, 42%	Varnaite and Raudoniene (2005)
Rice straw	*P. chrysosporium, Fusarium sp.89, Fusarium sp.82*	10	28.3%, 33.5%, 29.3%	Chang et al. (2012)
Corn stover	*I. lacteus*	28	25.48%	Song et al. (2013)
Rice straw	*Bacillus sp. CS-1, Lactic acid bacteria*	10	20%, 14.3%	Chang et al. (2014)
Wheat straw, corn stalk	*Microbial consortium XDC-2*	12	40.8%, 9.6%	Chang et al. (2014)

The preparation of biological products from biomass lignin has not been carried out on a large scale. A technical and economic analysis of the process of lignin production can help to optimize the cost factors of lignin, thereby reducing the technical risk and improving the competitiveness of biological refining.

In the process of separating the three major elements of lignocellulose, lignin will be chemically modified, and the degree of modification is closely related to the type of lignin (being more pronounced for sulfur-containing lignin than sulfur-free lignin). However, the structure of lignin is complex and intractable, which is the main challenge in the process of lignin separation and utilization. The biomass often needs prolonged biological treatment, and the cost of toxic wastewater treatment is very high. On the other hand, the process of microbial depolymerization of lignin is often accompanied by unsatisfactory crosslinking polymerization, which requires further in-depth research.

8.6 Conclusions and Future Trends

The existing chemical, biochemical, and thermochemical technologies are potential protocols for the conversion of biomass waste, such as lignocellulosic biomass and food waste, into chemicals, fuels, materials and energy. However, the collection of waste can be especially difficult because of its wide distribution and high transportation costs. Moreover, collected waste (especially food waste) cannot be stored for a long time, whereas a sustainable waste supply is required for large industries. In this context, small- or medium-scale plants may be more suitable for waste valorisation. On the other hand, the pretreatment methods (especially involved food waste) vary considerably depending on the substrate type, which would result in high cost. In some cases, the use of strong acids and bases for biomass waste hydrolysis is environmentally unfriendly. And in some perspectives, the conversion of biomass waste is limited by its intermediates. For example, 5-hydroxymethylfurfural (HMF) is an important intermediate for the conversion of lignocellulose to useful products, but the purification of HMF by a simple and cheap approach is difficult. Besides the production process of intermediates, the individual intermediate will be costly. Integration of existing technologies may be a cost-efficient and eco-friendly strategy to improve the economic value of biomass waste. However, such studies are still at the laboratory scale, and there is limited information about the up-scale conversion of biomass waste. In addition, owing to issues such as climate change, security of energy supply, and cost of investments and feedstock supply, the practical implementation of these proposed protocols remain great challenges. Lastly, future research should focus on biomass waste collection, pretreatment, and investigation of the feasibility of biomass waste valorisation, especially for large-scale production. Financial incentives or supports may be necessary to drive the commercial conversion of biomass waste to products.

Acknowledgements

The authors gratefully acknowledge financial support from the National Natural Science Foundation of China (Nos. 21676223, 21978248), the special fund for Fujian Ocean High-Tech Industry Development (No. FJHJF-L-2018-1), the Natural Science Foundation of Fujian Province of China (No.2019J06005), and the Energy Development Foundation of the College of Energy, Xiamen University (No. 2017NYFZ02).

References

Alkhatib, Y., Dewaldt, M., Moritz, S. et al. (2017). Controlled extended octenidine release from a bacterial nanocellulose/poloxamer hybrid system. *Eur. J. Pharm. Biopharm.* **112**: 164–176.

Bastidas-Oyanedel, J. and Schmidt, J.E. (2018). Increasing profits in food waste BiorefineryA techno-economic analysis. *Energies* **11** (15516).

Beckham, G.T., Johnson, C.W., Karp, E.M. et al. (2016). Opportunities and challenges in biological lignin valorisation. *Curr. Opin. Biotechnol.* **42**: 40–53.

Beltramino, F., Roncero, M.B., Vidal, T., and Valls, C. (2018). A novel enzymatic approach to nanocrystalline cellulose preparation. *Carbohydr. Polym.* **189**: 39–47.

Boakye-Boaten, N.A., Kurkalova, L., Xiu, S., and Shahbazi, A. (2017). Techno-economic analysis for the biochemical conversion of *Miscanthus x giganteus* into bioethanol. *Biomass Bioenergy* **98**: 85–94.

Bridgwater, A.V. (1994). Catalysis in thermal biomass conversion. *Appl. Catal., A* **116** (1–2): 5–47.

Cardona Alzate, C.A., Solarte Toro, J.C., and Gómez Peña, A. (2018). Fermentation, thermochemical and catalytic processes in the transformation of biomass through efficient biorefineries. *Catalysis Today* **302**: 61–72.

Carrero, J.J., Nakashima, A., Qureshi, A.R. et al. (2011). Protein-energy wasting modifies the association of ghrelin with inflammation, leptin, and mortality in hemodialysis patients. *Kidney Int.* **79** (7): 749–756.

Chang, A.J., Fan, J., and Wen, X. (2012). Screening of fungi capable of highly selective degradation of lignin in rice straw. *Int. Biodeterior. Biodegrad.* **72**: 26–30.

Chang, Y., Choi, D., Takamizawa, K., and Kikuchi, S. (2014). Isolation of *Bacillus* sp strains capable of decomposing alkali lignin and their application in combination with lactic acid bacteria for enhancing cellulase performance. *Bioresour. Technol.* **152**: 429–436.

Chuetor, S., Luque, R., Barron, C. et al. (2015). Innovative combined dry fractionation technologies for rice straw valorisation to biofuels. *Green Chem.* **17** (2): 926–936.

Corma, A., Iborra, S., and Velty, A. (2007). Chemical routes for the transformation of biomass into chemicals. *Chem. Rev.* **107** (6): 2411–2502.

Cui, S., Zhang, S., Ge, S. et al. (2016). Green preparation and characterization of size-controlled nanocrystalline cellulose via ultrasonic-assisted enzymatic hydrolysis. *Ind. Crops Prod.* **83**: 346–352.

Demichelis, F., Fiore, S., Pleissner, D., and Venus, J. (2018). Technical and economic assessment of food waste valorisation through a biorefinery chain. *Renewable Sustainable Energy Rev.* **94**: 38–48.

Diop, C.I.K. and Lavoie, J. (2017). Isolation of nanocrystalline cellulose: a technological route for valorizing recycled Tetra Pak aseptic multilayered food packaging wastes. *Waste Biomass Valorisation* **8** (1): 41–56.

Feng, Y., Zuo, M., Zeng, X. et al. (2018). Preparation of 5-hydroxymethylfurfural from glucose. *Prog. Chem.* **30** (2–3): 314–324.

Feng, Y., Li, M., Gao, Z. et al. (2019a). Development of betaine-based sustainable catalysts for green conversion of carbohydrates and biomass into 5-hydroxymethylfurfural. *ChemSusChem* **12** (2): 495–502.

Feng, Y., Yan, G., Wang, T. et al. (2019b). Synthesis of MCM-41-supported metal catalysts in deep eutectic solvent for the conversion of carbohydrates into 5-hydroxymethylfurfural. *ChemSusChem* **12** (5): 978–982.

Feng, Y., Zuo, M., Wang, T. et al. (2019c). Efficient synthesis of glucose into 5-hydroxymethylfurfural with SO_4^{2-}/ZrO_2 modified H^+ zeolites in different solvent systems. *J. Taiwan Inst. Chem. Eng.* **96**: 431–438.

Gumisiriza, R., Hawumba, J.F., Okure, M., and Hensel, O. (2017). Biomass waste-to-energy valorisation technologies: a review case for banana processing in Uganda. *Biotechnol. Biofuels* **10** (11).

Hassan, S.S., Williams, G.A., and Jaiswal, A.K. (2018). Emerging technologies for the pretreatment of lignocellulosic biomass. *Bioresour. Technol.* **262**: 310–318.

Hayes, D.J., Ross, J.R.H., Hayes, M.H.B., and Fitzpatrick, S. (2005). The biofine process: Production of levulinic acid, furfural, and formic acid from lignocellulosic feedstocks. In: *Biorefineries – Industrial Processes and Product* (eds. B. Kamm, P.R. Gruber and M. Kamm), 139–164. Wiley-VCH Verlag.

Islam, M.S., Kao, N., Bhattacharya, S.N., Gupta, R. (2017). An investigation between high and low pressure processes for nanocrystalline cellulose production from agro-waste biomass. *AIP Conference Proceedings, 32nd International Conference of the Polymer Processing Society* https://doi.org/10.1063/1.5016729

Jeon, W., Kim, Y., Hong, M. et al. (2018). Microcrystalline cellulose for delivery of recombinant protein-based antigen against erysipelas in mice. *Biomed. Res. Int.*: 7670505. https://doi.org/10.1155/2018/7670505.

Jeswani, H.K., Gujba, H., and Azapagic, A. (2011). Assessing options for electricity generation from biomass on a life cycle basis: environmental and economic evaluation. *Waste Biomass Valorisation* **2** (1): 33–42.

Karmee, S.K. (2016). Liquid biofuels from food waste: current trends, prospect and limitation. *Renewable Sustainable Energy Rev.* **53**: 945–953.

Kazharska, M., Ding, Y., Arif, M. et al. (2019). Cellulose nanocrystals derived from *Enteromorpha prolifera* and their use in developing bionanocomposite films with water-soluble polysaccharides extracted from *E. prolifera. Int. J. Biol. Macromol.* **134**: 390–396.

Khoo, R.Z., Chow, W.S., and Ismail, H. (2018). Sugarcane bagasse fiber and its cellulose nanocrystals for polymer reinforcement and heavy metal adsorbent: a review. *Cellulose* **25** (8): 4303–4330.

Li, C., Zhao, X., Wang, A. et al. (2015). Catalytic transformation of lignin for the production of chemicals and fuels. *Chem. Rev.* **115** (21): 11559–11624.

Liu, Z., Hao, N., Shinde, S. et al. (2019). Defining lignin nanoparticle properties through tailored lignin reactivity by sequential organosolv fragmentation approach (SOFA). *Green Chem.* **21** (2): 245–260.

Mats, K., Meine, N., Farès, C. et al. (2014). Fractionation of 'water-soluble lignocellulose' into C5/C6 sugars and sulfur-free lignins. *Green Chem.* **16**: 2454–2462.

Matsakas, L., Gao, Q., Jansson, S. et al. (2017). Green conversion of municipal solid wastes into fuels and chemicals. Electron. *J. Biotechnol.* **26**: 69–83.

Nitsos, C.K., Lazaridis, P.A., Mach-Aigner, A. et al. (2019). Enhancing lignocellulosic biomass hydrolysis by hydrothermal pretreatment, extraction of surface lignin, wet milling and production of cellulolytic enzymes. *ChemSusChem* **12** (6): 1179–1195.

Ogundare, S.A., Moodley, V., and van Zyl, W.E. (2017). Nanocrystalline cellulose isolated from discarded cigarette filters. *Carbohydr. Polym.* **175**: 273–281.

Pares, D., Angels Pelach, M., Toldra, M. et al. (2018). Nanofibrillated cellulose as functional ingredient in emulsion-type meat products. *Food Bioprocess Technol.* **11** (7): 1393–1401.

Ragauskas, A.J., Beckham, G.T., Biddy, M.J. et al. (2014). Lignin valorisation: improving lignin processing in the biorefinery. *Science* **344** (12468436185): 709.

Renders, T., Schutyser, W., Van den Bosch, S. et al. (2016). Influence of acidic (H_3PO_4) and alkaline (NaOH) additives on the catalytic reductive fractionation of lignocellulose. *ACS Catal.* **6** (3): 2055–2066.

Rohaizu, R. and Wanrosli, W.D. (2017). Sono-assisted TEMPO oxidation of oil palm lignocellulosic biomass for isolation of nanocrystalline cellulose. *Ultrason. Sonochem.* **34**: 631–639.

Sawatdeenarunat, C., Nguyen, D., Surendra, C. et al. (2016). Anaerobic biorefinery: current status, challenges, and opportunities. *Bioresour. Technol.* **215**: 304–313.

Siqueira, G., Bras, J., and Dufresne, A. (2010). Cellulosic bionanocomposites: a review of preparation, properties and applications. *Polymers-Basel* **2** (4): 728–765.

Sluiter, A., Hames, B., Ruiz, R. et al. (2008). *Determination of Structural Carbohydrates and Lignin in Biomass. Laboratory Analytical Procedure*. Golden, CO: NREL https://www.nrel.gov/docs/gen/fy13/42618.pdf (accessed 25 April 2020).

Song, L., Ma, F., Zeng, Y. et al. (2013). The promoting effects of manganese on biological pretreatment with Irpex lacteus and enzymatic hydrolysis of corn stover. *Bioresour. Technol.* **135**: 89–92.

Song, S., Zhang, J., Gozaydin, G., and Yan, N. (2019). Production of terephthalic acid from corn stover lignin. *Angew. Chem. Int. Ed.* **58** (15): 4934 4937.

Su, Y., Du, R., Guo, H. et al. (2015). Fractional pretreatment of lignocellulose by alkaline hydrogen peroxide: characterization of its major components. *Food Bioprod. Process.* **94**: 322–330.

Suhara, H., Kodama, S., Kamei, I. et al. (2012). Screening of selective lignin-degrading basidiomycetes and biological pretreatment for enzymatic hydrolysis of bamboo culms. *Int. Biodeterior. Biodegrad.* **75**: 176–180.

Sukyai, P., Anongjanya, P., Bunyahwuthakul, N. et al. (2018). Effect of cellulose nanocrystals from sugarcane bagasse on whey protein isolate-based films. *Food Res. Int.* **107**: 528–535.

Sun, J., Dutta, T., Parthasarathi, R. et al. (2016). Rapid room temperature solubilization and depolymerization of polymeric lignin at high loadings. *Green Chem.* **18** (22): 6012–6020.

Surov, O.V., Voronova, M.I., Rubleva, N.V. et al. (2018). A novel effective approach of nanocrystalline cellulose production: oxidation-hydrolysis strategy. *Cellulose* **25** (9): 5035–5048.

Tavakolian, M., Okshevsky, M., van de Ven, T.G.M., and Tufenkji, N. (2018). Developing antibacterial nanocrystalline cellulose using natural antibacterial agents. *ACS Appl. Mater. Interfaces* **10** (40): 33827–33838.

Tian, D., Hu, J., Bao, J. et al. (2017). Lignin valorisation: lignin nanoparticles as high-value bio-additive for multifunctional nanocomposites. *Biotechnol. Biofuels* **10** (192).

Tuck, C.O., Peréz, E., Horváth, I.T. et al. (2012). Valorisation of biomass: deriving more value from waste. *Science* **337** (6095): 695–699.

Ullah, H., Wahid, F., Santos, H.A., and Khan, T. (2016). Advances in biomedical and pharmaceutical applications of functional bacterial cellulose-based nanocomposites. *Carbohydr. Polym.* **150**: 330–352.

Varnaite, R. and Raudoniene, V. (2005). Enzymatic lignin degradation in rye straw by micromycetes. *Int. Biodeterior. Biodegrad.* **56** (3): 192–195.

Wei, X., Huang, T., Nie, J. et al. (2018). Bio-inspired functionalization of microcrystalline cellulose aerogel with high adsorption performance toward dyes. *Carbohydr. Polym.* **198**: 546–555.

Xu, T., Jiang, Q., Ghim, D. et al. (2018). Catalytically active bacterial nanocellulose-based ultrafiltration membrane. *Small* **14** (170400615).

Yan, G., Zhang, X., Li, M. et al. (2019). Stability of soluble dialdehyde cellulose and the formation of hollow microspheres: optimization and characterization. *ACS Sustainable Chem. Eng.* **7** (2): 2151–2159.

Zhai, Q., Li, F., Long, F. et al. (2019). Integrated separation process of CS sugars and phenolics from poplar wood using CO_2-assisted hydrolysis followed by hydrogenolysis. *ACS Sustainable Chem. Eng.* **7** (1): 526–536.

Zhang, Y.H.P., Zhu, Z., Rollin, J., and Sathitsuksanoh, N. (2010). Advances in cellulose solvent- and organic solvent-based lignocellulose fractionation (COSLIF). In: *Cellulose Solvents: For Analysis, Shaping and Chemical Modification*, ACS Symposium Series vol. 1033 (eds. T.F. Liebert, T.J. Heinze and K.J. Edgar), 365–379.

9

Life cycle Approaches for Evaluating Textile Biovalorisation Processes: Sustainable Decision-making in a Circular Economy

Karpagam Subramanian, Shauhrat S. Chopra, Cakin Ezgi, Xiaotong Li and Carol Sze Ki Lin

School of Energy and Environment, City University of Hong Kong, Kowloon Tong, Hong Kong, China

9.1 Introduction

Rapid urbanization and modernization drive many environmental changes, like pollution, threats to biodiversity, GHG emissions, and health consequences across local and global scales (Nagendra et al. 2018). It is a complex phenomenon that propels social, economic, and environmental footprints, compelling industries to achieve sustainable development in multiple dimensions (Pedro et al. 2019). Environmental problems associated with the production of conventional products using fossil fuels and consumer awareness of eco-friendly products, have also prompted the need for sustainable production approaches (González-García et al. 2016). Sustainable development was popularized as a concept by the Brundtland Commission (United Nations, 1987) to "meet the needs of the present without compromising the ability of future generations to meet their own needs."

Waste Valorisation: Waste Streams in a Circular Economy, First Edition.
Edited by Carol Sze Ki Lin, Guneet Kaur, Chong Li and Xiaofeng Yang.

Subsequently, the Rio Summit in 1992 introduced and emphasized a connection between society, economy, and environment as fundamental to sustainable development. Hence, sustainable production approaches that ensure social cohesion, environmental protection, and worldwide prosperity have to be developed.

The concept of biovalorisation, based on biowastes plays a key role in the transition from conventional products to bio-based products, and helps achieve sustainable development by addressing the three main dimensions: environmental sustainability, by converting waste to biofuels/virgin raw materials through valorisation; economic sustainability: production of value-added byproducts that can bring revenue when sold in various other industries/markets; and social sustainability: creation of job opportunities in involved industries and satisfying the growing customer demand for bio-based products (Fava et al. 2015). Recently, the demand for bio-based products derived from sustainable resources has increased in many sectors, such as food, fuel, textiles, and chemicals, as they help to improve ecosystem quality, achieve zero waste, and make a positive contribution to society and economy (González-García et al. 2018). Also, an increasing number of publications in the area of waste valorisation to produce bio-based products, indicating a growing trend, has been observed while searching in databases such as Scopus and Science Direct.

Now that the exploration and development of such biovalorisation techniques is growing, it is important to evaluate the efficiency of such technologies from a sustainability perspective. Life Cycle Assessment (LCA) has been widely used in the textile sector for analyzing the environmental impacts of various aspects of the supply chain (Muthu et al. 2012), but application of LCA to evaluate biovalorisation methods is still new. Few studies have evaluated the environmental and economic dimensions of biovalorisation techniques in the literature, and LCA is the most widely used tool to calculate potential impacts (Iribarren et al. 2016). Social dimensions of sustainability are completely neglected mainly due to the lack of: (i) standardized procedures to calculate social impacts, and (ii) understanding of the social mechanisms and interactions between the involved processes and stakeholders (Popovic et al. 2018; Subramanian et al. 2018).

The textile industry produces an enormous amount of waste globally (Hu et al. 2018). An increasing population, fast-changing fashion trends and the rise of purchasing power are all factors that have further stimulated worldwide clothes consumption and consequently, waste generation (Ling et al. 2019). Reuse is considered an ideal solution to this global problem, but it may not be a scalable solution; recycling might also have some environmental implications when the involved processes are driven by fossil fuel-derived energy (Sandin and Peters 2018). Hence, recently the textile sector has been paying attention to organic waste valorisation. The huge amount of textile waste generated can be a potential feedstock for biobased products, which can replace virgin material in the production process. Most of the textile wastes generated are a blend of cotton and polyester. The literature recommends polyester recycling, considering the kind of environmental impacts created, and also to reduce the depletion of petroleum resources. Additionally, 35–40% of textile waste, specifically, 95–99% of cotton, contains cellulose, which has numerous applications and is also a potential feedstock for bioethanol and biogas production in biorefineries (Shen et al. 2013). Recently, research on pretreatment and hydrolysis of cellulose, the biodegradable part in textile fabrics, has been carried out to convert it into fermentable glucose (Huang and Fu 2013). Byproducts such as glucose obtained during bioconversion also add to the economics. However, polyester in the cotton-based textiles inhibits bioconversion of textile waste, as it obstructs the enzymes from reaching the cotton during hydrolysis (Gholamzad et al. 2014). Hence, current research mainly focuses on optimizing the conditions for pretreatment and enzymatic hydrolysis, for effective and efficient bioconversion of cotton–polyester blends.

In this chapter, a textile biorecycling method used to recover polyethylene terephthalate (PET) fibers from a 50:50 cotton–polyester blend of waste textile fabrics is investigated from a sustainability perspective. The impacts are quantified using Life Cycle Assessment. In the analyzed biorecycling method, cellulose, the biodegradable part of the textile, is subjected to alkaline pretreatment and enzymatic hydrolysis under optimum conditions to convert it to glucose, while the remaining polyester, the non-biodegradable part, is recovered and re-spun into fibers. The polyester fiber recovered through the biorecycling method becomes virgin material for the next textile production cycle; glucose obtained as an additional byproduct has numerous applications in other markets. The analyzed textile biorecycling method also aligns with the circular economy principles of converting waste to resources. For instance, recently many textile industry retailers like H&M, Zara, Adidas, and Nike have started adopting such circular economy (CE) initiatives. The Hong Kong Research Institute of Textiles and Apparel (HKRITA) has been involved in research collaborations and development activities for transforming a fossil-based textile industry into a bio-based one (www.hkrita.com).

Though the biorecycling method can effectively recover polyester and convert cellulose in cotton into glucose, potential implications related to non-renewable energy consumption of the involved processes, production of the chemicals, and the cost of the enzymes used in the hydrolysis step might reduce the efficiency of the biorecycling method, and cannot be ignored (Ling et al. 2019). Hence, an evaluation of this biorecycling method is needed to understand its efficiency from a sustainability perspective. Previously, a techno-economic evaluation of the biorecycling method was carried out at a pilot scale, indicating the need to lower the enzyme usage, considering the huge cost involved (Kwan 2018). However, other possible advantages of biorecycling like moderate nonrenewable energy consumption, when compared with conventional recycling techniques, social aspects like job opportunities, improved health and safety, and organizational wellbeing are not explored in the literature (González-García et al. 2019). Similarly, LCA as a tool has been widely applied to assess various aspects of the textile supply chain, but the applicability of LCA to evaluate textile biovalorisation is less well explored (Neupane et al. 2013). Here, we adopt LCA methods to evaluate environmental and social implications of the textile biorecycling method. Inclusion of social dimensions along with environmental implications is mandatory to ensure safe and healthy working conditions for the involved workers and other related stakeholders (Lenzo et al. 2017).

This chapter describes the adoption of environmental LCA to identify potential environmental hotspots in the lifecycle of a biovalorisation process, taking bio-based PET as an example. Additionally, this work explores the possible social interaction associated with an integrated textile waste-based biorefinery and outlines an approach to evaluate social sustainability of biovalorisation processes based on social LCA. Such a comprehensive LCA framework that covers social and environmental aspects, along with a techno-economic evaluation, can guide businesses and other involved stakeholders to make informed decisions from a holistic perspective, rather than overemphasizing any one of the sustainability dimensions.

9.2 Literature Review

9.2.1 Circular Economy and Sustainable Development

Sustainability has started to gain more attention due to "wicked" environmental threats like global warming, soil acidification, biodiversity loss, and pollution (Duckett et al. 2016). The need for sustainable development and the impact of urbanization has been discussed in a variety of industries (Opoku 2019) involving the three pillars of sustainability: society, economics, and environment. Global sustainability can be achieved only by transforming the widely accepted and used linear economy model to a CE, which is restorative and regenerative by design. The CE aims to close the loops to regain resources back to the economy (Baydar et al. 2015). In order to do so, a new relationship between businesses and resources needs to be formed. Along with reduction of waste and efficient resource consumption, industries have to improve the safety of workers and customers, develop innovative products rather than conventional ones, and at the same time make a profit (Rugani et al. 2014). The main challenge while creating this new relationship is the lack of tools that support decision-making from a sustainability perspective (Traverso 2018). Environmental, social, and cost assessment methods that are easily applied commercially need to be developed to help in such decision-making processes, to facilitate economic growth while minimizing negative impacts on the environment and society.

9.2.2 Textile Industry – Sustainability Issues and Recycling

The textile industry is considered highly unsustainable due to the sheer volume of textiles that is produced and wasted, and concerns about disposal and management of textile waste are increasing (Pensupa et al. 2013). The fast-changing fashion trends and the increase in demand has resulted in an increase in the rate of textile production as well as the subsequent waste generation. As a consequence, more than 1.2 billion tonnes of carbon emissions are created every year through textile manufacturing processes, and most textile waste ends up in landfills (Zamani et al. 2015). The increasing amounts of textile waste, accompanied by decreasing landfill space in cities, has further prompted investigation into novel reuse and recycling methods (Sandin et al. 2011). Ideally, reuse of products is a more environmentally beneficial route as compare to recycling; however, the fast-changing fashion industry limits the reuse of clothing and encourages the development of recycling as the main solution to the textile waste problem.

9.2.2.1 Textile Recycling Routes

Currently, textile recycling processes are classified as mechanical, semi-mechanical, thermochemical, and chemical (Shen et al. 2010). Chemical recycling is usually carried out by either depolymerizing the polymers or dissolving them, whereas repolymerization of monomers or oligomers produces the recycled fibers. Mechanical recycling is executed by directly extruding polymers from PET flakes, followed by spinning and drawing filaments into fibers. Semi-mechanical methods, instead, pelletiz and melt-spin PET flakes into fibers, while later on ethylene glycol is usually added to increase the quality of the fiber (Shen et al. 2011). Thermal recycling produces flakes, pellets and chips mechanically, and melt-extrudes them into fibers. These recycling routes are not distinctively different from each other, which allows the methods to be referred to interchangeably. The environmental impacts of these methods are analyzed in the literature, which concludes that chemical recycling is the method with the highest environmental impact, and the mechanical method is the least impactful (Dahlbo et al. 2017).

There is now a pressing need to use sustainable raw materials in textiles, due to growing awareness and the need to reduce the environmental footprint. The concept of a textile-based biorefinery has been developed to produce raw materials and chemicals from textile wastes. Outside the conventional recycling methods, a few studies have investigated possible biovalorisation techniques that enable bioconversion of textile wastes. Cotton and sisal waste are bioconverted into biofuel by alcoholic fermentation and biogas technology, which has revealed the potential of using textile residues for energy production (Nzila et al. 2009). In the biorecycling method chosen to be analyzed, the biodegradable part of polyester–cotton textile waste is converted to a value-added byproduct, glucose syrup, while the non-biodegradable part is recovered to produce polyester fibers. Further to the rising interest in the development of textile biorefineries, sustainable development of these technologies is also essential to ensure efficient operation and encourage involved stakeholders to make investment decisions by understanding the tradeoffs (Budzianowski and Postawa 2016). Comparison of these novel technologies with conventional techniques (mechanical/chemical recycling) will also enable involved stakeholders to choose a sustainable technology for textile recycling.

9.2.2.2 Quantification of Sustainability in the Textile Industry

Sustainability in the textile industry can only be achieved with collaborative efforts of all the partners involved in the sector, from industry leaders to consumers and the government (Boström and Micheletti 2016). The traditional Triple Bottom Line (TBL) approach considers environment, society, and economy as the three pillars of sustainability, where each pillar carries equal importance and needs to be quantified. Environment and economy-related aspects of sustainability are mostly evaluated within the textile supply chain (Muthu et al. 2012). One of the most widely accepted tools for assessing environmental impact is LCA (Subramanian and Yung 2016). LCA considers the whole lifecycle of a product/service where raw material extraction, manufacturing, use, and disposal stages are all included.

LCA has been used to evaluate various textile product lifecycles and it is a quite well-established tool within the textile industry (Sandin and Peters 2018). Benchmarking of various textiles (cotton, polyester, nylon, and others) has been carried out using LCA (Van Der Velden et al. 2014). LCA was conducted to investigate environmental impacts of cotton/wood-based textile products and antibacterial T-shirts either ending the lifecycle with disposal at landfills or excluding the disposal/recycling stage from the process boundary (Moazzem et al. 2018). LCA has also been used to analyze the environmental impacts of the overall textile supply chain (Moazzem et al. 2018). Improved designs for the textile manufacturing process were developed based on LCA results (Zhang et al. 2018). Textile reuse and recycling techniques have been analyzed to evaluate overall environmental impacts, as well as single issues such as carbon footprint (Muthu et al. 2012; Sandin and Peters 2018; Zamani et al. 2015). Textile recycling using non-textile materials such as tires (Landi et al. 2018) and PET bottles (Shen et al. 2010, 2011) have all been previously analyzed in the literature using LCA. Application of recycled PET fiber in building thermal insulation was evaluated for its environmental implications based on the LCA approach (Ingrao et al. 2014). A European textile sorting center was evaluated based on material flow analysis and lifecycle inventory (Norup et al. 2019). The socioeconomic advantages of textile recycling were reviewed by Leal Filho et al. (2019). An environmental assessment of textile material recovery techniques by examining textile flows in Sweden was carried out by Youhanan (2013). A number of studies in the literature have used LCA for assessing environmental impacts associated with textile waste recycling, and the most studied material is polyester. However, LCA application to evaluate textile biovalorisation techniques is a new emerging tool. Some studies have evaluated the social aspects of the textile supply chain (Lenzo et al. 2017; Musaazi et al. 2015; Waas et al. 2014; Zamani et al. 2016).

Quantification of economic sustainability performance carries great importance in businesses that aim for financial profitability. Techno-economic analysis (TEA) is a frequently used method to determine the potential of a technology/product/service during its development stage, where results indicate the feasibility of the developing process/product. Investment decisions are based on TEA reports, where the areas that need to be improved are also revealed. Previously, a few studies in the literature have assessed the economic sustainability of textile procedures via TEA (Leal Filho et al. 2019).

The literature review clearly indicates that the social aspect of sustainability is often neglected, and there is a dearth of articles that quantify the environmental and social sustainability performance of textile biorefineries. LCA is a tool that can be used to evaluate multiple dimensions of sustainability, such as environment, society, and economics, by taking material and energy inputs/outputs, costs, benefits, and social aspects into consideration. Hence, this chapter focuses on the production of bio-based products that are beneficial to the environment and society.

9.3 Methods

9.3.1 Description of Environmental Assessment

Environmental Life Cycle Assessment (ELCA) is considered to be one of the best-established systems analysis tools to measure the environmental impacts of products/processes/services across the entire lifecycle, encompassing different stages from raw material extraction, through production, distribution, use, and disposal. It is widely used by researchers and policymakers for a holistic evaluation of the environmental dimensions of sustainability (Chopra et al. 2019; Chopra and Theis 2017; Falinski et al. 2018; Subramanian and Yung 2017; Yeo et al. 2019; Zaimes et al. 2015). As already mentioned, numerous studies have applied ELCA to analyze environmental aspects of various textile products, supply chain and recycling techniques following the ISO 14040 (2006) guidelines. Additionally, a few researchers have explored the applicability of ELCA to evaluate the environmental dimension of biorefineries (Moussa et al. 2016; Neupane et al. 2013; Smidt et al. 2016; Gilani and Stuart 2015). Hence, the applicability of ELCA for assessing environmental impacts of textile biorecycling is justified.

ELCA is conducted in four main phases based on ISO 2006 (Song et al. 2013): (i) goal and scope, (ii) Life Cycle Inventory (LCI), (iii) Life Cycle Impact Assessment (LCIA) and (iv) interpretation. The aim of the study, process flow, system boundaries, geographical limitations, functional unit used, and any other assumptions made, are included when defining the goal and scope in the first step. Collection of data regarding the inputs and outputs of the analyzed process needed to conduct an LCA study are discussed in the LCI step. Data can be collected from primary (site-specific) and secondary sources (literature sources, previous studies). In the third step of impact assessment, inventory data are used and the results are characterized into impact categories. It includes four main steps: classification, characterization, normalization, and weighting. Only classification and characterisation are included here – the optional steps of normalization and weighting are excluded. The characterization factors reported by the ReCiPe (H) method are considered in this analysis. The ReCiPe method was chosen for impact assessment as it can provide results at midpoint and endpoint level (Muthu et al. 2012). There are 18 midpoint categories into which the inventory results can be translated. The characterization factors are applied to the inventory results, and results are obtained in the corresponding indicator units. For example, results are obtained in kg CO_2 for the impact category of global warming potential (GWP). The midpoint results are represented as contribution percentages, the impact category that contributes the maximum environmental implications is represented as 100%, and the rest of the impact categories are calculated as relative percentages. The 18 midpoints are further grouped into three endpoints: human health, ecosystem quality, and resource depletion. Endpoint results are easily interpretable even by non-LCA experts, as they are represented using points. The following four basic and important midpoint impact categories are chosen for this study: global warming potential (GWP), human health, terrestrial acidification (TA), freshwater eutrophication (FE), and fossil resource scarcity (FRS). In the final interpretation step, the impact categories, the processes and materials that contribute to these impact categories can be identified and analyzed. Results help decision-makers to identify critical or key problematic processes in the analyzed system, and derive recommendations to mitigate the environmental impacts. SimaPro 8.5 software with the inbuilt Ecoinvent v3.5 database is used to implement LCA and derive results (Goedkoop et al. 2014).

9.3.2 Description of Social Assessment

Social Life Cycle Assessment (SLCA) is used to quantify social aspects of products/processes/systems through their entire lifecycle (Petti et al. 2018). SLCA is a new variant of LCA that analyses processes from a lifecycle perspective, hence it includes processing of raw materials, manufacturing, transportation, use, and disposal. Though human health is measured as an endpoint in ELCA, still SLCA varies from ELCA slightly due to its nature and objectives. The effect of production processes on the environment and consequently on human health is measured in ELCA, whereas effects of business activities and how they affect the human wellbeing of stakeholders are analyzed in SLCA (Subramanian and Yung 2018). Hence, SLCA can be used along with ELCA and LCC to assess social dimensions in sustainability assessments. Environmental consequences (impact categories) of the analyzed process are included in the inventory. Social consequences like forced labor, child labor, workers' rights, and discrimination in business activities are measured using social indicators in SLCA (Weidema 2018).

The social indicators need not necessarily be quantitative – they can be qualitative or semi-quantitative in nature. For instance, quantitative social indicators can be salary, number of working hours, percentage of child labor in an organization, women:men ratio in a business; qualitative indicators and representative data can be descriptions about cultural heritage in a region, or community engagement activities being conducted; while semi-quantitative indicators can be presence or absence of a child labor/technology development strategy in an organization. Overall, the social consequence or issue is represented by numbers or percentages for quantitative social indicators; yes/no answers for semi-quantitative indicators; and described in words for qualitative indicators. Due to the qualitative nature of the social data in most cases, measuring social aspects and deriving meaningful results can be complicated. UNEP/SETAC (2009) came up with guidelines based on the ISO 14040 framework, including the four main phases given in Section 9.3.1. In SLCA, human wellbeing is the main endpoint, and the effects of social pressures on this endpoint are measured using social indicators (Arcese et al. 2016). Organization and societal wellbeing are also considered as additional endpoints in this work to broaden the range of social aspects that possibly affect the involved stakeholders.

The measure of social aspects within sustainability assessment is emerging and is relatively very new at a technology level (Subramanian et al. 2018). Hence, the measurement of social impacts through the lifecycle of products from emerging technologies like biorefineries needs special attention. Human wellbeing directly measures social indicators like health and safety, working hours, discrimination; organizational wellbeing measures aspects like supplier relationships, fair competition; and societal wellbeing measures aspects like safe and healthy living conditions and access to resources.

There are many social indicators to measure a varied set of social aspects or issues proposed by the UNEP methodological sheets (Benoît-Norris et al. 2011). From these, social indicators that are relevant to the studied concept of biorefineries are selected and included here. The selected indicators linked to the corresponding endpoint can be found in Figure 9.1. These social indicators are linked to different social endpoints based on their

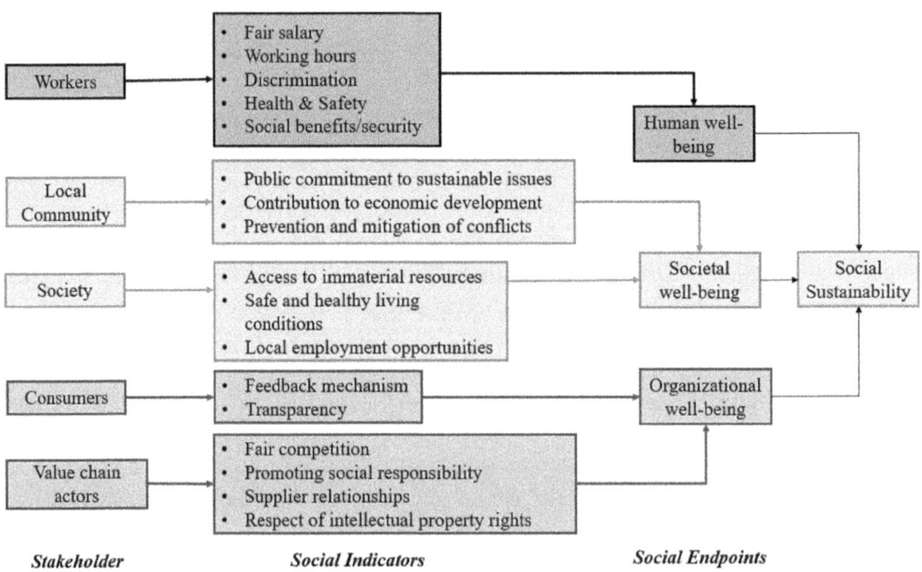

Figure 9.1 *A selected list of stakeholders, indicators, and endpoints.*

potential contribution to the social issues that affect the involved stakeholders like workers, the local community, and society. Finally, a social sustainability index can be derived by taking into account all three endpoints equally, with a set of indicators detailed below, that describe aspects like working hours, health, safety, transparency, and access to resources by the people involved in the bioproduct manufacture.

The following indicators are selected under each social endpoint to measure the wellbeing of stakeholders in biorefinery production and the effects on the local community and society where biorefineries are situated: workers, local community, society, and value-chain actors.

Human wellbeing

- Fair salary
- Working hours
- Discrimination
- Health and safety
- Social benefits/security

Organizational wellbeing

- Fair competition
- Promoting social responsibility
- Supplier relationships
- Respect for intellectual property rights
- Feedback mechanism
- Transparency

Societal wellbeing

- Safe and healthy living conditions
- Local employment opportunities
- Access to material resources
- Public commitment to sustainable issues
- Contribution to economic development
- Prevention and mitigation of conflicts
- Technology development

9.4 Case Study

9.4.1 Recovery of PET Fiber from Cotton–Polyester Blended Textile Waste

The textile waste biorecycling method analyzed in this work converts cotton–polyester blended textile waste into polyester and produces glucose as a byproduct. The polyester recovered is respun to produce polyethylene terephthalate (PET) fibers. The proposed biorecycling method can be considered as a circular economy initiative that brings resources back to the textile industry as a raw material. Environmental impacts of the biorecycling method are analyzed based on the LCA approach. Briefly, in the developed biorecycling method, cotton–polyester textile waste is crushed, then the crushed fabric is pretreated, and

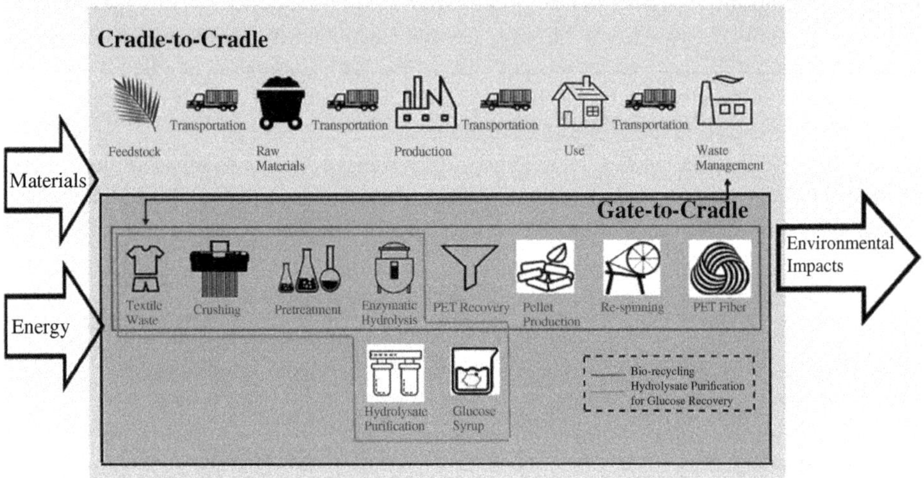

Figure 9.2 *Framework depicting the system boundaries of the textile biorecycling process.*

subjected to enzymatic hydrolysis to separate polyester and cotton. The recovered polyester is re-spun to PET fiber. In addition, the hydrolysate is purified to obtain glucose syrup as a value-added byproduct. Results demonstrate (i) the most impactful processes and subprocesses in the biorecycling method; (ii) the most detrimental materials and energy used in the impactful unit processes, (iii) the most affected impact category among the four major categories assessed (global warming potential (GWP), freshwater eutrophication (FE), terrestrial acidification (TA), fossil resource scarcity (FRS)).

All the related inputs and outputs of the studied recycling method are referred to the functional unit (FU), which is chosen as 1 kg of recovered PET fiber. Input is the 50:50 cotton–polyester blend waste textile fabric. The system boundary includes all unit processes like crushing, pretreatment, enzymatic hydrolysis, hydrolysate purification, and respinning (Figure 9.2). The use and end-of-life of the produced PET fiber, such as clothing, washing, and disposal of used clothes are not considered. The first life of the textile waste is not considered – a cutoff approach is used. Since the waste that enters as input into recycling is in any case considered waste, a burden-free approach is assumed.

9.4.2 System Description of the Biorecycling Method

A system description of the biorecycling method used to convert 50% cotton and 50% polyester waste fabric into polyester and glucose is given below, with brief explanations of each step (Figure 9.3). A much more detailed description can be found in Li et al. (2018).

Crushing: Insufficient surface area availability obstructs the abrasion of the biological components, which is needed to separate the biodegradable components from non-biodegradable ones. Therefore, a double shaft shredder was used to crush the input waste in order to increase the surface area, to enhance the effectiveness of the chemicals in pretreatment, and the efficiency of the enzymes used in the enzymatic hydrolysis step.

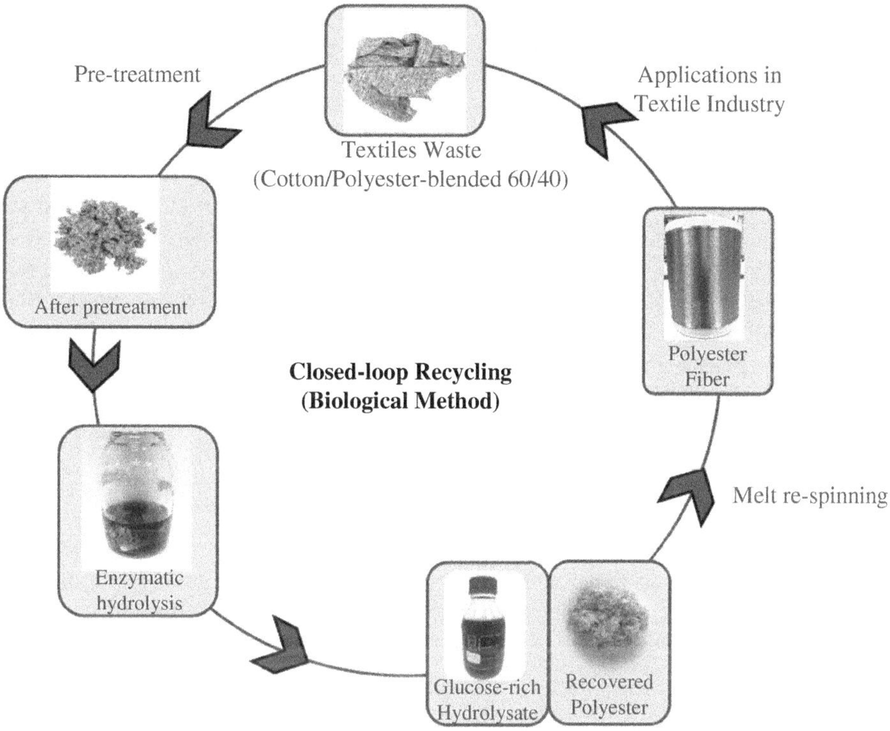

Figure 9.3 *System description of the biorecycling method.*

Pretreatment: In order to decrease the crystallinity of the cellulosic fibers, the crushed fabric was exposed to chemicals that are reusable. The crushed waste fabric was mixed with a solution of 12 w/v% urea and 7 w/v% NaOH and was stored in a freezer at −20°C for 6 h. The mixture was then taken out of the freezer and neutralized with HCl. Then, the sample was washed under tap water and dried in an oven at 40°C for 48 h.

Enzymatic hydrolysis: In order to separate polyester from cotton, the biodegradable component of the textile waste is hydrolyzed. The pretreated textile waste was mixed with sodium citrate buffer (50 mM, pH 5) at a solid loading of 3 w/v%. Cellulose from cotton, which is biodegradable, is hydrolyzed in a bioreactor with an enzyme solution made of 20 FPU/g cellulase and 10 U/g β-glucosidase at 50°C and pH 5.

PET recovery/purification of hydrolysate: The nonbiodegradable component of the textile waste was separated in a solid form by filtration and the recovered PET residue proceeds to the next step for re-spinning. Purification of glucose-rich hydrolysate solution was achieved by the addition of activated carbon at a loading of 15 w/w%. After the hydrolysate is purified from chemicals such as dye, activated carbon is removed by paper filtration. Then, for removal of ions, the filtered solution is passed through ion exchange columns. The columns were cleaned and regenerated with deionized waster, H_2SO_4 (50 w/w%) and NaOH (7 w/v%) for reusability purposes. Cleaning of resins resulted in a decrease in the efficiency of the glucose recovery method due to the glucose amount lost during washing.

Finally, glucose syrup is obtained by placing the hydrolysate solution in an evaporator. This glucose-rich syrup is a value-added byproduct and can be used in industries for production of bioplastics, biosurfactants (cleaning agents), and biochemicals.

Re-spinning: In the final step of the biorecycling method, the recovered PET residue is re-spun into PET fibers. Re-spinning the recovered PET into a fiber is needed for further weaving and knitting purposes, which can only be achieved if the fiber has a suitable tenacity. Therefore, the recovered PET residue is processed in three steps: granulation, solid-state polymerization (SSP), and melt spinning. In the granulating step, the filtered residue was washed and dried before granulation. The dried residue was then passed through the melting stages of granulation and cut into PET pellets. To increase the molecular weight of the pellets, the SSP process was applied. The recovered PET chips were heated at 230°C for 1800 min and mixed with PET bottle chips for the final step. The mixture was spun into PET fibers by the melt spinning process. The whole biorecycling method is designed to ensure the further processability of the final product. Therefore, the proposed method can be considered as a novel solution to the textile waste challenge.

9.4.3 Life Cycle Inventory

A reliable environmental assessment needs valid data. In this study, primary data specific to the analyzed biorecycling method are used for calculating the environmental impacts. All data are collected and converted according to the functional unit considered. Data related to all the inputs and outputs of the unit processes correspond to a lab-scale textile waste biorecycling process. Information from the lab is used to model the unit processes. An itemized cost report (ICR) and a materials and streams report (MSR) of the TEA, carried out by our research group previously, were used as the basis for creating the inventory. The original TEA report was generated based on textile biorecycling conducted at the laboratory scale for processing 10 MT of 50:50 cotton–polyester waste to produce PET fiber and glucose. Energy requirements of all the subprocesses involved in the unit processes, like heating, cooling, drying, and freezing, were all calculated from the TEA report according to the functional unit of this study. Machine specification, power consumption details, hours of usage, and so on, were also calculated from laboratory information.

Solid-to-liquid ratios and weight-to-volume ratios of the solutions used in the laboratory were used to determine stoichiometrically (based on molar mass) the mass of materials and chemicals (sodium hydroxide, urea, citric acid, and hydrochloric acid) used in the pretreatment and enzymatic hydrolysis steps. Beta-glucosidase was not found in the Ecoinvent database, hence only cellulase was used while modeling the enzymatic hydrolysis. Commercially, beta-glucosidase is used only when the cellulase enzyme does not contain enough beta-glucosidase; such a scenario was not assumed, and hence beta-glucosidase was excluded and replaced with cellulase while modeling. The TEA report was also used to determine the amount of activated carbon used in the hydrolysis purification step. Production processes associated with steam, tap water, electricity mix, activated carbon, sulfuric acid, cationic and anionic resins were all taken from Ecoinvent v3.5.

Polyester recovered in the enzymatic hydrolysis step was sent to Taiwan for melt re-spinning. Recovered polyester is mixed with 80% w/w PET bottle chips in the melt spinning process (80:20 ratio of PET bottle chips to recovered polyester sent for

Table 9.1 Life Cycle Inventory (LCI) of the biorecycling process used to produce 1 kg of PET fiber.

Process specification	Materials (kg)	Amount	Electricity (kWh)	Water (kg)	Steam (kg)
Crushing			0.54	—	—
Pretreatment	NaOH	0.68	10.09	1.46	—
	Urea	1.17			
	HCl	0.62			
Enzymatic hydrolysis	NaOH	0.10	0.08	16.24	3.54
	Citric acid	0.16			
	Cellulase	0.16			
	Beta glucosidase	0.01			
PET recovery/ purification of hydrolysate	Activated carbon	0.0002	0.0006	0.16	—
	H_2SO_4	0.0041			
	NaOH	0.0016			
	Monosphere 88 Resin (cationic)	0.12			
	Monosphere 66 Resin (anionic)	0.12			
Pellet production	—	—	10.91	12.02	—
Re-spinning	PET bottle chips –	0.8	0.10081301	0.84	7.3376

Source: Li et al. (2018), reproduced with permission of Springer.

re-spinning) to produce PET fiber. The PET fiber production process takes place in Taiwan and it includes three main processes: pellet production, melt spinning, and drawing processes. The TEA report was used to determine the electricity, water, and steam needed for re-spinning, again using the database Ecoinvent v3.5 for data related to production processes. Table 9.1 provides a summary of LCI for the biorecycling method used to produce 1 kg of PET fiber.

9.5 Results and Discussion

9.5.1 Environmental Sustainability of Bio-based PET Fiber

The characterization results per kg of valorized 50:50 cotton–polyester textile waste for the four selected midpoints is displayed in Figure 9.4. The relative contribution (%) for each step in the biorecycling towards the impact categories can be found in Figure 9.4. Alkaline pretreatment and melt spinning are identified as the main processes responsible for creating potential environmental impacts in the bio-based PET fiber recovery. Enzymatic hydrolysis is the third main step responsible for creating impacts. Among all the unit processes, pretreatment is the environmental hotspot, as it is responsible for the maximum contributions (100%) in all four selected impact categories. The rationale behind this huge environmental burden can be attributed to the electricity production (coal-powered generation) required in the freezing and drying for crystallinity reduction in the pretreatment step, together with the production of urea, a chemical required in alkaline pretreatment. Impacts can be traced back to ammonia liquid, commonly used in urea production. "Hard coal mine operation" (coal-powered electricity generation) and ammonia production also increase the contribution to GWP. Also, fossil resource scarcity is mainly impacted by coal and natural gas used for electricity generation.

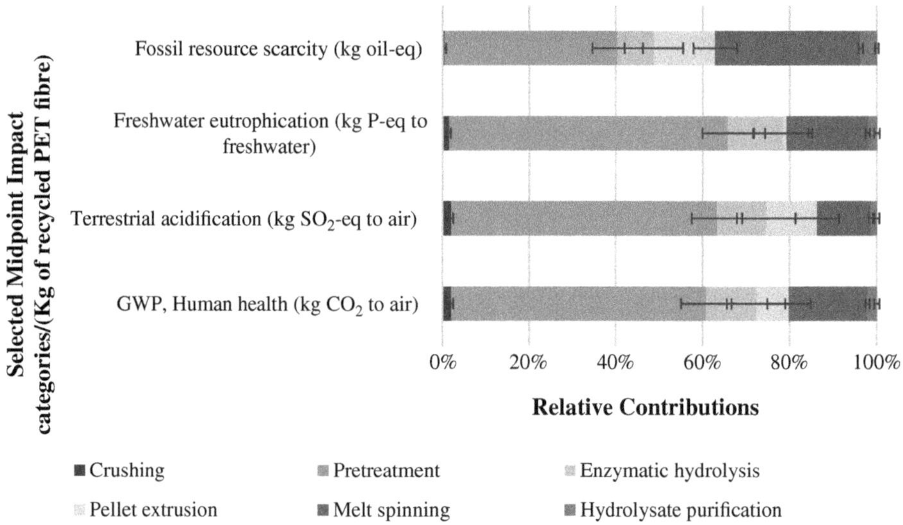

Figure 9.4 Contribution of impacts per unit process involved in the biorecycling process.

Melt spinning makes the second-highest contribution to all the four selected impacted categories. Production of electricity (oil and gas fired generation within electricity production) required in the granulation, SSP and re-spinning are considered as the main contributors to environmental impacts. Also, fossil resource scarcity is mainly affected by the melt spinning step with a contribution ratio higher than 80%, and pellet production with a contribution ratio of 36%. Impacts can be traced back to xylene production, a subprocess in the PET granulate, used in the production of PET bottles. Enzymatic hydrolysis makes the maximum contribution in affecting eutrophication potential. Impacts can be traced back to the production of citric acid in China, a chemical used in hydrolysis using the submerged fermentation process.

Pretreatment makes the maximum contribution on all four impact categories. Relative contributions from the melt spinning process are 31, 21, 29, and 84% of total GWP, TA, FE, and FRS respectively. Relative contributions from enzymatic hydrolysis are 19, 18, 20, and 21% of total GWP, TA, FE, and FRS. Thus, further optimization of textile biovalorisation techniques should focus on energy-efficient processes that demand less electricity and use chemicals that do not add to environmental impacts, or identify and use alternative organic chemicals and pretreatment agents that can be recycled. However, such replacement has to be again analyzed to check for a reduction in environmental impacts. The overall quality of the PET fiber recovered from hydrolysis has to be improved, to reduce the amount (%) of PET bottles being added in the melt spinning step, consequently reducing the associated environmental impacts. Though the final product is said to replace virgin PET fiber that can be used in the production cycle of textiles, rebound effects caused due to increased usage and economics have to be considered in the future.

Figure 9.5 presents the endpoint results for the production of 1 kg of PET fiber through the analyzed textile biorecycling method. Like the midpoint results, pretreatment caused the highest impact on all the three endpoints compared with the other steps in biorecycling

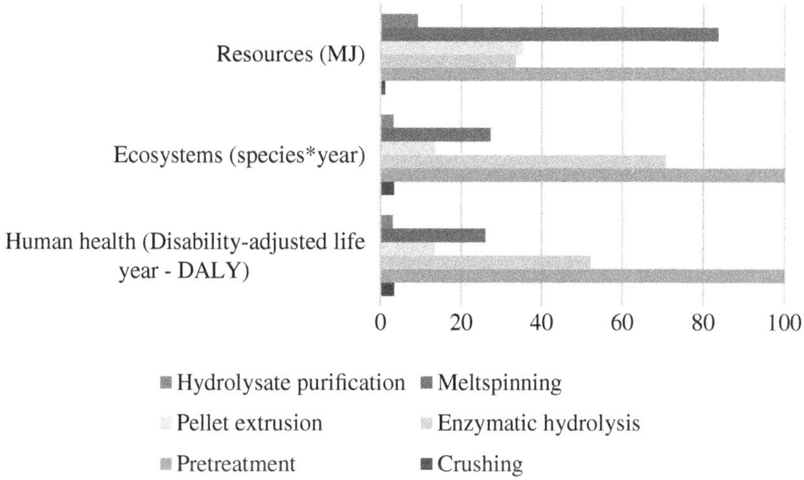

Figure 9.5 *Endpoint assessment results of textile biorecycling to produce 1 kg of PET fiber.*

(human health, ecosystems, and resource depletion respectively). Electricity production and ammonia within urea production are the major process/substance contributors to ecosystem quality and human health impacts. Process energy used includes use of coal and natural gas within electricity production, contributing the most to resources depletion. Enzymatic hydrolysis was the next most dominant step that affected ecosystem quality (70.7%) and human health (52.2%). This step caused comparatively lower impacts on mineral resources and fossil resources scarcity (33%). Melt spinning was the next most impactful step, and it majorly affected the resource depletion category (83.58%), but was comparatively less impactful on ecosystems and human health. Again, process energy used in electricity and steam production, as well as xylene used in PET bottle production, were responsible for the impacts. Like the midpoint characterization results, pellet production, hydrolysate purification, and crushing were found to be less critical in terms of environmental impacts.

9.5.2 Social and Economic Sustainability of Bio-based PET Fiber

An economic evaluation of PET recovery by the biological method was carried out based on the ICR prepared as a part of the techno-economic analysis conducted by our research group previously (Kwan 2018). It was done on a large scale. The annual production rate of the proposed plant at a 95% working capacity was approximately 170 000 MT PET fiber/year, manufactured from 83 000 MT 50:50 cotton–polyester textile waste/year. The three main stages of the process that were taken into consideration in the techno-economic analysis were enzymatic hydrolysis, purification of hydrolysate, and PET re-spinning. Expenses outside of material and utility costs, such as labor, wastewater treatment, logistics, maintenance, insurance, distribution, and marketing, were also taken into consideration. The enzymatic hydrolysis stage contributed 85% of the total material cost (US$1.08 billion) of the plant, mainly due to the high rate of cellulase enzyme consumption. PET bottle chips contributed 15% of the overall material cost. PET re-spinning and enzymatic hydrolysis stages generated 46% and 42% of the total labor costs (US$1 million), respectively. For 1 MT PET fiber production, raw materials cost approximately US$6300, 88% of the total cost. For recycling of 83 000 MT textile waste/year, the total generated cost was around US$1.22 million. The economic feasibility of a system can only be demonstrated through its benefits. The revenue generated by the biological method comes from the PET fiber (US$1450/kg) and glucose syrup (US$67/kg) obtained. The overall cost–benefit analysis showed a deficit of US$860 million per year (Kwan 2018). Raw material costs due to the commercial enzyme usage and the PET bottle chips added into the melt spinning affected the economics. Again, the PET bottle chips cannot be removed or reduced, as it firstly affects the final quality of the recovered fiber and secondly increases the environmental impacts. For now, the results from the economics analysis do not favor the textile biorecycling technique. However, future research should focus on reducing these deficits, for example, by optimizing the biorecycling method to recover polyester of improved quality that does not require addition of commercially bought PET bottle chips. Cellulase enzyme production that is economically feasible, hydrolysis reactor design, improvements in the process design to accommodate other kinds of textiles such as silk, nylon and wool, are a few other ways that can possibly reduce the raw material and operational costs.

Social sustainability results cannot be calculated as of now, since the process is currently conducted at the laboratory scale. More meaningful results for social dimensions of the biorefineries can be obtained from industry-scale biorefineries using the outlined approach. Social information for the selected indicators has to be gathered from the involved stakeholders through questionnaires and interviews. The collected social data has to be processed using social impact assessment methods to derive social sustainability results.

9.6 Conclusions and Future Trends

Huge amounts of textile waste are generated annually in Hong Kong. The Hong Kong government is taking initiatives to address this increasing textile waste problem. The most recent initiative is the development of an upcycling mill to recycle textile waste without affecting cost or quality. A few organizations such as the Hong Kong Research Institute of Textile and Apparel (HKRITA), in collaboration with clothing brands (H&M), are also developing sustainable solutions to manage textile waste effectively. One such direction is the proposed biological recycling method developed by our research group. Nowadays, such techniques are considered effective solutions to the global textile waste problem, as the organic waste valorisation aligns quite well with circular economy approaches. Few attempts have been made in the literature to evaluate the environmental and economic implications of such valorisation techniques; however, the social dimension is completely forgotten. Measuring the sustainability of such emerging technologies is emphasized (Falinski et al. 2018). It is also imperative for sustainability assessments to consider all three pillars. Thus, the preliminary environmental analysis conducted in this work is complemented by an approach to measure social sustainability of biorefineries. The social impact assessment approach includes multiple social impact categories that can measure human, social, and organizational wellbeing. The economics of the study are already analyzed and reported using TEA. Further research should focus on gathering real-time social data from biorefineries to validate the proposed social impact assessment approach. Together with environmental and economic impact assessment results, this can help stakeholders to make informed decisions.

Acknowledgement

The authors acknowledge the Hong Kong Research Institute of Textiles and Apparel (HKRITA), and Innovation and Technology Funding (ITP/002/19TP) from the Innovation and Technology Commission in Hong Kong. We are also grateful to the industrial partners and industrial sponsors, including the H&M Conscious Foundation, and H&M Hennes & Mauritz (Far East) Ltd. for providing textile waste.

References

Arcese, G., Lucchetti, M.C., Massa, I., and Valente, C. (2016). State of the art in S-LCA: integrating literature review and automatic text analysis. *Int. J. Life Cycle Assess.* **23**: 394–405. https://doi.org/10.1007/s11367-016-1082-0.

Baydar, G., Ciliz, N., and Mammadov, A. (2015). Life cycle assessment of cotton textile products in Turkey. *Res. Cons. Recycl.* **104**: 213–223. https://doi.org/10.1016/j.resconrec.2015.08.007.

Benoît-Norris, C., Vickery-Niederman, G., Valdivia, S. et al. (2011). Introducing the UNEP/SETAC methodological sheets for subcategories of social LCA. *Int. J. Life Cycle Assess.* **16**: 682–690. https://doi.org/10.1007/s11367-011-0301-y.

Boström, M. and Micheletti, M. (2016). Introducing the sustainability challenge of textiles and clothing. *J. Consum. Policy* **39**: 367–375. https://doi.org/10.1007/s10603-016-9336-6.

Budzianowski, W.M. and Postawa, K. (2016). Total Chain Integration of sustainable biorefinery systems. *Appl. Energy* **184**: 1432–1446. https://doi.org/10.1016/j.apenergy.2016.06.050.

Chopra, S.S. and Theis, T.L. (2017). Comparative cradle-to-gate energy assessment of indium phosphide and cadmium selenide quantum dot displays. *Environ. Sci. Nano* **4**: 244–254. https://doi.org/10.1039/c6en00326e.

Chopra, S.S., Bi, Y., Brown, F.C. et al. (2019). Interdisciplinary collaborations to address the uncertainty problem in life cycle assessment of nano-enabled products: Case of the quantum dot-enabled display. *Environ. Sci.*: 3256–3267. https://doi.org/10.1039/c9en00603f.

Dahlbo, H., Aalto, K., Eskelinen, H., and Salmenperä, H. (2017). Increasing textile circulation – consequences and requirements. *Sustainable Prod. Consumption* **9**: 44–57. https://doi.org/10.1016/j.spc.2016.06.005.

Duckett, D., Feliciano, D., Martin-Ortega, J., and Munoz-Rojas, J. (2016). Tackling wicked environmental problems: The discourse and its influence on praxis in Scotland. *Landscape Urban Plann* **154**: 44–56. https://doi.org/10.1016/j.landurbplan.2016.03.015.

Falinski, M.M., Plata, D.L., Chopra, S.S. et al. (2018). A framework for sustainable nanomaterial selection and design based on performance, hazard, and economic considerations. *Nat. Nanotechnol.* **13**: 708–714. https://doi.org/10.1038/s41565-018-0120-4.

Fava, F., Totaro, G., Diels, L. et al. (2015). Biowaste biorefinery in Europe: opportunities and research & development needs. *New Biotechnol.* **32** https://doi.org/10.1016/j.nbt.2013.11.003.

Gholamzad, E., Karimi, K., and Masoomi, M. (2014). Effective conversion of waste polyester-cotton textile to ethanol and recovery of polyester by alkaline pretreatment. *Chem. Eng. J.* **253**: 40–45. https://doi.org/10.1016/j.cej.2014.04.109.

Gilani, B. and Stuart, P.R. (2015). Life cycle assessment of an integrated forest biorefinery: Hot water extraction process case study. *Biofuels, Bioprod. Biorefin.* **9**: 677–695.

Goedkoop, M., Oele, M., Vieira, M. et al. (2014). *SimaPro Tutorial*. Netherlands: PRé https://www.pre-sustainability.com/download/SimaPro8Tutorial.pdf.

González-García, S., Gullón, B., Rivas, S. et al. (2016). Environmental performance of biomass refining into high-added value compounds. *J. Cleaner Prod.* **120**: 170–180. https://doi.org/10.1016/j.jclepro.2016.02.015.

González-García, S., Gullón, B., and Moreira, M.T. (2018). Environmental assessment of biorefinery processes for the valorisation of lignocellulosic wastes into oligosaccharides. *J. Cleaner Prod.* **172**: 4066–4073. https://doi.org/10.1016/j.jclepro.2017.02.164.

González-García, S., Gullón, P., and Gullón, B. (2019). Bio-compounds production from agri-food wastes under a biorefinery approach: exploring environmental and social sustainability. In: *Quantification of Sustainability Indicators in the Food Sector* (ed. S.S. Muthu), 25–53. Springer https://doi.org/10.1007/978-981-13-2408-6.

Hu, Y., Du, C., Leu, S.Y. et al. (2018). Valorisation of textile waste by fungal solid state fermentation: an example of circular waste-based biorefinery. *Resour. Conserv. Recycl.* **129**: 27–35. https://doi.org/10.1016/j.resconrec.2017.09.024.

Huang, Y.-B. and Fu, Y. (2013). Hydrolysis of cellulose to glucose by solid acid catalysts. *Green Chem.* **15**: 1095–1111. https://doi.org/10.1039/c3gc40136g.

Ingrao, C., Lo Giudice, A., Tricase, C. et al. (2014). Recycled-PET fibre based panels for building thermal insulation: environmental impact and improvement potential assessment for a greener production. *Sci. Total Environ.* **493**: 914–929. https://doi.org/10.1016/j.scitotenv.2014.06.022.

Iribarren, D., Martín-Gamboa, M., Mahony, T.O., and Dufour, J. (2016). Screening of socio-economic indicators for sustainability assessment: a combined life cycle assessment and data envelopment analysis approach. *Int. J. Life Cycle Assess.*: 202–214. https://doi.org/10.1007/s11367-015-1002-8.

Kwan, T.H. (2018). Integrated biorefinery strategies for valorisation of food and textile wastes. PhD thesis. City University of Hong Kong.

Landi, D., Gigli, S., Germani, M., and Marconi, M. (2018). Investigating the feasibility of a reuse scenario for textile fibres recovered from end-of-life tyres. *Waste Man.* **75**: 187–204. https://doi.org/10.1016/j.wasman.2018.02.018.

Leal Filho, W., Ellams, D., Han, S. et al. (2019). A review of the socio-economic advantages of textile recycling. *J. Cleaner Prod.* **218**: 10–20. https://doi.org/10.1016/j.jclepro.2019.01.210.

Lenzo, P., Traverso, M., Salomone, R., and Ioppolo, G. (2017). Social life cycle assessment in the textile sector: an Italian case study. *Sustainability* **9**: 2092. https://doi.org/10.3390/su9112092.

Li, X., Hu, Y., Du, C., and Lin, C.S.K. (2018). Recovery of glucose and polyester from textile waste by enzymatic hydrolysis. *Waste Biomass Valorisation* **10**: 3763–3772. https://doi.org/10.1007/s12649-018-0483-7.

Ling, C., Shi, S., Hou, W., and Yan, Z. (2019). Separation of waste polyester/cotton blended fabrics by phosphotungstic acid and preparation of terephthalic acid. *Polym. Degrad. Stab.* **161**: 157–165. https://doi.org/10.1016/j.polymdegradstab. 2019.01.022.

Moazzem, S., Daver, F., Crossin, E., and Wang, L. (2018). Assessing environmental impact of textile supply chain using life cycle assessment methodology. *J. Text. Inst.* **109**: 1574–1585. https://doi.org/10.1080/00405000.2018.1434113.

Moussa, H.I., Elkamel, A., and Young, S.B. (2016). Assessing energy performance of bio-based succinic acid production using LCA. *J. Cleaner Prod.* **139**: 761–769. https://doi.org/10.1016/j.jclepro.2016.08.104.

Musaazi, M.K., Mechtenberg, A.R., Nakibuule, J. et al. (2015). Quantification of social equity in life cycle assessment for increased sustainable production of sanitary products in Uganda. *J. Cleaner Prod.* **96**: 569–579. https://doi.org/10.1016/j.jclepro.2013.10.026.

Muthu, S.S., Li, Y., Hu, J.Y., and Ze, L. (2012). Carbon footprint reduction in the textile process chain: recycling of textile materials. *Fibers Polym.* **13**: 1065–1070. https://doi.org/10.1007/s12221-012-1065-0.

Nagendra, H., Bai, X., Brondizio, E.S., and Lwasa, S. (2018). The urban south and the predicament of global sustainability. *Nat. Sustain.* **1**: 341–349. https://doi.org/10.1038/s41893-018-0101-5.

Neupane, B., Halog, A., and Lilieholm, R.J. (2013). Environmental sustainability of wood-derived ethanol : a life cycle evaluation of resource intensity and emissions in Maine. *USA. J. Clean. Prod.* **44**: 77–84. https://doi.org/10.1016/j.jclepro.2012.11.039.

Norup, N., Pihl, K., Damgaard, A., and Scheutz, C. (2019). Evaluation of a European textile sorting centre: Material flow analysis and life cycle inventory. *Resour. Conserv. Recycl.* **143**: 310–319. https://doi.org/10.1016/j.resconrec.2019.01.010.

Nzila, C., Dewulf, J., Spanjers, H. et al. (2009). Valorisation of textile residues for energy production in Kenya. *Moi University 5th Annual International Conference, 4–8 August 2009, Eldoret, Kenya.*

Opoku, A. (2019). Biodiversity and the built environment: Implications for the Sustainable Development Goals (SDGs). *Resour. Conserv. Recycl.* **141**: 1–7. https://doi.org/10.1016/j.resconrec.2018.10.011.

Pedro, J., Silva, C., and Duarte, M. (2019). Land use policy integrating GIS spatial dimension into BREEAM communities sustainability assessment to support urban planning policies, Lisbon case study. *Land Use Policy* **83**: 424–434. https://doi.org/10.1016/j.landusepol.2019.02.003.

Pensupa, N., Jin, M., Kokolski, M. et al. (2013). A solid state fungal fermentation-based strategy for the hydrolysis of wheat straw. *Bioresour. Technol.* **149**: 261–267. https://doi.org/10.1016/j.biortech.2013.09.061.

Petti, L., Sanchez Ramirez, P.K., Traverso, M., and Ugaya, C.M.L. (2018). An Italian tomato "Cuore di Bue" case study: challenges and benefits using subcategory assessment method for social life cycle assessment. *Int. J. Life Cycle Assess* **23**: 569–580. https://doi.org/10.1007/s11367-016-1175-9.

Popovic, T., Barbosa-Póvoa, A., Kraslawski, A., and Carvalho, A. (2018). Quantitative indicators for social sustainability assessment of supply chains. *J. Cleaner Prod.* **180** https://doi.org/10.1016/j.jclepro.2018.01.142.

Rugani, B., Benetto, E., Igos, E. et al. (2014). Towards prospective life cycle sustainability analysis: exploring complementarities between social and environmental life cycle assessments for the case of Luxembourg's energy system. *Mater. Tech.* **102**: 605. https://doi.org/10.1051/mattech/2014043.

Sandin, G. and Peters, G.M. (2018). Environmental impact of textile reuse and recycling – a review. *J. Cleaner Prod.* **184**: 353–365. https://doi.org/10.1016/j.jclepro.2018.02.266.

Sandin, G., Peters, G., Pilgård, A. et al. (2011). Integrating sustainability considerations into product development: a practical tool for prioritising social sustainability indicators and experiences from real case application. In: *Towards Life Cycle Sustainability Management* (ed. M. Finkbeiner), 3–14. Springer https://doi.org/10.1007/978-94-007-1899-9.

Shen, L., Worrell, E., and Patel, M.K. (2010). Open-loop recycling: A LCA case study of PET bottle-to-fibre recycling. *Resour. Conserv. Recycl.* **55**: 34–52. https://doi.org/10.1016/j.resconrec.2010.06.014.

Shen, L., Nieuwlaar, E., Worrell, E., and Patel, M.K. (2011). Life cycle energy and GHG emissions of PET recycling: change-oriented effects. *Int. J. Life Cycle Assess.* **16**: 522–536. https://doi.org/10.1007/s11367-011-0296-4.

Shen, F., Xiao, W., Lin, L. et al. (2013). Enzymatic saccharification coupling with polyester recovery from cotton-based waste textiles by phosphoric acid pretreatment. *Bioresour. Technol.* **130**: 248–255. https://doi.org/10.1016/j.biortech.2012.12.025.

Smidt, M., Den Hollander, J., Bosch, H. et al. (2016). Life cycle assessment of biobased and fossil-based succinic acid. In: *Sustainability Assessment of Renewables-Based Products: Methods and Case Studies* (eds. J. Dewulf, S. De Meester and R.A.F. Alvarenga), 307–322. Wiley.

Song, Q., Wang, Z., Li, J., and Yuan, W. (2013). Life cycle assessment of desktop PCs in Macau. *Int. J. Life Cycle Assess.* **18**: 553–566. https://doi.org/10.1007/s11367-012-0515-7.

Subramanian, K. and Yung, W.K.C. (2016). Review of life cycle assessment on consumer electronic products: developments and the way ahead. *Crit. Rev. Env. Sci. Technol.* **46**: 1441–1497. https://doi.org/10.1080/10643389.2016.1245550.

Subramanian, K. and Yung, W.K.C. (2017). Life cycle assessment study of an integrated desktop device – comparison of two information and communication technologies: Desktop computers versus all-in-ones. *J. Cleaner Prod.* **156**: 828–837. https://doi.org/10.1016/j.jclepro.2017.04.089.

Subramanian, K. and Yung, W.K.C. (2018). Modeling Social Life Cycle Assessment framework for an electronic screen product – A case study of an integrated desktop computer. *J. Cleaner Prod.* **197**: 417–434. https://doi.org/10.1016/j.jclepro.2018.06.193.

Subramanian, K., Chau, C.K., and Yung, W.K.C. (2018). Relevance and feasibility of the existing social LCA methods and case studies from a decision-making perspective. *J. Cleaner Prod.* **171**: 690–703. https://doi.org/10.1016/j.jclepro.2017.10.006.

Traverso, M. (2018). Is social life cycle assessment really struggling in development or is it on a normal path towards harmonization/standardization? *Int. J. Life Cycle Assess.* **23**: 199–200. https://doi.org/10.1007/s11367-017-1387-7.

UNEP/SETAC. (2009). *Guidelines for Social Life Cycle Assessment of Products*. United Nations Environment Program. DTI/1164/PA. https://wedocs.unep.org/handle/20.500.11822/7912

United Nations. (1987). *Our Common Future*. Report of the World Commission on Environment and Development (Brundtland Commission). https://sustainabledevelopment.un.org/content/documents/5987our-common-future.pdf (accessed 28 April 2020).

Van Der Velden, N.M., Patel, M.K., and Vogtländer, J.G. (2014). LCA benchmarking study on textiles made of cotton, polyester, nylon, acryl, or elastane. *Int. J. Life Cycle Assess.* **19**: 331–356. https://doi.org/10.1007/s11367-013-0626-9.

Waas, T., Huge, J., Block, T. et al. (2014). Sustainability assessment and indicators: Tools in a decision-making strategy for sustainable development. *Sustainability* **6**: 5512–5534. https://doi.org/10.3390/su6095512.

Weidema, B.P. (2018). The social footprint – a practical approach to comprehensive and consistent social LCA. *Int. J. Life Cycle Assess.* **23**: 700–709. https://doi.org/10.1007/s11367-016-1172-z.

Yeo, J., Chopra, S.S., Zhang, L., and An, A.K. (2019). Life cycle assessment (LCA) of food waste treatment in Hong Kong: On-site fermentation methodology. *J. Environ. Manage.* **240**: 343–351. https://doi.org/10.1016/j.jenvman.2019.03.119.

Youhanan, L. (2013). Environmental assessment of textile material recovery techniques: Examining textile flows in Sweden. MSc thesis. Royal Institute of Technology Sweden (KTH).

Zaimes, G.G., Vora, N., Chopra, S.S. et al. (2015). Design of sustainable biofuel processes and supply chains: challenges and opportunities. *Processes* **3**: 634–663. https://doi.org/10.3390/pr3030634.

Zamani, B., Svanström, M., Peters, G., and Rydberg, T. (2015). A carbon footprint of textile recycling: A case study in Sweden. *J. Ind. Ecol.* **19**: 676–687. https://doi.org/10.1111/jiec.12208.

Zamani, B., Sandin, G., Svanström, M., and Peters, G.M. (2016). Hotspot identification in the clothing industry using social life cycle assessment – opportunities and challenges of input–output modelling. *Int. J. Life Cycle Assess.* **23**: 536–546. https://doi.org/10.1007/s11367-016-1113-x.

Zhang, Y., Kang, H., Hou, H. et al. (2018). Improved design for textile production process based on life cycle assessment. *Clean Technol. Environ. Policy* **20**: 1355–1365. https://doi.org/10.1007/s10098-018-1572-9.

10

Circular Waste-Based Biorefinery Development

Raffel Dharma Patria[1], Xiaotong Li[1], Huaimin Wang[1], Chenyu Du[2], Carol Sze Ki Lin[1] and Guneet Kaur[3,4]

[1] *School of Energy and Environment, City University of Hong Kong, Tat Chee Avenue, Kowloon Tong, Hong Kong*
[2] *School of Applied Sciences, University of Huddersfield, Huddersfield HD1 3DH, United Kingdom*
[3] *Department of Biology, Hong Kong Baptist University, Kowloon Tong, Hong Kong*
[4] *Department of Civil Engineering, York University, Toronto, Canada*

10.1 Introduction

The global sustainability goals of material and energy efficiency, environmental protection, and food security have become the key drivers for the production of bio-based alternatives. This has resulted in increased promotion and development of the concept of the biorefinery. According to the International Energy Agency (IEA), a biorefinery is defined as "a sustainable processing of biomass into a spectrum of marketable products and energy" (Morais and Lukasik 2013). This essentially implies the conversion of biomass into a plethora of products ranging from bulk products such as bioenergy and/or platform chemicals, to specialty products including fine chemicals, flavors, and fragrances. A biorefinery can be considered analogous to current oil refineries, which produce multiple fuels and chemicals from petroleum (Cherubini 2010). However, with this perspective, it is important to consider that the variety of both raw materials and products is one of the most important

Waste Valorisation: Waste Streams in a Circular Economy, First Edition.
Edited by Carol Sze Ki Lin, Guneet Kaur, Chong Li and Xiaofeng Yang.

characteristics of a biorefinery. If broadly understood, a biorefinery should process all kinds of biomass, for example, organic residues, energy crops, and other agricultural wastes, to produce numerous products such as fuel, chemicals, materials, power, food, and animal feed. In fact, maximum processing of raw materials and minimum production of residues will give maximum sustainability for a biorefinery (Demirbas 2009; Moncada et al. 2013). This indeed can be achieved by an integrated biorefinery, which, as opposed to a single productive process (stand-alone biorefinery), combines several biomass conversion processes and equipment to produce chemicals, materials, energy, and food, for greater profitability and environmental sustainability.

One of the most important aspects of a biorefinery is the sourcing of raw materials for processing into high-value products. There has been a sequential appearance and utilization of first-generation (1G) (edible vegetable oils, rice, wheat, cane), second-generation (2G) (non-food crops, forestry residues, biomass resources) and third-generation (3G) (microalgae) feedstocks, for reasons such as higher cost efficiency, environmental considerations, and better renewability (Moncada et al. 2016). In recent years, the utilization of wastes including agricultural wastes, industrial wastes, food wastes, textile wastes, and so on, as feedstocks has attracted a great deal of attention for the development of so-called "waste biorefineries" (Venkata Mohan 2014b). The major compelling reasons for such a concept are surplus availability of biogenic waste, high (residual) nutrient value of wastes, and significant environmental and economic prospects. Municipal solid waste (MSW) is one of the most abundant organically-bound carbon sources produced each year throughout the world. Current annual MSW generation amounts to 1.3 billion tonnes and is projected to increase to 2.2 billion tonnes by 2025 (Sadef et al. 2016). Increasing urbanization and economic development will be the major contributors to increasing MSW amounts. MSW comprises electronic wastes, construction and demolition waste, medical wastes, and wastes generated from households, offices, shops, agricultural and industrial activities.

Although the MSW composition differs for various countries, food waste as a component of MSW has lately received great attention worldwide. According to the Food and Agricultural Organization, roughly a third of the total food produced for human consumption is wasted globally (Ong et al. 2018). The staggering amounts of food waste generated have enormous chemical potential due to the functionalized molecules stored in them. Consequently, in recent years, the food waste-based biorefinery has been proposed to convert the widely available food waste into platform chemicals, fuels, and materials (Lin et al. 2013). The idea of harvesting high value-added products from food waste streams is particularly attractive since it facilitates the recovery of nutrients, which would otherwise have been lost, and aligns with the concept of sustainable development. Thus, the valorisation scheme for food waste has been endorsed by developed and emerging economies (European Commission 2012). The greatest development in food waste-based transformation into high-value products has been observed in the European Union, which has supported multinational projects and innovation networks such as COST actions FP1303, FP1003, FP0602 (BIOBIO), and TD1203 (EuBis) (COST 2017). Different bioconversion processes leading from food waste to platform chemicals such as ethanol, butanol, biopolymers (polyhydroxybutyrate), microalgal biomass, and so on, have been reported in recent years. Thus, the valorisation of food waste in the context of waste-based biorefinery development for high-value products is a lucrative option.

Another type of MSW that has not only caught the attention of researchers, but also corporate leaders, is textile waste. Recent decades have witnessed the generation of enormous quantities of textile waste emerging from the fiber, textile and clothing manufacturing process, commercial service and consumption. Around 32 kg per capita of textile wastes are discarded each year as a global average, which is expected to increase further due to a general increase in household purchasing power (Hu et al. 2018). Famous fashion brands such as Zara and H&M (2016) have launched an initiative of "closing the loop" by exploring solutions to production, recycling and reuse of textiles in an environmentally, economically and socially sustainable way. Researchers, on the other hand, are investigating methods to utilize textile waste within a biorefinery system to produce valuable products. With respect to a potential feedstock for a biorefinery, the key ingredient in textile waste is cellulose, which accounts for 35–40% of textile (Shen et al. 2013). The general idea is conversion of available cellulose to glucose using chemical, biological or thermal techniques, and subsequent utilization of this glucose as substrate for bioconversions to ethanol, succinic acid, and other useful products.

Biorefinery models which incorporate the utilization of waste as a bioresource, support the transition to a circular bioeconomy. The prospect of material and energy recovery using waste resources also contributes towards waste remediation and management (Venkata Mohan 2014a). Utilization of waste resources is not new to mankind; however, in the present times, burgeoning interest and effort for effective exploitation of organic wastes through biorefinery also aims at achieving both economic and environmental sustainability. Despite the overall benefits of a biorefinery, its design approach is still an important issue. This is mainly due to various aspects such as raw materials, technologies, products, transformation processes, social aspects, policies and laws that complicate the biorefinery system (Moncada et al. 2016). A simple example is the meager consideration of aspects of food security in the biorefinery design, in the context of developing countries or regions with food scarcity. In such a scenario, the claimed benefits from the biorefinery approach are not anticipated and require the development of patterns and design approaches for local solutions (Douglas 1988; Fallot et al. 2008). This also leads to an important understanding that the assumption of availability of raw material is not always appropriate, and must consider important restrictions in local regions. For example, the availability of lignocellulosic materials in Brazil or Colombia might be much greater than in European countries, and therefore the obvious problem to solve will be to decide which raw material(s) to use in order to effectively produce the desired range of products through the same platforms (Bentsen and Felby 2012). A similar situation can be applied to a waste-based biorefinery depending on food waste, industrial waste, textile wastes, and so on. Hence, the availability of a specific type of waste, possible products which could be derived from it, local regulations and policies, and market acceptance for waste-derived products should be the deciding factors for biorefinery design. Finally, the economic competitiveness of the biorefinery with existing processes should also be demonstrated.

This chapter discusses some key aspects of circular waste-based biorefinery development, including the recirculation of resources in terms of their reuse and recycling, as well as requirements and challenges for the transition from a linear economy to a circular economy. It further describes case studies of food waste and textile waste-based biorefineries, particularly in the context of Hong Kong, for the production of high value-added products such as chemicals and materials.

10.2 Transitioning from Current Linear to Stronger Circular Economy Models

The current global sustainability challenges of climate change, biodiversity loss, material and energy scarcity, and pollution, are all interrelated. In order to solve these challenges, we need to design and adopt holistic approaches that integrate various components of human and natural systems to effectively identify and propose suitable solutions (Liu et al. 2015). The concept of sustainable development is the most appropriate way to fulfill the above requirement. Sustainable development can be defined as a process that considers health and environment, social conditions, research and innovation in interdisciplinary and transdisciplinary studies, as well as economic aspects, within the ecological and resource limits of our planet (Blum et al. 2017). This paradigm includes all areas where energy, chemicals and materials are produced, used, integrated into products, recycled, disposed of, and released to the environment. The latter transition thus demands a closed loop structure for resource circulation which maintains the value of products, materials and resources in the economy for as long as practically possible, while also contributing to conservation of energy and protection of the environment (Kaur et al. 2018). This indeed is the concept of the circular economy.

10.2.1 Integration of Circular Economy and Sustainable Development

The circular economy concept was launched in 2010 and can be defined as a "system that replaces the linear, 'take-use-dispose' process and emphasizes on restorative and regenerative industrial design." As highlighted recently by the Ellen MacArthur Foundation, the circular economy model can be realized via use of renewable energy, nontoxic manufacture routes for chemicals and materials, and development of bio-benign products, to enable their efficient reuse and return to the biosphere (Ellen MacArthur Foundation 2016). Furthermore, in addition to the technical aspects, another key aspect of the circular economy is social and economic development, which indeed aligns it with the overall attainment of sustainable development. The latter aspects are fulfilled by increased participation of policymakers, and incentives to businesses and people overall (Preston 2012).

The circular economy concept underpins the green chemistry and bio-processing approach which involves the use of (bio-)benign renewable resources, including waste resources in benign reaction media, to develop processes and products in an environmentally friendly manner (Sheldon 2016). With respect to the economic component of the circular economy concept, the utilization of waste biogenic resources as chemical feedstock is indeed an opportunity. As discussed in Section 10.1, abundant availability and low-cost origins of waste-based feedstock is an important factor for development of a circular economy. However, it should be borne in mind that other key factors such as biodegradability, recyclability, and consumer acceptance of waste-derived and/or recycled products are also equally vital to the success of a circular economy. Thus, it is imperative that the circular economy approach accounts for development of renewable and greener processes with high production efficiency and productivity, while utilizing resources efficiently, generating reduced waste flows, and creating economic opportunities for people.

10.2.2 Requirements for Transition to a Circular Economy

The transition from the current linear to a stronger circular economy needs integrated, hybrid models that consider resource management, creation of sustainable solutions for services, products and businesses which address the problems of environmental pollution, scarcity and/or recovery of resources. This implies that the key requirements of a circular economy are sourcing of products, design of production routes, and reuse/recycling schemes (Clark et al. 2016).

In the context of a bio-based economy, a major concern for the adoption of a circular economy is exploring the role of renewable resources. In recent years, bio-based sourcing has attracted a lot of attention worldwide, while there is increasing pressure to replace fossil-fuel based feedstocks with bio-based, renewable feedstocks (Tuck et al. 2012). At this point, it is important to understand that renewable and bio-based sourcing of products is only one of the contributing factors to sustainable development goals. If other associated factors are suitably considered, maximum benefit could be achieved. For example, the sourcing of bio-based resources requires no significant externalities, but application of regenerative agricultural principles and conservation of land use and biodiversity (Sherwood et al. 2017). As outlined in Section 10.1, the transition from 1G to 2G to 3G feedstocks for production of industrially important commodities that has been witnessed in recent decades is a manifestation of this crucial requirement. The latter fulfills the need to overcome the challenges of the fuel versus food dilemma, efficient land use and energy consumption, renewability and other social and economic factors to make a significant contribution towards advancement of the circular economy.

The prospect of utilizing organic wastes has been emphasized in recent times within the circular economy concept. In fact, the circular economy proposal encompasses the solution to the current waste management problem that plagues various countries throughout the world (Lin et al. 2013). Waste valorisation is a concept related to waste management that has existed for a long time. However, the main conversion processes were often limited to physicothermal processes for producing energy and/or fuel from waste. The rapid depletion of natural and primary resources and the search for more socially, environmentally and economically viable options have led to increased global efforts towards transition to a closed loop (bio)economy. The circular economy proposal can be inspired by the biorefinery system which is viewed as a holistic approach involving valorisation of widely sourced (bio)feedstocks to produce high value-added products via green and sustainable production chains (Zuin 2016). Here, the use of biogenic organic wastes fulfills the requirement of quantity and quality of feedstock, and establishment of associated robust technologies for transformation to useful products.

In addition to feedstock, the second important requirement for a successful biorefinery-based circular economy is design and implementation of robust technological schemes for achieving high-efficiency conversion yields. An integrated biorefinery platform that combines raw materials and produces bioenergy and chemical platforms through transformation of organic material via transformation routes, integrated processes and equipment, is the most appropriate platform for maximizing the potential of a circular economy (Yue et al. 2014). In this respect, the biorefinery platform should support the integrated production of several products such as energy, platform chemicals, and biopolymers. The more common route for transformation of bio-based (biomass, biowaste) resources is thermochemical processes including gasification, pyrolysis, combustion, and biochemical processes such as pretreatment, hydrolysis, chemical and biological conversions (Moncada et al. 2016).

A relevant example is the production of biofuel, which involves a combination of several biochemical processes. The first stage of pretreatment subjects the biomass to a variety of mechanical and/or physical processes to fractionate the lignocellulosic complex. This is followed by chemical (via acid treatment) or biological (via enzymes) processes to allow the release of glucans. A further purification step generally follows before the sugar stream is utilized for conversion to useful products. Subsequently, fermentation with microorganisms or catalyst-based reactions is used to convert sugar to biofuel (Saucedo-Luna et al. 2011). Thus, all the above-mentioned processes are required to be integrated in a robust and efficient manner in a biorefinery to allow successful development of a bio-based circular economy. Furthermore, the design of transformation routes and processes should be suitable to allow handling of more than one kind of feedstock, subject to seasonal and/or regional availability (Moncada et al. 2016).

For the fermentation route for production of chemicals and fuels, process intensification via the development of hybrid and integrated systems such as in-situ product removal (ISPR) techniques is considered to provide several sustainable benefits. ISPR allows real-time coupling of upstream (fermentation) processes utilizing second- and third-generation waste feedstocks, and downstream (product recovery or separation) processes such as adsorption, pervaporation, reactive extraction, ion-exchange, electrodialysis, and so on. Product enrichment, high productivities, use of concentrated feedstocks, reduced wastewater generation and lower energy consumption are some of the benefits achieved via ISPR processes (Kaur and Elst 2014; van Hecke et al. 2014). These contribute significantly towards achievement of the objectives of a circular and sustainable economy, and therefore increased implementation of ISPR processes for fermentation-based production schemes in the biorefinery platform cannot be ruled out.

Finally, another important concern for fulfilling the requirement of a circular economy is increase of market opportunities, promotion of reuse and recycling, and application of corporate social responsibility (Blum et al. 2017). With respect to businesses, emphasis on product lifecycle and circularity during development of new products, services, business models and corporations would certainly be a step in the right direction. A recent example is the production of renewable polyester, polyethylene furanoate (PEF), from plant-based sugars by a Dutch company Avantium. PEF is projected as a replacement for conventional petroleum-based polyethylene terephthalate (PET) because of its superior properties, which make it suitable for wider applications (Kaur et al. 2018). In another example, several initiatives have been taken by leading fashion companies such as Zara and H&M for collection of used textiles/fabrics at their stores, and production of new fabrics using recycled textiles and green production processes with environmentally acceptable chemicals and reduced energy consumption. These initiatives contribute significantly toward improved waste management and sustainable production schemes, while creating more job opportunities and social acceptance of recycle products (H&M 2016; Inditex 2016).

While creating new products and/or service models using bio-based or recycled materials, other important factors concerning the circular economy are product functionality, recyclability and/or biodegradability (Lieder and Rashid 2016). The first one can be easily understood from the above example of PEF, which is an alternative to a long-existing product on market, PET. In this case, a new product has been created which has new properties and functionalities, and therefore appropriate strategies for its introduction into the market are needed. This is indeed applicable to all products (and services) developed via the circular economy approach. The newer products need to demonstrate their functional superiority, in addition to their bio-based sourcing, and compete with the existing products. Another ensuing concern is the question of recyclability or biodegradability. For the circular economy, the former is more advantageous since it retains the product in the recirculation loop of resources, while biodegradation converts it into innocuous yet essentially low-value elements such as water and carbon dioxide.

It is worthwhile emphasizing that the role of government policies and regulations is imperative to provide a concrete framework for closing the loop of product lifecycles, while creating opportunities for people. This aspect has been focused on by the European Union Action Plan, which lays great emphasis on development of quality patterns for secondary raw materials, improving durability and recyclability of products, and revising regulations for recognition of bio or waste derived products in the market (European Commission 2015). These measures provide economic initiatives for businesses to launch sustainable products in the market, while also supporting increased research and development efforts in this direction. Such regulations also lay greater responsibility on producers/manufacturers for both products/processes, and costs and risks associated with the waste generated. Overall, such a concerted initiative helps in the efficient and successful transition to a circular economy where the interests of society are considered along with the capacity of our planet. This is expected to lead to advancement of a closed-loop economy from the current extract–process–consume–dispose linear economy.

10.3 Case Study 1: Circular Textile Waste-based Biorefinery for Production of Chemicals, Materials, and Fuels

10.3.1 Need for a Circular Textile Waste-based Biorefinery

There is an increasing demand for textile products due to the growth of the world's population (Hu et al. 2018; Pensupa et al. 2017). In 2016, around 95 million tons of textile was produced at the global level, which was roughly a fourfold increase compared with the production volume in 1975 (Statista 2016). However, current textile production is an environmentally harmful process, since a considerable amount of resource is consumed, including chemicals, water, and oil, as well as the generation of wastewater and greenhouse gas emissions (Claudio 2007; Pensupa et al. 2017). Furthermore, the enlargement of the textile industry results in the growth of textile waste generation. The amounts of textile waste produced in China and the UK are 26.0 and 1.7 million tons per year, respectively (Hu et al. 2018). Conventional textile waste management methods are landfill and incineration. In the United States, more than 16 million tons of textile waste were generated in 2014, of which approximately 10.46 million tons were sent to landfills and 3.14 million tons were combusted with energy recovery (EPA 2016). At this stage, only 14–15% of textile waste is recycled, although 95% of it can be recycled (Hu et al. 2018). Additionally, the major application of recycled textile waste is to be reused as some low-value product, such as apparel, wiping rags, home insulation, and building materials (Sandin and Peters 2018). According to the concept of the circular biorefinery, the abundant utilization of textile waste and production of high value-added products through sustainable and green biotransformation routes must be considered (Kaur et al. 2018). Therefore, the case of a circular textile waste-based biorefinery is discussed.

10.3.2 Circular Textile Biorefinery

Based on the raw materials for textiles, there are three common types of textile fibers, including cellulose fiber obtained from plants, protein fiber from animals, and synthetic fibers from artificial synthesis (Ghaly et al. 2014). Cotton fibers are collected from the plant *Gossypium* genus. As one of main raw materials among different types of textile fiber, cotton accounts for a large share of the textile market, accounting for roughly a third of the global market (Alay et al. 2016). The annual production of cotton is around 25 million tons (Pensupa et al. 2017), and a considerable amount of cotton-based textile waste is generated owing to fast fashion. Typical cotton fibers are composed of 88–96% cellulose, and the rest of the components are wax, protein and pectin materials. Before use in textile applications, cotton needs to be scoured and bleached. After these treatments, the minor components are removed and the cellulose content is up to 99% (Jeihanipour and Taherzadeh 2009). As the major component that could be used for the production of glucose-rich hydrolysate, cotton from textile waste is regarded as a promising carbon source for the production of value-added chemicals and fuels via bioconversion (Hong et al. 2012; Jeihanipour et al. 2010a; Kiel et al. 1981; Kuo et al. 2014; Li et al. 2012; Mahalakshmi et al. 2011; Opwis and Gutmann 2012). However, the crystallinity index and degree of polymerization of cotton is relatively high compared with other typical cellulosic materials, as presented in Table 10.1 (Jeihanipour and Taherzadeh 2009), which is one of most significant factors affecting the efficiency of biotransformation (Pensupa et al. 2017).

10.3.2.1 Pretreatment Methods

In order to have an efficient enzymatic hydrolysis, pretreatment is crucial to disrupt inter- and intra-hydrogen bonds and produce cellulose in an amorphous state. After the pretreatment step, a decrease in crystallinity is observed, and it is easier for cellulose to react with cellulase. Various pretreatment approaches could contribute to different levels of structural change and enhance the efficiency of subsequent enzymatic hydrolysis, as summarized in Table 10.2. The conventional approach is chemical pretreatment, including acid, alkali, and ionic liquid. Acid pretreatment is the most widely applied method for cellulose fiber, which can contribute to the decomposition of the structure of cellulose. Cellulose in amorphous

Table 10.1 *Physical properties of typical cellulosic materials.*

Substrate	Crystallinity index	Degree of polymerization
Cotton	0.81–0.95	1000–3000
Wood pulp	0.5–0.7	500–1500
Avicel	0.5–0.6	300
Filter paper	0.45	750

Table 10.2 Current pretreatment methods used in textile waste valorisation.

Substrate	Pretreatment	Glucose yield (%)	Reference
White 40/60 polyester/ cotton blend	NaOH/urea (7/12 wt%) at −20°C for 1 h	91	Gholamzad et al. (2014)
Cotton liner	12 wt% NaOH at 0°C for 3 h	99.1	Jeihanipour and Taherzadeh (2009)
Used jeans	85% phosphoric acid at 50°C for 7 h	79.2	Shen et al. (2013)
100% cotton linters	5 g/l $Na_2S_2O_4$ and Na_2CO_3 solution 100°C for 1 h and treated by 85% phosphoric acid 50°C, 100 rpm for 2 h	90	Kuo et al. (2014)
Red T-shirt		80	
Blue polyester/cotton (40/60) blended shirt		60	
100% cotton T-shirts	[AMIM]Cl (ionic liquid) 110°C for 90 min	95	Hong et al. (2012)
Orange 50/50 polyester/cotton blend	85% N-methylmorpholine-N-oxide and n-propyl galate 120°C for 2 h	94.4	Jeihanipour et al. (2010a)

phase can be hydrolyzed from polysaccharide to short saccharide chains, cellobioses and monosaccharide. Meanwhile, more reducing ends and non-reducing ends are released in crystalline regions. The sugar yield from acid pretreatment is related to the acid dosage, operating temperature, and reaction time. Another widely used approach is alkali pretreatment. A swelling effect is the key function of alkali treatment and results in more amorphous phase in the cellulose structure through destroying the inter- and intra-hydrogen bonds. Ionic liquid is also considered as a useful pretreatment method due to its capability to dissolve cellulose. However, some properties of the solvent limit its application, such as high production costs, toxicity, and flammability.

10.3.2.2 Textile Waste Biorefinery

One of the important products from cotton-based textile waste is glucose, hydrolyzed from cotton via enzymes. Glucose is a monosaccharide, which can serve as a carbon source, and is then fermented into value-added products, such as ethanol and methane. Table 10.3 summarizes the latest studies on textile waste valorisation. Many previous works have demonstrated the feasibility of biotransformation from textile waste to ethanol (Kuo et al. 2014; Mahalakshmi et al. 2011). For example, Kuo and Mahalakshmi have reported the production of ethanol from textile waste hydrolysate by microorganisms (Kuo et al. 2014; Mahalakshmi et al. 2011). Hydrogen has also been produced from textile waste-based materials by *Clostridium butyricum* and *Klebsiella oxytoca* (Li et al. 2012). Furthermore, Hong et al. investigated the possibility of bacterial cellulose production using ionic liquid-treated textile waste from *Gluconacetobacter xylinus*. Results suggested that glucose-rich hydrolysate from textile waste could be a high-quality carbon source (Hong et al. 2012). Biogas production via anaerobic digestion is an alternative way to use textile waste. Jeihanipour et al. (2010b) reported that 415 ml of methane were generated from each gram of cellulose. The above examples illustrated that inexpensive and renewable sugars from cotton-based textile waste can be used for production of high value-added products.

Table 10.3 *Recent studies on textile waste valorisation.*

Feedstock	Products	Microorganism	Results	Reference
Cotton linter	Methane	*Saccharomyces cerevisiae*	415 ml/g	Jeihanipour et al. (2010b)
Glucose (cotton)	Ethanol	*Saccharomyces cerevisiae*	0.47 g/g	Mahalakshmi et al. (2011)
Textile wastewater	Hydrogen	*Clostridium butyricum and Klebsiella oxytoca*	1.37 mol/mol	Li et al. (2012)
Glucose (cotton)	Ethanol	*Zymomonas mobilis*	0.43 g/g	Kuo et al. (2014)
Glucose (cotton)	Ethanol	*Saccharomyces cerevisiae*	50 g/g cellulose	Jeihanipour et al. (2010a)
Cotton/polyester	Cellulase	*Aspergillus niger*	1.18 FPU/g	Hu et al. (2018)
Cotton-based textile	Bacterial cellulose	*Gluconacetobacter xylinus*	10.8 g/l	Hong et al. (2012)

Another important high-value component is polyester fiber, which is the largest-volume fiber consumed at a global level and accounts for 50% of the overall natural and synthetic fiber market. The average annual growth rate is nearly 7%, as observed in overall global demand for polyester fiber since 1990 (IHS 2016). However, polyester production requires a large amount of energy, and 1 kg of polyester fiber is responsible for more than 1.5 times the carbon dioxide emissions during the full lifecycle, compared with cotton fiber production (Alay et al. 2016). In order to reduce non-renewable resource input and negative impact on the environment, some yarn manufactures started to produce recycled polyester from used PET (polyester) bottles, which was used by many well-known companies, such as H&M, Nike, C&A, Adidas, and others (Ellen MacArthur Foundation 2017). However, there are not yet any clothing-to-clothing recycling operations at a large scale in the world. One of the key limitations is material blends, where different fibers are combined to give improvements in appearance, comfort, and performance. Polyester blended with cotton is a most popular blending choice in garment production. However, it is more difficult to capture material value through recycling when the textile is blended. An ongoing project entitled "Textile waste recycling by biological method" in our laboratory is focused on the circular textile waste-based biorefinery, as presented in Figure 10.1. This project developed a novel bioprocess using textile wastes for the recovery of glucose and synthetic fibers for textile application. The first step was pretreatment using freezing sodium hydroxide and urea to break down the highly ordered crystalline structure and make the fibers easily accessible to enzymes. In the subsequent enzymatic hydrolysis, enzymes (i.e., cellulase and β-glucosidase) were used to degrade the biodegradable textiles, such as cotton, into glucose, while the non-biodegradable textiles, such as polyester, remained intact and were separated in solid form by centrifugation. In view of the high cost of commercial enzymes, the feasibility of cellulase production through solid-state fermentation using textile waste was investigated in this project. Hu et al. (2018) reported that a complete system of cellulosic enzymes was produced using *Aspergillus niger*, and the cellulase activity harvested was 1.18 FPU/g after nine days' cultivation. Furthermore, the highest β-glucosidase activity was obtained to date, reaching 1731 U/g. This fungal enzyme solution was successfully applied in textile waste hydrolysis with comparable efficiency to commercial cellulase (Hu et al.

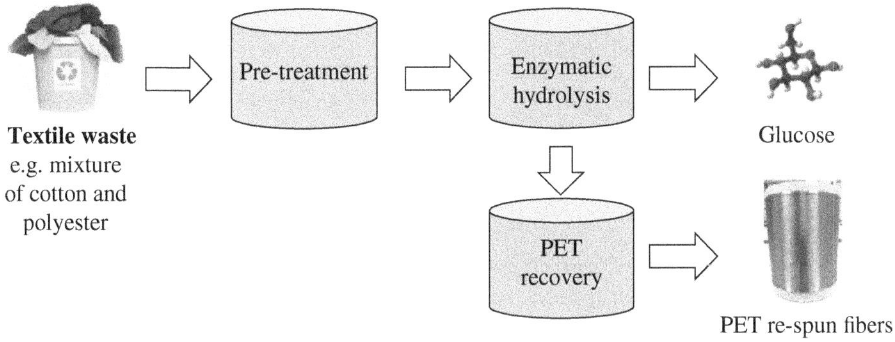

Figure 10.1 Process flow diagram of the utilization of textile waste for glucose recovery and PET recycling.

2018). After enzymatic hydrolysis, recovered polyester from textile wastes was re-spun successfully for further textile application by melt spinning. Melt spinning was a commonly used method for polyester production (Fambri et al. 1997). Washing was not required during spinning since no solvent was used and thereby no wastewater was generated. Therefore, this process was environmentally friendly and met the requirement for green chemistry (Militky 2009). This project successfully achieved the goals of capturing the embodied value of cotton and polyester fibers, which is an important contribution towards the development of a circular textile waste-based biorefinery, as illustrated in Figure 10.2.

10.4 Case Study 2: Circular Food Waste-based Biorefinery for Production of Chemicals, Materials, and Fuels

In recent years, around a third of food is wasted globally, resulting in 1.3 billion metric tonnes of food waste per year. Research also shows that most of the food wastes are produced by developed regions in the world, with annual food waste per capita in North America and Oceania reaching almost 300 kg/year (Food and Agriculture Organization 2018). Much of this could be recycled and processed properly to produce useful materials via the biorefinery concept. Some countries have started raising concerns over insufficient landfill space and environmental issues caused by current food waste disposal methods. In 2012, food waste constituted 36% of MSW in Hong Kong (3337 tonnes/day) (Environmental Protection Department 2017). Thus, reducing or processing food waste into useful energy and materials would be an attractive solution to environmental management.

In many developed countries such as the United States (US), food waste is also a major environmental problem. In 2014, around 38 million tonnes of food waste was generated in the US, which is the largest food waste contributing country in the world. This amounts to around 40% of the total food produced by the US farming sector. In addition, the cost of loss/processing this food waste is approximately US$165 billion/year, according to the Natural Resources Defense Council (NRDC) (Gunders 2017). By generating food waste, not only does the Government have to provide the necessary landfill space, but also resources are lost due to the production and packaging of food that eventually goes to landfill. Twenty

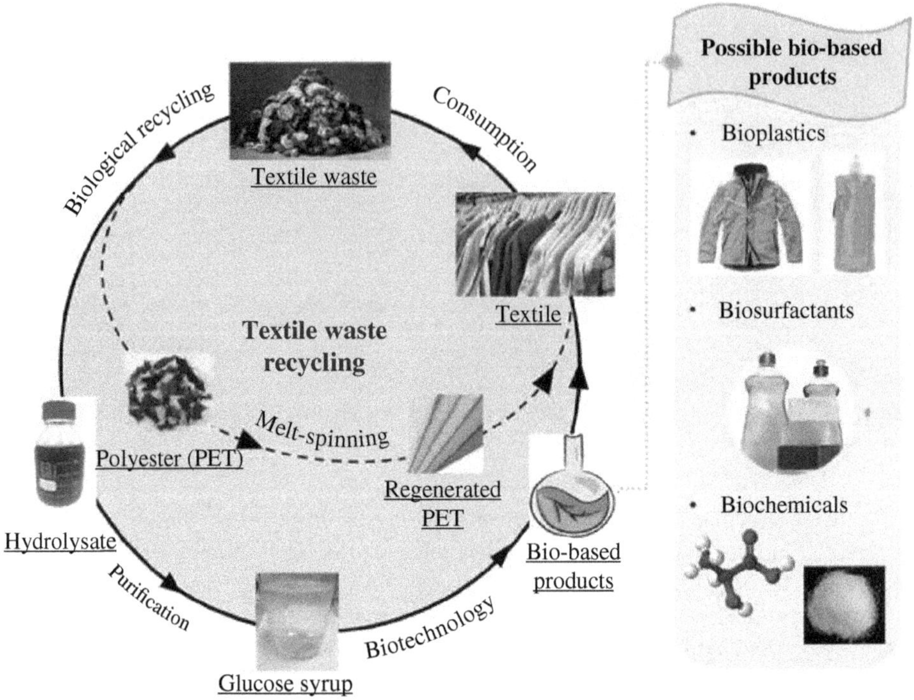

Figure 10.2 *Circular textile waste-based biorefinery for bio-based products and polyester recycling.*

five percent of fresh water in the US is used in food production, the majority of which becomes wastewater after usage. In addition, food waste thrown into landfill is also a major issue for environment since it emits methane, which is a greenhouse gas that is 20 times more harmful than CO_2 (Lendon 2012).

While concerns over global food waste generation have been raised, solutions and efforts to resolve this issue have been rather limited. Two possible solutions are proposed: (i) preventing generation of food waste; (ii) upcycling food waste. While it is better to prevent generation of food waste, it is unavoidable that people worldwide will throw away spoiled or unappetizing food. Food waste can also be generated through poor handling or storage in harvest, post-harvest, and in the supply chain (Lipinski et al. 2013). Therefore, designing feasible recycling or processing methods for food waste seems to be a feasible alternative solution.

Another advantage of processing food waste is that we can produce useful materials that were originally produced using edible food. This will also avoid competition for food resources. Figure 10.3 exemplifies three possible bioconversion routes for food waste, of which we discuss two below. Section 10.4.1 describes the circular bioconversion of food waste into polyethylene furanoate (PEF). Section 10.4.2 discusses the circular bioconversion of food waste into biosurfactant.

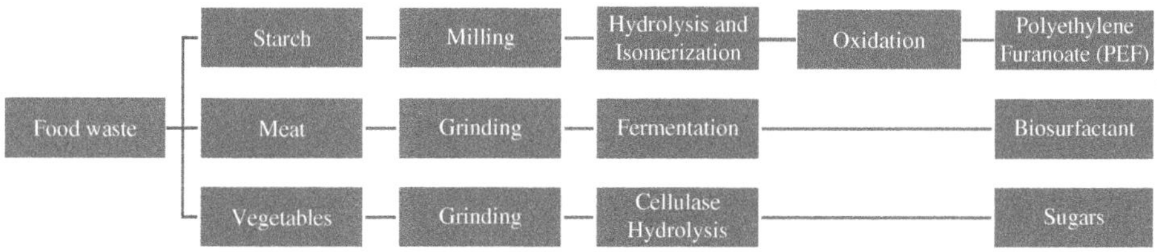

Figure 10.3 *Bioconversion routes for food waste.*

10.4.1 Circular Bioconversion of Food Waste into Polyethylene Furanoate (PEF)

This section will focus on one of the examples suggested in Figure 10.1, production of PEF from food waste. Currently, a project funded by the Innovation Technology Commission in Hong Kong entitled "Conversion of Food Waste into Bio-based Polyethylene Furanoate (PEF) Yarn" is being conducted by our group in the School of Energy and Environment at the City University of Hong Kong. The idea is analogous to the corn-derived high-fructose corn syrup (HFCS). The aim is to produce PEF via food waste-derived fructose. This will potentially reduce the diversion of produced corn toward materials such as PEF and avoid competition with food. Section 10.4.1.1 briefly describes the advantages of PEF compared with PET. Then, Section 10.4.1.2 explains the methodology of PEF production from food waste. Lastly, Section 10.4.1.3 underlines the circular economy of transforming food waste into PEF plastic bottles.

10.4.1.1 What is Polyethylene Furanoate (PEF)?

People worldwide have begun to realize the importance of green products to minimize global warming. Therefore, many government laboratories and universities are now investing their funds into research projects that involve optimization of green product processing. One such project is PEF production using food waste. Bio-based PEF is an alternative to polyethylene terephthalate (PET), potentially used as a precursor for the manufacture of beverage bottles. PEF is a green material with better properties than PET, for example, since it provides six times stronger protection against oxygen and three times stronger protection against carbon dioxide (Reinshagen 2012). This is an important characteristic in a beverage bottle. For example, usage of PEF in a fruit juice bottle would potentially reduce the amount of oxygen scavenger required in the production process. Another example is usage of PEF in a soft drink bottle, which could delay reduction in the quality of the soft drink due to loss of carbon dioxide through diffusion from the plastic bottle. Furthermore, PEF also has a lower melting temperature and higher glass-transition temperature. In addition, it is a cheaper solution than many other green plastic bottles on the market currently (Reinshagen 2012) and compared to PET, production of PEF could reduce non-renewable energy use by 40–50% and greenhouse gas emissions by 45–55% (Eerhart et al. 2012).

Examples of several companies producing bioplastics are readily available. Corbion, a Dutch company that focuses on biopolymers, began production of PEF bioplastics in 2016 (Corbion 2020). Another Dutch company named Avantium, specializing in renewable bio-based products, is now focusing on PEF production (Avantium 2020). Avantium succeeded in securing a €25M Horizon 2020 grant from the Bio-based Industries Joint Undertaking. The company intended to acquire a total funding of €150M by the end of 2020 and planned to open its PEF plant by 2023 (Cowen 2019). Due to its several advantages, PEF plastic is useful for production of water or carbonated soft-drinks bottles, replacing the currently used PET bottles. The methodology of producing PEF from food waste will be discussed in Section 10.4.1.2.

10.4.1.2 Production of PEF from Food Waste

Production of PEF from food waste begins with the selection of food waste feedstock, then processing the feedstock into fructose, and lastly using this fructose as a feedstock in the production of PEF. The production process is summarized in Figure 10.2.

Feedstock selection: Choosing the most abundant and obtainable type of food waste for efficient PEF production Availability of resources has been one of the primary concerns in the production of useful materials (e.g., PEF) through waste-based feedstock. Therefore, it is important to select the most abundant food waste as feedstock for PEF production. In general, food waste can be classified into many categories, such as fruit, vegetables, meat and fish, cereal, dairy and eggs, bakery, and beverage wastes (Parfitt et al. 2010). Fruit and beverage wastes can be used as feedstocks since they contain glucose and fructose. Similarly, bakery and cereal wastes can also be used since these can be hydrolyzed into glucose through chemical or enzymatic processes. Vegetable waste can also be used by converting cellulose into glucose using thermochemical treatment and/or enzymatic hydrolysis, with cellulase as biocatalyst. The produced glucose will subsequently be isomerized into fructose through isomerization processes.

Utilization of different waste-based feedstocks would affect the fructose concentrations in the PEF production process. In order to choose the most abundant food waste for PEF production, different types need to be compared while considering distinct locations and phases of food waste generation (i.e., agricultural, post-harvest, and consumption phases). For example, it might be better to use bakery, vegetable, or fruit waste in Western countries since they do not produce or eat much cereal (Parfitt et al. 2010). On the other hand, it might be more suitable to use cereal or vegetable wastes in Eastern countries due to their higher availability in these regions. There is a great amount of cereal waste in Asia owing to the high consumption of rice. Besides, there are also plenty of vegetable wastes in Asia, mainly due to large losses in the agricultural (harvesting) and post-harvest phases (Food and Agriculture Organization 2013). On the worldwide scale, Asia contributes around 51% of the total food waste produced globally (Lipinski et al. 2013).

A technical report by the FAO suggests that the cereal waste from the consumption phase in industrialized Asia (e.g., Japan, Korea, Hong Kong) is the third largest food waste contributor, comprising more than 3% of the total food wastage worldwide. The second largest food waste contributor is vegetable waste in industrialized Asia from the consumption and agricultural production phases (Food and Agriculture Organization 2013). In this case, it might be better to use vegetable waste since it is one of the largest contributors to food waste. However, vegetable waste involves greater processing demands since it naturally contains cell walls that need to be broken down by physical or thermochemical treatments (e.g., grinding and torrefaction). On the other hand, cereal waste from the consumption phase can be hydrolyzed by adding glucoamylase without requiring rigorous physical treatment, and therefore it is more efficient to use this cereal waste. Cereal waste mainly consists of glucose, which improves the fructose productivity from the process. Hence, in order to optimize the efficiency and availability of feedstocks for PEF production, cereal waste from the consumption phase in industrialized Asia is selected as the primary feedstock in this chapter. The cereal waste used will be primarily rice, since it is the most abundant cereal waste in Asia.

Processing food waste into fructose as feedstock for PEF production The next stage in PEF production is producing its feedstock (fructose) from food waste through enzymatic hydrolysis, followed by several purification processes. Firstly, food waste is collected from several restaurants, and a pretreatment process is performed. Pretreatment of food waste is essential for the whole process. There are two ways to pretreat the food waste. One way is to mill and grind the food waste. This pretreatment will greatly increase efficiency of the hydrolysis process. Another way is to manually select and remove meat, vegetable matter and other unneeded materials in the food waste, leaving behind only rice or other starch components. Although this pretreatment is inefficient and strenuous, it eliminates many unnecessary impurities that could hinder the process and decrease the quality of final product (i.e., fructose). In laboratory-scale experiments, the latter pretreatment process is preferred since it produces a higher fructose yield. However, as it is too inefficient for large-scale production, milling and grinding pretreatment is performed.

The second step is hydrolysis of food waste using a paddle mixer. Since the final target is fructose, the starch components in food waste are hydrolyzed by adding amylase, glucoamylase, and water into the paddle mixer. With only milling and grinding pretreatment before this hydrolysis reaction, pieces of meat and vegetable remain inside the hydrolysate. Therefore, the hydrolysate must be centrifuged, clarified, and then filtered to separate meat, vegetables, and other remaining solids. After this step, high glucose concentrations will be obtained in the hydrolysate.

Lastly, the remaining steps after the hydrolysis process are isomerization and purification processes. The isomerization process converts glucose into fructose with a maximum efficiency of 50%. The purification process includes centrifugation, filtration, adsorption, ion-exchange chromatography, and ligand-exchange chromatography with the simulated moving bed (SMB) technique. Throughout all these processes, several impurities are removed, such as unwanted remaining solids, colorants, cations and anions. The final chromatography with the SMB technique is necessary to separate fructose from the glucose–fructose mixture. Then, the remaining glucose is recycled and isomerized again to produce fructose. Through all these downstream processes, high-purity fructose syrup is obtained.

Processing fructose into PEF yarn The fructose produced from food waste will be used as a feedstock for PEF production. In the conventional production process, three steps are involved in the production of PEF yarn from fructose. The first step is dehydration of fructose. In this step, fructose is converted into 5-hydroxymethylfurfural (HMF) by acid/base catalyst. It is important to note that humin can often be produced as a byproduct in this step as a result of side-reactions of fructose/HMF with acid/base catalyst (Motagamwala et al. 2018). Several other impurities can also be formed through these side-reactions, such as formic acid and levulinic acid (van der Klis et al. 2017). Therefore, prevention of this side-reaction or purification steps in downstream processes are of paramount importance in order to ensure the quality of PEF yarn.

The second step is oxidation of HMF with Pt/C as a catalyst to produce furan-2,5 -dicarboxylic acid (FDCA). FDCA is the monomer needed to produce PEF yarn. In the oxidation reaction of HMF, several intermediate chemicals are produced, specifically diformylfuran (DFF), 5-hydroxymethyl-2-furancarboxylicacid (HMFCA) and 5-formylfuran-2-carboxylicacid (FFCA). However, incomplete oxidation of HMF toward FDCA can occur using this commonly used method. This will generate small quantities of intermediates within the final product, namely DFF and HMFCA, resulting in another complex purification process or termination of chain growth during polymerization processes in the downstream step. Some research projects have tried to tackle this problem, such as finding other methods to produce FDCA.

The last step is purification and polymerization of FDCA to produce PEF yarn. Several chromatography processes are performed here in order to purify FDCA. Then, PEF yarn is produced through a polymerization process with FDCA and ethylene.

In summary, the production of PEF yarn from food waste starts from processing food waste into fructose, which is then converted into FDCA. The FDCA is subsequently polymerized with ethylene to form PEF yarn. The flowchart of this overall process is shown in Figure 10.4 (Yu et al. 2018).

10.4.1.3 Circular Economy of Processing Food Waste into PEF Plastic Bottles

The concept of the circular economy has been widely recognized worldwide. A definition by the Ellen MacArthur Foundation suggests that the circular economy is a restorative and regenerative system that aims to redefine products and services to design waste out, while minimizing negative impacts (Ellen MacArthur Foundation 2016). In other words, the circular economy proposes that the most important factors to be considered in product and process design are renewable resources and recyclability (Kaur et al. 2018). The circular economy consists of several aspects to form a loop.

The opposite of a circular economy is a linear economy. The linear economy is a straightforward system that involves the extract–consume–dispose concept rather than recycling. A typical example of a linear economy is the current plastic packaging value chain, which represents 26% of total plastics. In the plastic packaging industry, the feedstock and manufacturing process utilizes non-renewable resources of petroleum, depleting 6% of worldwide oil production (Kaur et al. 2018). In addition, only 5% of the product is currently recycled, while the other 95% is lost after first use (Ellen MacArthur Foundation 2016). The non-recycled plastic goes to landfill and causes environmental issues such as emission of greenhouse gases. This case exemplifies the importance of a circular economy, which in contrast emphasizes the concept of renewable sources. It should be pointed out that renewability is not the same as sustainability. The sustainability concept usually underlines the economic component of the system. For example, using edible food crops for production of biofuel is not a sustainable option since it creates a "food versus fuel" controversy; however, it is a renewable option since food crops can be renewed and grown infinitely as long as there is still enough arable land to do so.

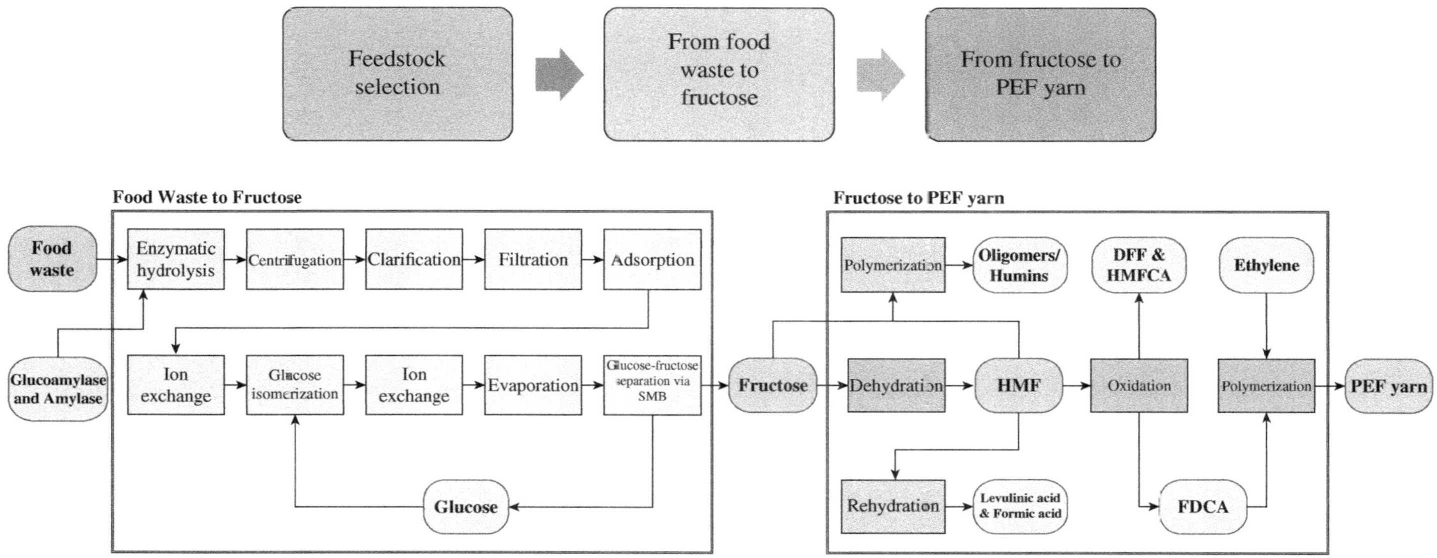

Figure 10.4 Flowchart of overall production of PEF yarn from food waste.

Other important concepts are biodegradability, recyclability, and potential reuse applications (Kaur et al. 2018). These three terms have different meanings; each of them serves as an important part of the circular economy. Biodegradability is the ability of a material to be broken down into simple and low-value elements through microorganism activity, such as water and carbon dioxide. It is important to note that biodegradability is not the same as a bio-based product. A bio-based product is composed of biological products or renewable agriculture materials; however, it is not necessarily biodegradable. For example, bioplastic made from polyhydroxybutyrate (PHB) is biodegradable, while bioplastic from bio-PET is not biodegradable. Recyclability is the ability of a material to be reused without changing its essential form. One common example is recycling PET bottles into other bottles or plastic sheet. Lastly, potential reuse application means that the material can be reused again for the same application. For example, plastic bags obtained in the supermarket can be reused as plastic bags. In the circular economy, recycling and reuse pathways are the preferred options since they retain the material in a closed-loop resource circulation economy (Kaur et al. 2018).

One good example of the circular economy is production of PEF plastic bottles from food waste. PEF is a suitable material that emphasizes the concept of the circular economy by using renewable sources (food waste). It has been explained earlier in this chapter that food waste is renewable due to its sheer availability worldwide and high nutritional content (Kaur et al. 2018). Furthermore, since PEF bottles also have high product functionality, they can reliably replace PET bottles.

Although PEF is not biodegradable, it can be recycled properly with current recycling technology. Currently, plastic is recycled by collecting it, sorting it based on the polymer in each plastic, then chipping it into pellets and melting it. Realistically, PEF plastic could be recycled if there was a systematic recycling process proposed by government. The European PET Bottle Platform (EPBP) gave interim approval for the recyclability of PEF plastics in the European bottle recycling market up to 2% of the total volume of the recycling stream (Barrett 2020). EPBP also mentioned that it would help in establishing systematic recycling processes that involve incentives for consumers to return or dispose of PEF bottles in similar manner as they dispose of PET bottles. Figure 10.5 shows the circular economy of PEF plastic from food waste.

In conclusion, production of PEF from food waste is a promising project that conforms to the circular economy concept. It provides renewable resources (food waste) as well as systematic recycling and reuse processes.

10.4.2 Circular Bioconversion of Food Waste into Biosurfactant

In recent years, interest in microbial surfactants has significantly increased for several reasons. First, biosurfactants are regarded as alternatives to chemically synthesized surfactants because of their low toxicity and biodegradability. Second, biosurfactants possess similar properties to chemical surfactants, covering diverse applications ranging from biotechnology to environmental cleanup. In spite of having broad potential applications in industries, uncompetitive production of biosurfactants has been reported compared with chemical surfactants (Campos et al. 2013) due to expensive raw materials, high recovery costs, and low yields obtained. The aim of this section is to provide an overview of biosurfactants, their properties, production and downstream processing for further research.

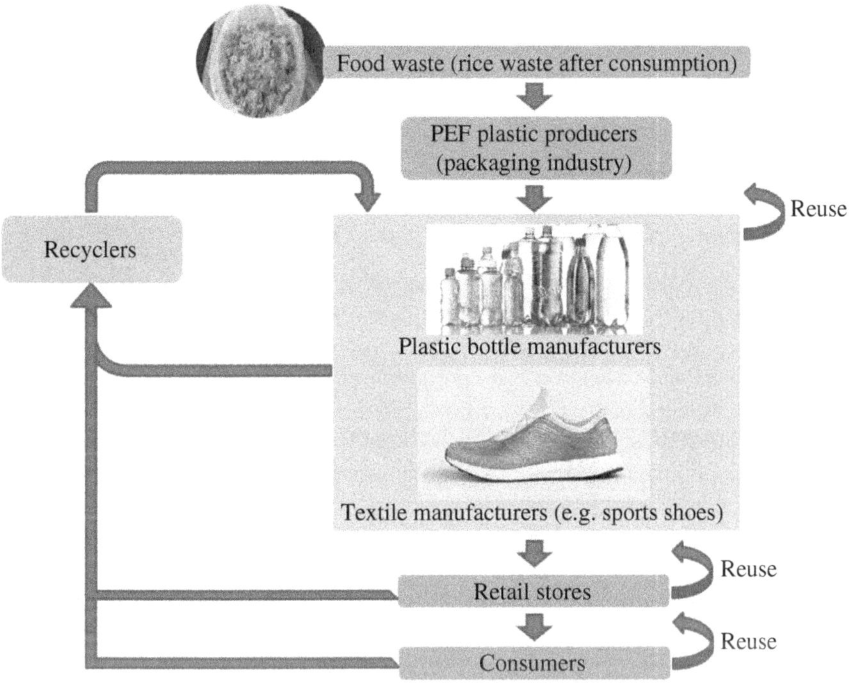

Figure 10.5 Circular economy of PEF plastic bottles from food waste.

10.4.2.1 Biosurfactants

Biosurfactants are surface-active agents naturally produced by microorganisms (fungi, yeast, bacteria) and are amphiphilic compounds with both hydrophilic and hydrophobic properties. The hydrophilic head is composed of carbohydrate, amino acid, peptide anions, cations, and polysaccharides, and the hydrophobic tail comprises saturated or unsaturated fatty acids (Georgiou et al. 1992). This unique structure allows the microbial surfactants to reduce surface and interfacial tension between two phases, for example at liquid–liquid, liquid–solid or liquid–air interfaces (Motevasel 2014). This important property of microbial surfactant, which enhances the interaction between two media in different phases, has a wide variety of uses, such as solubilization, emulsification, demulsification, foaming, wetting, and so on.

Biosurfactants are classified according to their microbial origin and molecular structure (Sharma et al. 2014). Glycolipid surfactants are the most studied molecules and can be divided into four categories: rhamnolipids, sophorolipids, trehalolipids, and mannosylerythritol lipids. Among them, sophorolipids are the most promising microbial surfactants, produced by non-pathogenic species in very high yields and with simple recovery methods (de Oliveira et al. 2015).

Sophorolipids (SLs) are glycolipid biosurfactants produced on microbial cell surfaces or secreted as extracellular products. SLs were first identified in 1961 as being produced by yeast species *Candida apicola* (Gorin et al. 1961). Later, several yeast species were found and reported to synthesize SLs, such as *Starmerella bombicola, Rhodotorula bogoriensis, Wickerhamiella domercqiae,* and *Candida batistae*. With an optimized fermentation process, sophorolipid production by microbes increases to a very high yield of over 400 g/l (Van Bogaert et al. 2007). The promising yield indicates that sophorolipid is an attractive alternative for industrial and commercial production and an environmentally friendly alternative to petroleum-based surfactants.

10.4.2.2 Medium Optimization for Sophorolipid Production

Biosurfactants are produced by microorganisms during the stationary phase and can be regarded as secondary metabolites. Hydrophilic carbon sources are used as the primary energy source for growth of microbes, and are incorporated into the structure of biosurfactants. Glucose is the most common hydrophilic source for sophorolipid production. Several types of carbohydrates can be used as substrate, such as fructose, mannose, saccharose, raffinose, maltose, lactose, sucrose, and glycerol. Optimal synthesis for biosurfactants occurs when both hydrophilic and hydrophobic carbon substrates are present in the fermentation medium (Cooper and Paddock 1984). The presence of lipophilic sources can result in higher productivity, and yeast extract is also important for growth and biosurfactant production.

10.4.2.3 Waste Feedstock as Substrate for Sophorolipid Production

Economy is always a hindrance for microbial technology, especially biosurfactant production. The reason for limited applications of biosurfactants is the high cost of substrates, which accounts for nearly 60% of the total operating expenses. Therefore, researchers have focused on alternative substrates to replace the conventional raw materials to produce surfactants. Table 10.4 summarizes various waste substrates for biosurfactant production by different microorganisms and yields.

Different wastes have been reviewed and targeted as potential feedstocks for biosurfactant production, and two common features can be found from Table 10.4. First, organic waste materials are exploited as substrates to synthesize surfactants by various microorganisms. Second, these substrates are mainly byproducts from the food processing chain, and therefore it demonstrates the possibility of utilizing food waste as feedstocks for transformation into valuable products. For example, cassava flour wastewater and potato processing effluent are waste materials rich in carbohydrate content in the form of starch or reducing sugar. Wastes from dairy and sugar industries provide a balance of nutrients to support good microbial growth. Whey is a byproduct of cheese production and achieved high yields of SLs up to 422 g/l by combining whey concentrate and rapeseed oil (Daniel et al. 1999). Molasses is a co-product of sugar production with high sugar content, and has been successfully used as a substrate to produce biosurfactants.

Waste stream utilization is the main strategy for the fermentation process along with efficient waste management. Food waste is a promising source of feedstocks which may be utilized because of their valuable compositions. Starch, proteins and lipids are three major constituents of food waste (Paritosh et al. 2017). Selection of suitable food waste in terms of balance of hydrophilic carbon and lipids composition helps to support microbial growth and produces the desired biosurfactant yields. Using food waste as cheaper substrates is of great benefit to lower the production costs of biosurfactants along with better waste management.

Table 10.4 *Summary of inexpensive feedstocks for biosurfactant production using various microbes.*

Waste feedstock	Producer	Fermentation mode	Biosurfactant titer (g l^{-1})	Time (h)	Productivity (g/l-h)	Reference
Deproteinized whey concentrate. rapeseed oil, waste cell oil	*Candida bombicola*	Fed batch	422	410	1.03	Daniel et al. (1999)
Glucose, sugarcane molasses, soybean oil	*Candida albicans*	Fed batch	108.7	102	1.07	Yang et al. (2012)
Waste cooking oil, glucose	*Candida bombicola*	Fed batch	55.6	240	0.23	Maddikeri et al. (2015)
Sweet sorghum bagasse, corn stover hydrolysate, yellow grease	*Candida bombicola*	Fed batch	52.1	168	0.31	Samad et al. (2017)
Sugarcane molasses, dried milk powder, ghee, soybean oil	*S. bombicola* NRRL Y-17069	Batch	38.76	192	0.20	Daverey et al. (2009)
Deproteinized whey concentrate, glucose, oleic acid	*Candida bombicola*	Batch	33	192	0.17	Daverey and Pakshirajan (2010)
Sunflower oil cake waste, motor oil	*C. bombicola NRRL Y-17069*	Batch	26.4	96	0.275	Rashad et al. (2014)
Potato processing waste	*Bacillus subtilis*	Batch	0.9	48	0.019	Noah et al. (2005)
Cassava wastewater, frying oil	*Pseudomonas aeruginosa*	Batch	0.66	120	0.005	Costa et al. (2009)
Cashew apple juice pomace	*Bacillus subtilis*	Batch	0.44	72	0.007	de Oliveira et al. (2013)

10.4.2.4 Optimization of the Culture Conditions for Cell Growth and Sophorolipid Production

Fermentation is carried out in a vessel for the growth of microorganisms known as a fermenter. The main function of the fermenter is to provide a suitable environment in which microorganisms can efficiently produce the desired products under controlled conditions. Since the operating conditions of the fermenter strongly affect the fermentation process, several parameters should be carefully controlled in order to provide the optimum environment for the desired product. Temperature, agitation rate and aeration speed, pH, and nutrient availability are of primary importance. Optimization of the fermentation process is important for reducing production costs and improving productivity.

Temperature The optimal temperature for biosurfactant production has been reported in the range of 25–30°C with no significant difference in the yields between the temperature. However, fermentation carried out at 25°C results in lower biomass growth and higher glucose consumption compared with fermentation at 30°C (Casas and García-Ochoa 1999). In the fermenter, addition of heat is provided by use of internal heating coils, and removal of heat is achieved by circulating deionized water through the water jacket. A temperature probe is placed directly in the fermentation chamber for monitoring temperature regularly.

Dissolved oxygen Oxygen is necessary for all aerobic fermentation. The dissolved oxygen level in the fermenter is a very important factor for the fermentation process since it affects microbial growth and the overall efficiency of biosurfactant production. A dissolved oxygen probe is used to detect oxygen levels in the fermenter. Aeration rate and agitation speed are two major parameters controlling dissolved oxygen levels in the fermenter. Excessive airflow rate and agitation speed reduces biosurfactant production due to foam formation (Shaligram and Singhal 2010).

pH Microorganisms grow faster in specific pH, and therefore it is important to control pH to grow the targeted microorganisms for biosurfactant production. The optimal pH value for biosurfactant production can vary depending on the microbial producer and type of biosurfactant. The optimal pH for SL production is 3.5 (Cooper and Paddock 1984). A pH probe is inserted into the fermenter for measurement, and an automatic pH control system is used to maintain a desired pH during the fermentation.

10.4.2.5 Downstream Processing of Biosurfactant

High capital investment is needed for downstream processing to recover and purify microbial surfactants from fermentation broths. Feedstocks from the waste stream may reduce raw material costs, but it can also lead to expensive treatment and lower productivity due to the presence of impurities or additional complex purification processes. Efficient recovery of the biosurfactants plays an important role in evaluating the feasibility of a production process on a commercial scale. There are several conventional methods for the recovery of biosurfactants, such as organic solvent extraction, phase separation, foam separation, and ultrafiltration (see Table 10.5). The purification step involves crystallization and chromatography. The choice of recovery methods depends on the type of biosurfactant.

Table 10.5 Summary of downstream processing of biosurfactants.

Methods	Principles
Solvent extraction	Solubility of molecules in organic solvents
Centrifugation	Separation of surfactant from cells in the aqueous phase by centrifugation
Ultrafiltration	Above the critical micelle concentration, micelle form and are retained by the polymeric membrane because of their large molecular size
Foam fractionation	Foam surface activity
Chromatography	Affinity to adsorbent surface of the stationary phase based on size, charge and polarity

Solvent extraction Organic solvent extraction is a common recuperation technique for biosurfactants on a laboratory scale. It involves two phase extractions. Polar solvents such as ethyl acetate, alcohols, and methyl-tert-butyl ether (MTBE) are used to extract biosurfactants and dissolve hydrophobic impurities (Kuyukina et al. 2001). When a lipidic carbon source supplements the culture broth as a co-substrate, a second extraction with non-polar solvent, such as n-pentane or n-hexane, is performed to remove the lipidic residue. According to Karadi et al. (2014), using chloroform and methanol in a 1 : 1 ratio was found to be the most suitable solvent for recovery of biosurfactant. However, the drawback of this method is that it uses a large amount of solvent and thus results in increasing production costs. The toxicity of the solvent also needs to be considered.

Centrifugation Physical separation is the preferred method for large-scale and industrial applications. For high yields of biosurfactants in the product-rich phase, products can be isolated from the fermentation medium by centrifugation. Centrifugation allows the separation of heavier molecules from fermenting (production) cells in the aqueous phase. However, the disadvantage is increased contamination with cells, since biosurfactants are heavier than the product phase but the transfer is hindered because of the high viscosity of the product phase (Kosaric and Sukan 2014).

Foam fractionation Foam fractionation is a chemical process in which hydrophobic molecules are preferentially separated from a liquid solution using rising columns of foam. Surfactants get concentrated at a gas–liquid interface, leading to the formation of bubbles through the liquid and on the interface, resulting in foam formation (Willenbacher et al. 2015). A simple foam column is inserted through the fermentation vessel headplate to collect foam product. Employment of foam fractionation by continuously removing the product can result in good yield values (Cooper et al. 1981).

Membrane ultrafiltration Ultrafiltration can be used for the purification of biosurfactants from supernatant by utilizing semipermeable membranes to separate molecules of different sizes. Above the critical micelle concentration, micelle formed are retained by the polymeric membrane due to their larger size (Jauregi et al. 2013). Under hydraulic pressure, small molecules and impurities are forced through a membrane of very fine pore size. Two-step ultrafiltration can improve the recovery of biosurfactant (Isa et al. 2007).

Chromatography Chromatography is a process for separation and identification as well as purification of individual compounds in a mixture by passing solution through a column. The separation principle is that components in the mixture have different affinities to adsorb onto a stationary phase or dissolve in a solvent, due to its size, charge, and polarity of solutes. All chromatographic methods require a stationary phase and a mobile phase. Different interactions with the stationary phase and mobile phase have different traveling rates. Molecules with more affinity towards the stationary phase migrate down the chromatographic column at a slower rate compared with the molecules with more affinity towards the mobile phase. The mobile phase must be degassed to eliminate the formation of air bubbles.

Compared with other chromatographic separation methods, high-performance liquid chromatography (HPLC) is the most precise approach for biosurfactant identification and quantification (Heyd et al. 2008). This sophisticated method allows coupling with different detection systems. Mass spectrometry provides the detailed structure and composition of the mixture. Evaporative light scattering detection (ELSD) is capable of detecting almost all compounds that are less volatile than the mobile phase. ELSD is useful for detection of compounds with low light absorbance such as biosurfactants. This property is not available with UV detector systems (Ribeiro et al. 2012). The signal intensity of the ELSD is proportional to the quantity of the target substance. A refractive index detector (RI) detects compounds with limited or no UV absorption such carbohydrates or sugars. Another advantage of HPLC is its high sample capacity of up to 120 samples per run.

10.5 Conclusions and Future Trends

The chapter discussed the concept of the circular economy through the biorefinery model. The production of high value-added products such as platform chemicals, materials, energy, and fuel via the valorisation of inexpensive and renewable resources and/or organic wastes (including food waste, agricultural waste, industrial waste, etc.) in a biorefinery has been considered as one of the important factors contributing to the development of a circular economy. The establishment of robust chemical, thermal or biological transformation routes, the ability to utilize multiple feedstocks, and generation of more than a single product in an integrated biorefinery, are factors which could lead to increased overall profitability. It has also been demonstrated in this chapter that while renewability is one aspect of circularity, other underlying factors such as market acceptance and penetration of new bio-based and/or waste-derived products, government regulations to promote the development of such products, along with creation of market opportunities, businesses and jobs for people, and acceptance of corporate social responsibility are also crucial. The case studies of food waste and textile waste transformations presented in this chapter illustrate how the concerted action of researchers, government and industry can facilitate the transition from a linear to a closed-loop economy.

Acknowledgements

Appreciation and gratitude are expressed to the Hong Kong Research Institute of Textiles and Apparel (HKRITA), the Pou Chen Corporation, Innovation and Technology Fund (ITF),

Bio-Based Europe Power Plant (BBEPP), the Industrial Technology Research Institute (ITRI) Taiwan, the International College Hong Kong (ICHK), Novozymes, and professors, research supervisors, research fellows, and other supportive parties in the City University of Hong Kong. Without them, we would not have been able to produce this book chapter.

References

Alay, E., Duran, K., and Korlu, A. (2016). A sample work on green manufacturing in textile industry. *Sustainable Chem. Pharm.* **3**: 39–46. https://doi.org/10.1016/j.scp.2016.03.001.

Avantium (2020). YXY Technology. https://www.avantium.com/technologies/yxy/ (accessed 23 April 2020).

Barrett, A. (2020). PET and PEF. *Bioplastics News*. https://bioplasticsnews.com/2020/03/31/pet-pef/ (accessed 28 April 2020).

Bentsen, N.S. and Felby, C. (2012). Biomass for energy in the European Union – a review of bioenergy resource assessments. *Biotechnol. Biofuels* **5**: 1–10.

Blum, C., Bunke, D., Hungsberg, M. et al. (2017). The concept of sustainable chemistry: key drivers for the transition towards sustainable development. *Sustainable Chem. Pharm.* **5**: 94–104. https://doi.org/10.1016/j.scp.2017.01.001.

Campos, J.M., Montenegro Stamford, T.L., Sarubbo, L.A. et al. (2013). Microbial biosurfactants as additives for food industries. *Biotechnol. Progr.* **29** (5): 1097–1108.

Casas, J. and García-Ochoa, F. (1999). Sophorolipid production by Candida bombicola: medium composition and culture methods. *J. Biosci. Bioeng.* **88** (5): 488–494.

Cherubini, F. (2010). The biorefinery concept: using biomass instead of oil for producing energy and chemicals. *Energy Convers. Manage.* **51**: 1412–1421.

Clark, J.H., Farmer, T., Herrero-Davila, L., and Sherwood, J. (2016). Circular economy design considerations for research and process development in the chemical sciences. *Green Chem.* **18**: 3914–3934.

Claudio, L. (2007). Waste couture: environmental impact of the clothing industry. *Environ. Health Perspect.* **115** (9): A449–A454.

Cooper, D.G. and Paddock, D. (1984). Production of a biosurfactant from Torulopsis bombicola. *Appl. Environ. Microbiol.* **47** (1): 173–176.

Cooper, D., Macdonald, C., Duff, S., and Kosaric, N. (1981). Enhanced production of surfactin from *Bacillus subtilis* by continuous product removal and metal cation additions. *Appl. Environ. Microbiol.* **42** (3): 408–412.

Corbion. (2020). The future of plastics is biobased. https://www.corbion.com/bioplastics/fdca (accessed 23 April 2020).

COST. (2017). About COST. European Cooperation in Science and Technology. http://www.cost.eu/about_cost (accessed 23 April 2020).

Costa, S.G., Lépine, F., Milot, S. et al. (2009). Cassava wastewater as a substrate for the simultaneous production of rhamnolipids and polyhydroxyalkanoates by *Pseudomonas aeruginosa*. *J. Ind. Microbiol. Biotechnol.* **36** (8): 1063–1072.

Cowen, L. (2019). Bioplastics consortium granted €25M EU funding. https://www.labiotech.eu/industrial/bioplastics-avantium-horizon2020-grant/ (accessed 27 April 2020)

Daniel, H.-J., Otto, R., Binder, M. et al. (1999). Production of sophorolipids from whey: development of a two-stage process with *Cryptococcus curvatus* ATCC 20509 and *Candida bombicola* ATCC 22214 using deproteinized whey concentrates as substrates. *Appl. Microbiol. Biotechnol.* **51** (1): 40–45.

Daverey, A. and Pakshirajan, K. (2010). Sophorolipids from *Candida bombicola* using mixed hydrophilic substrates: production, purification and characterization. *Colloids Surf., B* **79** (1): 246–253.

Daverey, A., Pakshirajan, K., and Sangeetha, P. (2009). Sophorolipids production by *Candida bombicola* using synthetic dairy wastewater. *Int. J. Environ. Sci. Eng.* **1** (4): 173–175.

Demirbas, A. (2009). Biorefineries: current activities and future developments. *Energy Convers. Manage.* **50**: 2782–2801.

Douglas, J.M. (1988). *Conceptual Design of Chemical Processes*. McGraw-Hill.

Eerhart, A.J.J.E., Faaij, A., and Patel, M. (2012). Replacing fossil based PET with biobased PEF: Process analysis, energy and GHG balance. *Energy Environ. Sci.* **5**: 6407–6422.

Ellen MacArthur Foundation. (2016). *The New Plastics Economy: Rethinking the Future of Plastics*. http://www.ellenmacarthurfoundation.org/publications (accessed 23 April 2020).

Ellen MacArthur Foundation. (2017). *A New Textiles Economy: Redesigning Fashion's Future*. http://www.ellenmacarthurfoundation.org/publications (accessed 23 April 2020).

Environmental Protection Department. (2017). *Problems & Solutions: Food Waste Challenge*. Environmental Protection Department of Hong Kong. www.epd.gov.hk/epd/english/environmentinhk/waste/prob_solutions/food_waste_challenge.html (accessed 23 April 2020).

EPA. (2016). Advancing Sustainable Materials Management: 2014 Fact Sheet. https://www.epa.gov/sites/production/files/2016-11/documents/2014_smmfactsheet_508.pdf (accessed 23 April 2020).

European Commission. (2012). *Innovating for Sustainable Growth: A Bioeconomy for Europe. Communication from the Commission to the European Parliament, the Council, the European Economic and Social Committee and the Committee of the Regions*. http://ec.europa.eu/research/bioeconomy/pdf/official-strategy_en.pdf (accessed 23 April 2020).

European Commission. (2015). Closing the loop: Commission adopts ambitious new Circular Economy Package to boost competitiveness, create jobs and generate sustainable growth. http://europa.eu/rapid/press-release_IP-15-6203_en.htm (accessed 23 April 2020).

Fallot, A., Saint-André, L., Laclau, J. et al. (2008). Biomass sustainability, availability and productivity. *Rev. Met. Paris* **106** (10): 410–418. https://doi.org/10.1051/metal/2009072.

Fambri, L., Pegoretti, A., Fenner, R. et al. (1997). Biodegradable fibres of poly (L-lactic acid) produced by melt spinning. *Polymer* **38** (1): 79–85.

Food and Agriculture Organization (FAO) (2013). *Food Wastage Footprint: Impacts on Natural Resources*. FAO. http://www.fao.org/3/i3347e/i3347e.pdf (accessed 23 April 2020).

Food and Agriculture Organization (FAO). (2018). SAVE FOOD: Global Initiative on Food Loss and Waste Reduction. http://www.fao.org/save-food/resources/en/ (accessed 23 April 2020).

Georgiou, G., Lin, S.-C., and Sharma, M.M. (1992). Surface-active compounds from microorganisms. *Nat. Biotechnol.* **10** (1): 60.

Ghaly, A., Ananthashankar, R., Alhattab, M., and Ramakrishnan, V. (2014). Production, characterization and treatment of textile effluents: a critical review. *J. Chem. Eng. Process Technol.* **5** (1): 1–18.

Gholamzad, E., Karimi, K., and Masoomi, M. (2014). Effective conversion of waste polyester–cotton textile to ethanol and recovery of polyester by alkaline pretreatment. *Chem. Eng. J.* **253**: 40–45.

Gorin, P., Spencer, J., and Tulloch, A. (1961). Hydroxy fatty acid glycosides of sophorose from *Torulopsis magnoliae*. *Can. J. Chem.* **39** (4): 846–855.

Gunders, D. (2017). *Wasted: How America Is Losing Up to 40 Percent of Its Food from Farm to Fork to Landfill*. United States: National Resources Defense Council (NRDC).

H&M. (2016). *The H&M Group Sustainability Report*. https://sustainability.hm.com/content/dam/hm/about/documents/en/CSR/Report%202016/HM_group_SustainabilityReport_2016_CircularAndRenewable_en.pdf (accessed 23 April 2020).

van Hecke, W., Kaur, G., and De Wever, H. (2014). Advances in in-situ product recovery (ISPR) in whole cell biotechnology during the last decade. *Biotechnol. Adv.* **32**: 1245–1255.

Heyd, M., Kohnert, A., Tan, T.-H. et al. (2008). Development and trends of biosurfactant analysis and purification using rhamnolipids as an example. *Anal. Bioanal.Chem.* **391** (5): 1579–1590.

Hong, F., Guo, X., Zhang, S. et al. (2012). Bacterial cellulose production from cotton-based waste textiles: enzymatic saccharification enhanced by ionic liquid pretreatment. *Bioresour. Technol.* **104**: 503–508.

Hu, Y., Du, C., Leu, S.-Y. et al. (2018). Valorisation of textile waste by fungal solid state fermentation: an example of circular waste-based biorefinery. *Resour. Conserv. Recycl.* **129**: 27–35. https://doi.org/10.1016/j.resconrec.2017.09.024.

IHS. (2016). *Chemical Economics Handbook: Polyester Fibers*. https://ihsmarkit.com/products/polyester-fibers-chemical-economics-handbook.html (accessed 23 April 2020).

Inditex. (2016). *Inditex Annual Report 2016*. https://www.inditex.com/documents/10279/319575/Inditex+Annual+Report+2016/6f8a6f55-ed5b-41f4-b043-6c104a305035 (accessed 23 April 2020).

Isa, M.H.M., Coraglia, D.E., Frazier, R.A., and Jauregi, P. (2007). Recovery and purification of surfactin from fermentation broth by a two-step ultrafiltration process. *J. Membr. Sci.* **296** (1–2): 51–57.

Jauregi, P., Coutte, F., Catiau, L. et al. (2013). Micelle size characterization of lipopeptides produced by *B. subtilis* and their recovery by the two-step ultrafiltration process. *Sep. Purif. Technol.* **104**: 175–182.

Jeihanipour, A. and Taherzadeh, M.J. (2009). Ethanol production from cotton-based waste textiles. *Bioresour. Technol.* **100** (2): 1007–1010.

Jeihanipour, A., Karimi, K., Niklasson, C., and Taherzadeh, M.J. (2010a). A novel process for ethanol or biogas production from cellulose in blended-fibers waste textiles. *Waste Manage. (Oxford)* **30** (12): 2504–2509.

Jeihanipour, A., Karimi, K., and Taherzadeh, M.J. (2010b). Enhancement of ethanol and biogas production from high-crystalline cellulose by different modes of NMO pretreatment. *Biotechnol. Bioeng.* **105** (3): 469–476.

Karadi, R.V., Lokesh, K.N., Channarayappa, M.V. et al. (2014). Standardization of biosurfactant enrichment process by factorial design and elucidating its physico-chemical and structural characteristics. *J. Biochem. Tech.* **3** (5): 161–166.

Kaur, G. and Elst, K. (2014). Development of reactive extraction systems for itaconic acid fermentation: a step towards ISPR of itaconic acid. *RSC Adv.* **4**: 45029–45039.

Kaur, G., Uisan, K., Ong, K.L., and Lin, C.S.K. (2018). Recent trends in green and sustainable chemistry; waste valorisation: rethinking plastics in a circular economy. *Curr. Opin. Green Sustainable Chem.* **9**. 30–39.

Kiel, H., Guvrin, R., and Henis, Y. (1981). Citric acid fermentation by Aspergillus Niger on low sugar concentrations and cotton waste. *Appl. Environ. Microbiol.* **42** (1): 1–4.

van der Klis, F., van Haveren, J., van Es, D.S., and Bitter, J.H. (2017). Synthesis of furandicarboxylic acid esters from nonfood feedstocks without concomitant levulinic acid formation. *ChemSusChem* **10** (7): 1460–1468. https://doi.org/10.1002/cssc.201700051.

Kosaric, N. and Sukan, F.V. (2014). *Biosurfactants: Production and Utilization – Processes, Technologies, and Economics*, vol. **159**. CRC Press.

Kuo, C.-H., Lin, P.-J., Wu, Y.-Q. et al. (2014). Simultaneous saccharification and fermentation of waste textiles for ethanol production. *BioResources* **9** (2): 2866–2875.

Kuyukina, M.S., Ivshina, I.B., Philp, J.C. et al. (2001). Recovery of Rhodococcus biosurfactants using methyl tertiary-butyl ether extraction. *J. Microbiol. Methods* **46** (2): 149–156.

Lendon, B. (2012). 40% of US food wasted, report says. *Cable News Network (CNN)*. http://news.blogs.cnn.com/2012/08/22/40-of-u-s-food-wasted-report-says (accessed 23 April 2020).

Li, Y.-C., Chu, C.-Y., Wu, S.-Y. et al. (2012). Feasible pretreatment of textile wastewater for dark fermentative hydrogen production. *Int. J. Hydrogen Energy* **37** (20): 15511–15517.

Lieder, M. and Rashid, A. (2016). Towards circular economy implementation: a comprehensive review in context of manufacturing industry. *J. Cleaner Prod.* **115**: 36–51.

Lin, C.S.K., Pfaltzgraff, L.A., Herrero-Davila, L. et al. (2013). Food waste as a valuable resource for the production of chemicals, materials and fuels. Current situation and global perspective. *Energy Environ. Sci.* **6** (2): 426–464.

Lipinski, B., Hanson, C., Waite, R., et al. (2013). *Reducing Food Loss and Waste. Creating a Sustainable Food Future, Installment Two*. World Resources Report, Food Loss and Waste Protocol. https://www.wri.org/publication/reducing-food-loss-and-waste (accessed 23 April 2020).

Liu, J., Mooney, H., Hull, V. et al. (2015). Systems integration for global sustainability. *Science* **347**: 12588320–12588329.

Maddikeri, G.L., Gogate, P.R., and Pandit, A.B. (2015). Improved synthesis of sophorolipids from waste cooking oil using fed batch approach in the presence of ultrasound. *Chem. Eng. J.* **263**: 479–487.

Mahalakshmi, M., Angayarkanni, J., Rajendran, R., and Rajesh, R. (2011). Bioconversion of cotton waste from textile mills to bioethanol by microbial saccharification and fermentation. *Ann. Biol. Res.* **2** (3): 380–388.

Militky, J. (2009). The chemistry, manufacture and tensile behaviour of polyester fibers. In: *Handbook of Tensile Properties of Textile and Technical Fibres* (ed. A.R. Bunsell), 223–314. Woodhead Publishing.

Moncada, J., El-Halwagi, M.M., and Cardona, C.A. (2013). Techno-economic analysis for a sugarcane biorefinery: Colombian case. *Bioresour. Technol.* **135**: 533–543.

Moncada, B.J., Aristizábal, M.V., and Cardona, A.C.A. (2016). Design strategies for sustainable biorefineries. *Biochem. Eng. J.* **116**: 122–134.

Morais, A.R.C. and Lukasik, R.M. (2013). Green chemistry and the biorefinery concept. *Sustainable Chem. Processes* **1**: 18–19.

Motagamwala, A.H., Won, W., Sener, C. et al. (2018). Toward biomass-derived renewable plastics: production of 2,5-furandicarboxylic acid from fructose. *Sci. Adv.* **4** (1) https://doi.org/10.1126/sciadv.aap9722.

Motevasel, M. (2014). A study of surface biosurfactants applications on oil degradation. *Am. J. Oil Chem. Technol.* **2** (9).

Noah, K.S., Bruhn, D.F., and Bala, G.A. (2005). Surfactin production from potato process effluent by *Bacillus subtilis* in a chemostat. *Appl. Biochem. Biotechnol.* **121–124**: 465–473.

de Oliveira, D.W.F., França, Í.W.L., Félix, A.K.N. et al. (2013). Kinetic study of biosurfactant production by *Bacillus subtilis* LAMI005 grown in clarified cashew apple juice. *Colloids Surf., B* **101**: 34–43.

de Oliveira, M.R., Magri, A., Baldo, C. et al. (2015). Sophorolipids a promising biosurfactant and it's applications. *Int. J. Adv. Biotechnol. Res.* **6**: 161–174.

Ong, K.L., Kaur, G., Pensupa, N. et al. (2018). Trends in food waste valorisation for the production of chemicals, materials and fuels: case study south and Southeast Asia. *Bioresour. Technol.* **248** (Part A): 100–112.

Opwis, K. and Gutmann, J.S. (2012). Generation of biogas from textile waste waters. *Chem. Eng.* **27**: 103–108. https://doi.org/10.3303/CET1227018.

Parfitt, J., Barthel, M., and Macnaughton, S. (2010). Food waste within food supply chains: quantification and potential for change to 2050. *Philos. Trans. R. Soc. Lond. B Biol. Sci.* **365** (1554): 3065–3081. https://doi.org/10.1098/rstb.2010.0126.

Paritosh, K., Kushwaha, S.K., Yadav, M. et al. (2017). Food waste to energy: an overview of sustainable approaches for food waste management and nutrient recycling. *Biomed. Res. Int.* **2017**: 1–19.

Pensupa, N., Leu, S.-Y., Hu, Y. et al. (2017). Recent trends in sustainable textile waste recycling methods: current situation and future prospects. *Top. Curr. Chem.* **375** (5): 76. https://doi.org/10.1007/s41061-017-0165-0.

Preston, F. (2012). *A Global Redesign? Shaping the Circular Economy*. Chatham House. https://www.chathamhouse.org/sites/default/files/public/Research/Energy,%20Environment%20and%20Development/bp0312_preston.pdf (accessed 23 April 2020).

Rashad, M., Al-Kashef, A., Nooman, M., and El-din-Mahmoud, A. (2014). Co-utilization of motor oil waste and sunflower oil cake on the production of new sophorolipids by Candida bombicola NRRL Y-17069. *Res. J. Pharm. Biol. Chem. Sci.* **5** (4): 1515–1528.

Reinshagen, P. (2012). Why PEF is better than PET. Bio Based Press. https://www.biobasedpress.eu/2012/07/why-pef-is-better-than-pet (accessed 23 April 2020).

Ribeiro, I.A., Bronze, M.R., Castro, M.F., and Ribeiro, M.H.L. (2012). Optimization and correlation of HPLC-ELSD and HPLC–MS/MS methods for identification and characterization of sophorolipids. *J. Chromatogr. B* **899**: 72–80. https://doi.org/10.1016/j.jchromb.2012.04.037.

Sadef, Y., Nizami, A.S., Batool, S.A. et al. (2016). Waste-to-energy and recycling value for developing integrated solid waste management plan in Lahore. *Energy Sources Part B* **11** (7): 569–579.

Samad, A., Zhang, J., Chen, D. et al. (2017). Sweet sorghum bagasse and corn Stover serving as substrates for producing sophorolipids. *J. Ind. Microbiol. Biotechnol.* **44** (3): 353–362.

Sandin, G. and Peters, G.M. (2018). Environmental impact of textile reuse and recycling – A review. *J. Cleaner Prod.* **184**: 353–365.

Saucedo-Luna, J., Castro-Montoya, A.J., Martinez-Pacheco, M.M. et al. (2011). Efficient chemical and enzymaticsaccharification of the lignocellulosic residue from Agave tequilana bagasse toproduce ethanol by Pichia caribbica. *J. Ind. Microbiol. Biotechnol.* **38**: 725–732.

Shaligram, N.S. and Singhal, R.S. (2010). Surfactin – a review on biosynthesis, fermentation, purification and applications. *Food Tech. Biotechnol.* **48** (2): 119–134.

Sharma, S.K., Mulligan, C.N., and Mudhoo, A. (2014). *Biosurfactants: Research Trends and Applications*. CRC Press.

Sheldon, R.A. (2016). Green chemistry and resource efficiency: towards a green economy. *Green Chem.* **18**: 3180–3183.

Shen, F., Xiao, W., Lin, L. et al. (2013). Enzymatic saccharification coupling with polyester recovery from cotton-based waste textiles by phosphoric acid pretreatment. *Bioresour. Technol.* **130**: 248–255.

Sherwood, J., Clark, J., Farmer, T.J. et al. (2017). Recirculation: a new concept to drive innovation in sustainable product design for bio-based products. *Molecules* **22**: 1–17.

Statista. (2016). Worldwide production volume of chemical and fibers from 1975 to 2016. The Statistics Portal. https://www.statista.com/statistics/263154/worldwide-production-volume-of-textile-fibers-since-1975 (accessed 23 April 2020).

Tuck, C.O., Pérez, E., Horvath, I.T. et al. (2012). Valorisation of biomass: deriving more value from waste. *Science* **337**: 695–699.

Van Bogaert, I.N., Saerens, K., De Muynck, C. et al. (2007). Microbial production and application of sophorolipids. *Appl. Microbiol. Biotechnol.* **76** (1): 23–34.

Venkata Mohan, S. (2014a). Reorienting waste remediation towards harnessing bioenergy: a paradigm shift. In: *Industrial Wastewater Treatment, Recycling and Reuse* (eds. V.V. Ranade and V.M. Bhandari), 235–281. Butterworth-Heinemann and Elsevier.

Venkata Mohan, S. (2014b). Sustainable waste remediation: a paradigm shift towards environmental biorefinery. *Chem. Eng. World* **49** (12): 29–35.

Willenbacher, J., Rau, J.-T., Rogalla, J. et al. (2015). Foam-free production of Surfactin via anaerobic fermentation of *Bacillus subtilis* DSM 10T. *AMB Express* **5** (1): 21.

Yang, X., Zhu, L., Xue, C. et al. (2012). Recovery of purified lactonic sophorolipids by spontaneous crystallization during the fermentation of sugarcane molasses with *Candida albicans* O-13-1. *Enzyme Microb. Technol.* **51** (6–7): 348–353.

Yu, I.K.M., Ong, K.L., Tsang, D.C.W. et al. (2018). Chemical transformation of food and beverage waste-derived fructose to hydroxymethylfurfural as a value-added product. *Catal. Today* https://doi.org/10.1016/j.cattod.2018.01.011.

Yue, D., You, F., and Snyder, S.W. (2014). Biomass-to-bioenergy and biofuel supply chainoptimization: overview, key issues and challenges. *Comput. Chem. Eng.* **66**: 36–56.

Zuin, V.G. (2016). Circularity in green chemical products, processes and services: innovative routes based on integrated eco-design and solution systems. *Curr. Opin. Green Sustainable Chem.* **2**: 40–44.

Index

Waste Valorisation: Waste Streams in a Circular Economy, First Edition.
Edited by Carol Sze Ki Lin, Guneet Kaur, Chong Li and Xiaofeng Yang.